Physics-Based Computational Methods for Nonlinear Framed Structures and Plates/Shells

Physics-Based Computational Methods for Nonlinear Framed Structures and Plates/Shells

Yeong-Bin Yang
Chongqing University, China

Der-Shen Yang
Tongji University, China

Shyh-Rong Kuo
National Taiwan Ocean University, Taiwan

World Scientific

NEW JERSEY • LONDON • SINGAPORE • BEIJING • SHANGHAI • TAIPEI • CHENNAI

Published by

World Scientific Publishing Co. Pte. Ltd.
5 Toh Tuck Link, Singapore 596224
USA office: 27 Warren Street, Suite 401-402, Hackensack, NJ 07601
UK office: 57 Shelton Street, Covent Garden, London WC2H 9HE

Library of Congress Control Number: 2025003791

British Library Cataloguing-in-Publication Data
A catalogue record for this book is available from the British Library.

PHYSICS-BASED COMPUTATIONAL METHODS FOR NONLINEAR FRAMED STRUCTURES AND PLATES/SHELLS

ISBN 978-981-98-0821-2 (hardcover)
ISBN 978-981-98-0822-9 (ebook for institutions)
ISBN 978-981-98-0823-6 (ebook for individuals)

For any available supplementary material, please visit
https://www.worldscientific.com/worldscibooks/10.1142/14183#t=suppl

Desk Editors: Kannan Krishnan/Amanda Yun

Typeset by Stallion Press
Email: enquiries@stallionpress.com

Foreword

I have had the distinct privilege of knowing Professor Y.B. Yang for over four decades, dating back to his time as a graduate student at Cornell University. Throughout these years, I have witnessed his profound and sustained contributions to the field of structural engineering, particularly in nonlinear analysis. His latest book, *Physics-Based Computational Methods for Nonlinear Framed Structures and Plates/Shells*, exemplifies his unceasing commitment to advancing both the theories and practical applications of mechanics.

Professor Yang has always sought to tackle complex engineering challenges with clarity and insight. His earlier work, *Theory and Analysis of Nonlinear Framed Structures* (1994), made significant strides in establishing the mathematical underpinnings of nonlinear structural analysis. In this new volume, he shifts his focus to make nonlinear analysis more intuitive and accessible, emphasizing the physical meanings of governing principles. By bridging the gap between theoretical complexity and practical understanding, the book serves as a vital resource for students, researchers, and practitioners working on nonlinear structural systems.

The text addresses key challenges in the field, including the physical justification of nonlinear finite elements, the integration of three-dimensional moment rotational behaviors, and the reliable tracing of post-buckling paths. What distinguishes this work is Professor Yang's ability to distill these complex topics into accessible insights without sacrificing solution accuracy. The introduction of the "rigid body rule" as an effective tool to verify various aspects of the analysis exemplifies his innovative approach, offering readers a fresh perspective on some of the most difficult problems in nonlinear analysis.

This book is more than an academic treatise; it is a practical guide that balances theoretical depth with real-world applications. Researchers and students alike will find it an invaluable tool for tackling both foundational and advanced aspects of nonlinear analysis. Its clarity, rigor, and practicality ensure its enduring relevance and impact on structural engineering.

Having known Professor Yang for so many years, it is deeply gratifying to see his continued influence on the next generation of engineers, just as he inspired me in the early days of his career. I am confident that this book, like his previous works, will leave a lasting legacy in the field and serve as a cornerstone for further development in structural engineering.

S. Kitipornchai
University of Queensland

Preface

Unlike our previous book that depends heavily on mathematics (Yang and Kuo, *Theory and Analysis of Nonlinear Framed Structures*, 1994, Prentice-Hall), we choose to bring closer the distance to authors by deleting some mathematically intricate parts, while stressing more on the physical meaning of the conceptual flows. The nonlinear analysis of structures seems not new to the academe. But in fact, there are many issues that remain not well answered, e.g.,

(1) How can a nonlinear finite element be physically justified?
(2) A moment differs from a force. How to accommodate the three-dimensional rotational behavior of a moment in a nonlinear analysis?
(3) How to update the initial nodal forces existing at each incremental step?
(4) Can we devise a path-tracing method to trace faithfully the multi-critical points of a post-buckling response?

In this book, we will try to answer all the above questions by adopting a physically justified approach, i.e., the universally valid rigid body rule. Focus is placed more on the physical interpretation of each issue involved in the nonlinear analysis of structures, ranging from the theory, element formulation, quality testing, and benchmarking, to the implementation of path-tracing schemes. We believe that the book in the present format is more friendly to researchers, engineers, graduate students, and senior undergraduate students.

This book is based entirely on the *stiffness or displacement formulation* of the finite element method (FEM) for its popularity in engineering practice. While the FEM has stridden in the past over half a century and has given us the impression that it can be used to solve almost any structural problem. In fact, this is only true for linear or static problems. For *nonlinear* or *dynamic* problems, not all the parts of the computational procedures have come to the state of maturity as that for linear and static problems.

For *nonlinear structures*, the deformations or strains of the constituting members of the structure under loadings are such large that they should be included in establishing the equations of equilibrium in analysis. Aside from laborious mathematical derivations for analysis of nonlinear structures presented in the previous version, we think a more efficient approach is to focus on the *physical aspect* involved, by using the *rigid body rule* to interpret the structural nonlinear behavior, to make the entire formulation more simplified, elegant, and physically meaningful. This is what will be undertaken in preparing this book.

The structures that are considered in the book are mainly *framed structures*, except for the minor adventure to plates and shells. A framed structure is composed of members that are long in comparison with their cross-sectional dimensions, such as width and depth. Members of this sort are called slender members and are conventionally represented by one-dimensional line elements in the finite element analysis. It is assumed that the shearing deformation on the cross-sectional plane is small and can be neglected. Four categories of framed structures are considered in this book, i.e., plane trusses, space trusses, plane frames, and space frames, which appear in the order of increasing complexity. Each of these categories represents a class of structures with specific characteristics. For the plane and space frames, the joints are assumed to be *rigid-connected*, while for the plane and space trusses, they are *pinned-connected*. While no consideration will be made for *semi-rigid joints*, the semi-rigid or elastic joints can be easily accommodated by inserting springs between the connected members to represent the joint deformation effect.

For nonlinear structures, both the *linear* and *nonlinear parts* of the strains should be considered, say, in the virtual work formulation. The linear and nonlinear strains we know from elasticity are good for elastic or deformable bodies. However, in the presence of the

Bernoulli-Euler kinematic hypothesis (of plane cross sections remaining plane) for beams or the extended Kirchhoff hypothesis for plates, the linear strains remain good, but *the nonlinear strains may produce some high-order stiffnesses that cannot be physically justified.* This means that the finite element formulation for the beams or plates based on the displacement approach (i.e., by relating strains to displacements and by relating member forces to stresses using the virtual work formulation) is not straightforward. It is only for trusses, for which no kinematic hypothesis is needed, that the stiffness terms of all orders can be perfectly interpreted by physics. Obviously, the conventional kinematic hypothesis has introduced some constraints to the deformation of the elastic bodies, especially on the nonlinear strain parts.

Linear analysis for finding the deformations, stresses, and forces of structural members can be conducted using the elastic stiffness $[k_e]$ alone. In a virtual work formulation, the *elastic stiffness* $[k_e]$ can be derived from the (virtual) strain energy that involves the linear strains of the structural element. This is the part that is considered well established for the structural members considered, including the truss element, beam element, and the triangular 3-node plate element considered herein. Most of the elastic or linear stiffnesses $[k_e]$ have been tested to be physically sound, especially by the *patch test*, in that for a patch of elements subjected to a rigid displacement (or rotation) field, no strains or stresses should be induced on the elements, if they are originally stress-free. Such a test has been adopted in the early stage of finite element development to ensure the quality of a newly formulated element. The formulation of the elastic stiffnesses $[k_e]$ for structural elements has been the subject of courses on FEMs or matrix structural analysis. The readers are assumed to have been acquainted with this part, which will not be covered herein.

As for the *buckling* or *geometric nonlinear analysis* of structures, we need to bring in the *geometric stiffness* $[k_g]$, which can be derived from the (virtual) potential energy that involves the *nonlinear strains* and *initial stresses*. The initial stresses (or forces) existing on the element have been recognized as the source of instability for the element. However, due to the adoption of the kinematic hypothesis (i.e., the Bernoulli-Euler hypothesis), the initial stresses coupled with the nonlinear strains in the virtual work sense will produce some high-order (stiffness) terms that cannot be physically interpreted, especially from the point of rigid rotations. Given that the elastic stiffnesses $[k_e]$ for structural elements are rigid-qualified as a prerequisite, we will focus entirely in this book on the derivation of *rigid-qualified* geometric stiffness $[k_g]$ for various structural elements.

In essence, the capability of an initially stressed element (and particularly its geometric stiffness $[k_g]$) to displace correctly under the rigid rotation is the first requirement for the nonlinear element to be acceptable. In this regard, we have to call on the *rigid body rule*, which states that if we impose a *rigid rotation field* on an initially stressed (and equilibrated) finite element, then the element should remain in equilibrium with no change in the magnitude of the acting forces after the rigid rotation. This rule provides the criterion for us to retain or discard the confusing, physically unjustified high-order terms brought by the kinematic hypothesis in the mathematical formulation for structural members The ability to exhibit rigid body rule is the bottom line that should be obeyed by each nonlinear finite element.

Among the framed structures considered, the *truss* is of the simplest form in that it is composed of two-force members. Unlike the beam, a truss member does not need any kinematic hypothesis to describe its cross-sectional behavior. Because of this, all the stiffness terms derived from the nonlinear strains in the virtual work expression conform perfectly to the rigid body rule and are physically justified.

In fact, the rigid body rule goes beyond what it means. Firstly, as was pointed above, it can be used to test the quality of a newly derived nonlinear element, especially its associated geometric stiffness $[k_g]$. Secondly, for an element initially stressed by nodal forces, the rigid body rule can be employed to update the nodal forces initially acting at the end of each iteration subjected to rigid rotations, see Chapter 3 and Section 7.3. Thirdly, for an element equilibrated by a set of initial nodal forces, the rigid body rule can be utilized to derive the geometric stiffness $[k_g]$ based on the nodal degrees of six rigid displacements and rotations in the three-dimensional space, as will be demonstrated in Section 8.4. Finally, concerning the rigid-rotation behavior, the geometric stiffness $[k_g]$ for the three-node *triangular plate element* can be obtained by simply assembling the geometric stiffnesses $[k_g]$ of the rigid elements that form the three sides of the triangle, see Section 8.6. It is from the point of rigid body rule that the geometric nonlinear analysis of plates/shells is covered in this book. All the above features on the rigid body rule will be demonstrated throughout the book.

The outlines of the book are as follows: Chapter 1 presents the fundamentals of mechanics, such as strains and stresses, and the principle of virtual works, that form the backbone of the theoretical formulations in the following chapters. Chapter 2 summarizes the linear analysis as a prerequisite for nonlinear analysis. In Chapter 3, the most important *rigid body rule* is introduced and adopted to verify the rationality of the plane beam element derived.

In Chapter 4, the truss element is derived. Since no kinematical hypothesis has been employed, the truss element serves as a perfect example for demonstrating the rigid body rule, in that all the nonlinear terms involved can be physically explained. In Chapter 5, the nonlinear theory of space frames is rigorously derived along with the inclusion of the induced moments resulting from rotations of the nodal moments. The governing equations derived in this chapter are confirmed to pass the rigid body rule. Based on the theory of Chapter 5, the stiffness matrices are derived for the space frame element in Chapter 6 and applied to the buckling analysis of a number of simple frames.

Chapter 7 is devoted to the incremental-iterative analysis of structures involving the post-buckling analysis. Firstly, the procedure involving the predictor, corrector, and geometry updating that was not adequately presented previously is thoroughly discussed. Secondly, the *generalized displacement control* (GDC) method enhanced by the *general stiffness parameter* (GDP) is demonstrated to be the most effective tool for tracing the post-buckling response of structures involving multiple critical points.

In Chapter 8, the rigid body rule is employed first to derive the geometric stiffness matrix of a rigid beam. Then, by assembling three rigid beam elements that form the boundary of a triangular plate element (TPE), the geometric stiffness matrix for the TPE is derived and used in the post-buckling analysis of a number of plate/shell problems.

To offer an independent reference for comparing with the results presented by the finite element approach, analytical solutions are derived in Chapter 9 for some two-member frames based on the governing equations, and boundary and continuity conditions.

The senior (first) author returned to his alma mater, the National Taiwan University (NTU), for the teaching/research career since he got his Ph.D. degree from Cornell University in 1984. During the thirty years of service at the NTU, he is lucky to have so many brilliant students for the Master's and Ph.D. programs. In 2014, he retired from the NTU and moved to Chongqing University in China. Through his teaching and research career, it is amazing that he continues to receive some very brilliant students under his supervision at Chongqing University. In general, most of the former students have contributed in certain form to the completion of this book. Particularly, those students who have contributed directly to the book can be identified from the authorship of the papers that have been included as parts of the book. To add one more, the senior author's graduate student Mr. Bo Pang has assisted in preparing some of the figures and analysis results for Chapter 3. The senior author is indebted to all the students for their contribution either in general or in particular.

About the Authors

Yeong-Bin Yang received his PhD from Cornell University in 1984. He is a member of the Chinese Academy of Engineering and the European Academy of Sciences and Arts and a foreign member of the Austrian Academy of Sciences. Currently, he is the Honorary Dean of the School of Civil Engineering, Chongqing University, China, Professor Emeritus of National Taiwan University (NTU), and Editor-in-Chief of the *International Journal of Structural Stability and Dynamics* (IJSSD).

Dr Yang was President of the National Yunlin University of Science and Technology (YunTech), Taiwan, President of the Asian-Pacific Association of Computational Mechanics (APACM), and Chairman of the East Asia-Pacific Conference on Structural Engineering and Construction (EASEC). He was also President of the Association of Computational Mechanics Taiwan (ACMT), the Institute of Engineering Education Taiwan (IEET), and the Chinese Institute of Civil and Hydraulic Engineering.

Dr Yang received a number of awards, including the Lifetime Achievement Medal, American Society of Civil Engineers (ASCE) Greater China Section (2023); Zienkiewicz Medal, Asian Pacific Association for Computational Mechanics (APACM) (2022); Li-Guo-Hao Bridge Innovation Medal, Tongji University, China (2022); Nishino Medal, EASEC (2016); Tsuboi Award, International Association for Shell and Spatial Structures (IASS) (2008); Munro Prize, Engineering Structures (2003); Outstanding Scientist, Nobel Laureate Dr Y T Lee's Foundation (1998); Distinguished Research Award, National Science Council, Taiwan (1988–1998); among others.

Der-Shen (Ted) Yang is a researcher and educator with a PhD in Structural Engineering from the University of Queensland, Australia, where he pioneered methods for analyzing structural dynamics under moving loads. His academic journey includes a master's degree from Cornell University, USA, and substantial research and teaching roles at Monash University, Australia, where he has integrated structural health monitoring, signal processing, and machine learning in his interdisciplinary work. His research has led to widely cited publications on topics like vehicle-bridge interaction and infrastructure health monitoring, earning recognition through Australia's "Global Talent" program. With a unique combination of academic and industry experience — beginning as a structural engineer at China Construction Design International — he is committed to bridging research and practical applications. In 2025, he joins Tongji University as an associate professor.

Shyh-Rong Kuo received his PhD from National Taiwan University in 1991. He is currently Dean of the College of Engineering at National Taiwan Ocean University (NTOU) and was Chairman of the Department of River and Marine Engineering at NTOU. Globally, he was most reputed for his works on the stability of curved beams, out-of-plane buckling or torsional buckling of structural frames, three-dimensional formulation of nonlinear beam elements, and various structural vibration problems.

Acknowledgments

Parts of the materials presented in this book have been revised from the book (Yang and Kuo, 1994) or papers published previously by the authors and co-workers. Efforts have been undertaken to update, digest, and rewrite the materials acquired from each source, such that a unified and progressive style of presentation can be achieved throughout the book. We like to express our appreciation to the publishers of the following papers and book authored by us, that have formed parts of the book:

References

Kuo, S.R., Yang, Y.B., and Chou, J.H. (1993). Nonlinear analysis of space frames with finite rotations, *J. Struct. Eng.*, ASCE, 119(1), 1–15.

Leu, L.J. and Yang, Y.B. (1990). Effects of rigid body and stretching on nonlinear analysis of trusses, *J. Struct. Eng.*, ASCE, 116(10), 2582–2598.

Yang, Y.B. and Chiou, H.T. (1987). Rigid body motion test for nonlinear analysis with beam elements, *J. Eng. Mech.*, ASCE, 113(9), 1404–1419.

Yang, Y.B. and Kuo, S.R. (1991). Out-of-plane buckling of angled frames, *Int. J. Mech. Sci.*, 33(1), 55–67.

Yang, Y.B. and Kuo, S.R. (1991). Buckling of frames under various torsional loadings, *J. Eng. Mech.*, ASCE, 117(8), 1681–1697.

Yang, Y.B. and Kuo, S.R. (1992). Frame buckling analysis with full consideration of joint compatibilities, *J. Eng. Mech.*, ASCE, 118(5), 871–889.

Yang, Y.B. and Kuo, S.R. (1994). *Theory and Analysis of Nonlinear Framed Structures*, Prentice-Hall, Englewood Cliffs, New Jersey, USA.

Yang, Y.B., Kuo, S.R., and Wu, Y.S. (2002). Incrementally small-deformation theory for nonlinear analysis of structural frames, *Eng. Struct.*, 24(6), 783–798.

Yang, Y.B., Lin, S.P., and Chen, C.S. (2007). Rigid body concept for geometric nonlinear analysis of 3D frames, plates and shells based on the updated Lagrangian formulation, *Comp. Meth. Appl. Mech. & Eng.*, 196(7), 1178–1192.

Yang, Y.B. and Shieh, M.S. (1990). Solution method for nonlinear problems with multiple critical points, *AIAA J.*, 28(12), 2110–2116.

Contents

List of Figures

List of Tables

List of Symbols

A general guideline for the notation used in the text is given in Section 1.2. The following is a list of the symbols used throughout each chapter. All the symbols are defined when they first appear in the text. The left superscript of a symbol denotes the configuration in which the quantity occurs, while the left subscript denotes the configuration to which the quantity is measured. The left subscripts may be dropped in the cases where the reference configuration can be easily identified. Quantities that are generated during each incremental step are denoted with no left superscripts. Matrices, vectors, and rows are enclosed by brackets [ ], braces { }, and the symbols $\langle\ \rangle$, respectively.

A	area of member cross-sections
${}^0A,\ {}^1A,\ {}^2A$	surface area at C_0, C_1 and C_2
$[a],\ [\bar{a}]$	antisymmetric matrices
C_0	initial undeformed configuration of the structure
C_1	last deformed (known) configuration of the structure
C_2	current deformed (unknown) configuration of the structure
${}_0C_{ijkl},\ {}_1C_{ijkl}$	incremental constitutive tensor with respect to C_0 and C_1
CSP	current stiffness parameter
$[D]$	diagonal matrix
${}^0ds,\ {}^1ds,\ {}^2ds$	length of an infinitesimal element at C_0, C_1 and C_2

E	elastic modulus
e_{ij}	linear strain (linear analysis)
e_{ijk}	permutation symbol
${}_1e$	stretch of the truss element
${}_0e_{ij}$	linear part of Green strain increment tensor (also known as infinitesimal strain tensor referred to C_0)
${}_1e_{ij}$	linear part of the updated Green strain increment tensor (also known as infinitesimal strain tensor referred to C_1)
${}_2e_{ij}$	linear part of Euler strain tensor (also known as infinitesimal strain tensor referred to C_2)
$\{{}^1\bar{e}_y\}$, $\{{}^1\bar{e}_z\}$	averaged unit vectors (Appendix B)
F	force action
F_{xl}, F_{xn}	linear and nonlinear parts of force increment of truss element
$\vec{F}_D$	direct forces forming the external moment
${}^1F$	initial member force at C_1
${}^2F$	member force at C_2
f	single-valued function for material law
f_i	body forces (linear analysis)
$\{f\}$	element force vector
${}^2f_i$	body force per unit volume at C_2
${}^1_1f_i$, ${}^2_1f_i$	body forces at C_1 and C_2, but referred to C_1
${}^1_0f_i$, ${}^2_0f_i$	body forces at C_1 and C_2, but referred to C_0
$\{{}^1f\}$ (or $\{{}^1_1f\}$)	initial element force vector at C_1 and referred to C_1
$\{{}^2f\}$	element force vector at C_2 but referred to C_1
$\{{}^2_2f\}$	element force vector at C_2 and referred to C_2
$\{\Delta f\}$	element force increment vector
G	shear modulus
GSP	general stiffness parameter
I_y, I_z	moments of inertia about y- and z-axis
$[I]$	unit matrix
i $(=x/L)$	normalized axial coordinate
J	torsional constant
${}^2_0J$, ${}^2_1J$	Jacobian determinant from C_0 to C_2 and from C_1 to C_2

l, m, n	direction cosines
$\bar{K}$	Wagner coefficient
$[K]$	structural stiffness matrix
$[K_e]$	structural linear or elastic stiffness matrix
$[K_g]$	structural geometric stiffness matrix
$[K_j]$	joint moment matrix for structure
$[K_{gh}^{stv}]$	integral matrix (Appendix A)
$[k]$ (or $[k_e]$)	element linear or elastic stiffness matrix
$[k_g]$	element geometric stiffness matrix
$[k_g]_{r.b.}$	geometric stiffness matrix for the rigid beam element
$[k_g]_{TPE}$	geometric stiffness matrix for the triangular plate element
$[k_i]$	induced moment matrix
$[k_j]$	joint moment matrix for element
$\Delta[k_g]$	sum of $[s_2]_{eq}$ and $[s_3]_{eq}$
L	length of the element
${}^2L$	length of element at C_2
$[L]$	lower triangular matrix
M	moment action
M_{cr}, $M_{0,cr}$	critical moment
$\{M\}$, $\{\vec{M}\}$	moment vectors
${}^1M$	initial member moment at C_1
${}^2M$	member moment at C_2
${}^1\vec{M}$	external moment vector
ΔM	induced moment at a nodal point
n_j	components of the direction cosines of the unit normal ν to the surface
$\vec{n}$	unit vector
$\vec{n}_a$	unit direction vector
$\{n_1\}$, $\{n_3\}$	interpolation functions
P_{cr}, P_{cy}, P_{cz}	critical loads
$\{P\}$	structural load vector
$\{\hat{P}\}$	reference loads for incremental-iterative analysis
$\{{}^1P\}$	structural load vector at C_1
$\{{}^2P\}$	structural load vector at C_2
R	external virtual work (linear analysis)
R	radius (Chapter 8)

$[R]$	rotation matrix
$\{R_{j-1}^i\}$	unbalanced forces for the current iteration (Chapter 7)
${}^2R$ (or ${}^2_1R$)	external virtual work at C_2, but referred to C_1
${}^1_1R$, ${}^2_1R$	external virtual work at C_1 and C_2 with reference to C_1
${}^1_0R$, ${}^2_0R$	external virtual work at C_1 and C_2 with reference to C_0
$\vec{r}$, $\vec{r}'$	position vectors (Appendix B)
${}^1\vec{r}_{CD}$	lever arm of the external moment
${}^2S_u$	surface area with prescribed displacements
${}^2S_\tau$	surface area with prescribed tractions
${}_0S_{ij}$	Kirchhoff stress increment tensor
${}_1S_{ij}$	updated Kirchhoff stress increment tensor
${}^1_0S_{ij}$, ${}^2_0S_{ij}$	second Piola–Kirchhoff stress tensor at C_1 and C_2
ΔS	arc length
$[s]$	symmetric matrix
$[s_1]$, $[s_2]$, $[s_3]$	higher order matrices of truss element
$[s_1]_{eq}$, $[s_2]_{eq}$, $[s_3]_{eq}$	equivalent stiffness matrices of $[s_1]$, $[s_2]$ and $[s_3]$
T_{cr}, $T_{0,cr}$	critical torque
$[T]$	transformation matrix
t_i	surface tractions (linear analysis)
${}^1_1t_i$, ${}^2_1t_i$	initial tractions at C_1 and C_2 referred to C_1
${}^1_0t_i$, ${}^2_0t_i$	initial tractions at C_1 and C_2 referred to C_0
${}^2t_i$ (or ${}^2_2t_i$)	surface tractions on ${}^2S_\tau$ at C_2
U_b	element axial elongation (Appendix B)
$\{U\}$	structural displacement vector
δU	virtual strain energy
$\{\Delta U\}$	structural displacement increment (Chapter 7)
$\{\Delta U_j^i\}$	total force increment for the current iteration (Chapter 7)
$\{\Delta \hat{U}_j^i\}$	displacements corresponding to reference loads $\{\hat{P}\}$ (Chapter 7)
$\{\Delta \bar{U}_j^i\}$	displacements corresponding to unbalanced forces $\{R_{j-1}^i\}$ (Chapter 7)
u_i	displacement increments from C_1 to C_2
u, v, w	centroidal displacements of the beam

$u_r,\ v_r,\ w_r$	rigid displacements
${}^1u_i,\ {}^2u_i$	displacements at C_1 and C_2
$\{u\}$	element displacement vector
$\{u\}_r$	element rigid displacement vector
$\Delta u,\ \Delta v,\ \Delta w$	relative displacements of two nodes of a truss element
$\Delta\vec{u}_a$	displacement increments of node A (Appendix B)
${}^0V,\ {}^1V,\ {}^2V$	volume at C_0, C_1 and C_2
δV	virtual potential energy
$V.W.I.$	virtual work increment
X_{ij}	$= X_i - X_j$ (Chapter 8)
${}^0x_i,\ {}^1x_i,\ {}^2x_i$	coordinates at C_0, C_1 and C_2
${}^1\bar{x}, {}^1\bar{y}, {}^1\bar{z}$	C_1 axes fixed in space
${}^2\bar{x}, {}^2\bar{y}, {}^2\bar{z}$	C_2 axes fixed in space
${}^i\vec{x},\ {}^i\vec{y},\ {}^i\vec{z}$	element axes at configuration C_i (Appendix B)
$({}^iX_a,\ {}^iY_a,\ {}^iZ_a)$	coordinates of node A at configuration C_i
Y_{ij}	$=Y_i - Y_j$ (Chapter 8)
α	torsional parameter
${}^i\vec{\alpha}_a,\ {}^i\vec{\beta}_a,\ {}^i\vec{\gamma}_a$	section axes of node A at C_i
$[\Gamma]$	transformation matrix
γ_{ij}	engineering shear strains
$[\gamma]$	rotation matrix
$\delta(\cdot)$	virtual or variational quantity
δ_{ij}	Kronecker delta
${}_0\epsilon_{ij}$	Green strain increment tensors from C_1 to C_2
${}_1\epsilon_{ij}$ (or ${}^2_1\epsilon_{ij}$)	updated Green strain increment tensor
${}_2\epsilon_{ij}$	Euler strain tensor
${}^1_0\epsilon_{ij},\ {}^2_0\epsilon_{ij}$	Green–Lagrange strain tensors at the C_1 and C_2 (referred to C_0)
${}_0\eta_{ij}$	nonlinear part of Green strain increment tensor
${}_1\eta_{ij}$	nonlinear part of the updated Green strain increment tensor
$\eta,\ \zeta$	embedded axes of cross-section
$\Theta_{xa},\ \Theta_{ya},\ \Theta_{za}$	natural rotations at node A (Appendix B)
θ	rotational angle
θ_r	rigid rotation
$\theta_{xr},\ \theta_{yr},\ \theta_{zr}$	rigid rotations

$\Delta\vec{\theta}_a$	rotation increments of node A (Appendix B)
Λ_j^i	load factor for the current iteration (Chapter 7)
λ	eigenvalue
λ_{cr}	critical load factor
λ_j^i	load increment parameter for ith increment and jth iteration
${}^i\vec{\xi}_a,\ {}^i\vec{\eta}_a,\ {}^i\vec{\zeta}_a$	reference axes of node A at configuration C_i (Appendix B)
${}^0\rho,\ {}^1\rho,\ {}^2\rho$	specific mass at C_0, C_1, and C_2
${}^1\tau_{ij},\ {}^2\tau_{ij}$	Cauchy (or Euler) stresses at C_1 and C_2
ν	unit normal to the surface
ν	Poisson's ratio in numerical examples
$[\Phi]$	transformation matrix
ϕ_a	magnitude of rotation vector of node A (Appendix B)
${}_2\omega_{ij}$	skey-symmetric part of the infinitesimal strain at C_2

List of Abbreviations

CPU	central processing unit
CSP	current stiffness parameter
DOF	degree of freedom
GDC	generalized displacement control (method)
GSP	general stiffness parameter
PiCj	predictor i combined with corrector j as defined in Section 8.9
QT-1	first-kind quasitangential (moment)
QT-2	second-kind quasitangential (moment)
r.b.	rigid beam
ST	semitangential (moment)
TL	total Lagrangian (formulation)
TPE	triangular plate element
TRIC	plate element derived by Argyris et al. (1979)
V.W.I.	virtual work increment
UL	updated Lagrangian (formulation)

Chapter 1

Introduction

1.1 Background

This chapter provides a complete review of the strains, stresses, and reference configurations that are essential to the geometric nonlinear analysis of framed structures based on the updated Lagrangian (UL) formulation. A *framed structure* is said to consist of members that are long in comparison with their cross-sectional dimensions, such as width and depth. Members of this sort are called *slender elements* and are conventionally represented by line elements in the finite element analysis. A truss is considered a special form of the framed structure. A framed structure is characterized by the presence of axial forces, shear forces, moments, and torques on each member, but for a truss, only the axial forces will be present. The joints of framed structures are assumed to be *rigid-connected*, and those of the trusses are *pinned-connected*.

The nonlinear analysis of structures differs from the linear analysis in that the deformations of the structure are not small, such that a distinction needs to be made between the current state and initial state, particularly with respect to the definitions of strains and stresses and the calculation of element forces. This will require the loading on the structure to be applied in a number of incremental steps, each of which can be described in the linearized sense with the structure subjected to initial nodal forces at the beginning of each incremental step. From the point of computation, the buckling of structures can be regarded as a special case of geometric nonlinear analysis that has only a single incremental step from the configuration of zero loads to the configuration under the buckling loads. For this reason, the theory will be presented only in the incremental-iterative form that is suitable for geometric nonlinear analysis of elastic structures in this book.

In the geometric nonlinear analysis of elastic structures, the *UL formulation* is known to be more efficient than the total Lagrangian (TL) formulation. Therefore, it is the formulation that will be exclusively adopted in this book. In this regard, only the physical quantities that are essential to the UL formulation, such as strains, stresses, and their transformation between two configurations, will be presented in this chapter. Throughout the book, an attempt will be made to present all the necessary materials and derivations in a brief, but sufficient manner for the understanding of the nonlinear structural analysis. Moreover, emphasis will be placed on the *physical interpretation* or *justification* of the mathematical derivations. The material presented in this chapter is basically an updated version of the one presented in Chapter 1 of Yang and Kuo (1994).

1.2 Notation and Nomenclature

All numerical procedures for nonlinear analysis have to start from the basic mechanics principles. In this chapter, the fundamental mechanics equations and relevant variables to be used in later chapters for nonlinear analysis will be summarized. Prior to this, the notation and nomenclature for a typical nonlinear analysis will be presented in this section.

Both *Eulerian formulation* and *Lagrangian formulation* have been adopted for describing the motion of solid bodies. In the Eulerian formulation, the static and kinematic variables are referred to the current deformed configuration at each load step, also known as the moving coordinates, while in the Lagrangian formulation, all the static and kinematic variables are referred to the initial or a previously known configuration. The Lagrangian formulation is particularly suitable for the step-by-step nonlinear analysis of the solid body, in which we are interested in the deformation of each point of the body in the loading process. In contrast, the Eulerian formulation has been widely adopted in the analysis of fluid mechanics problems,

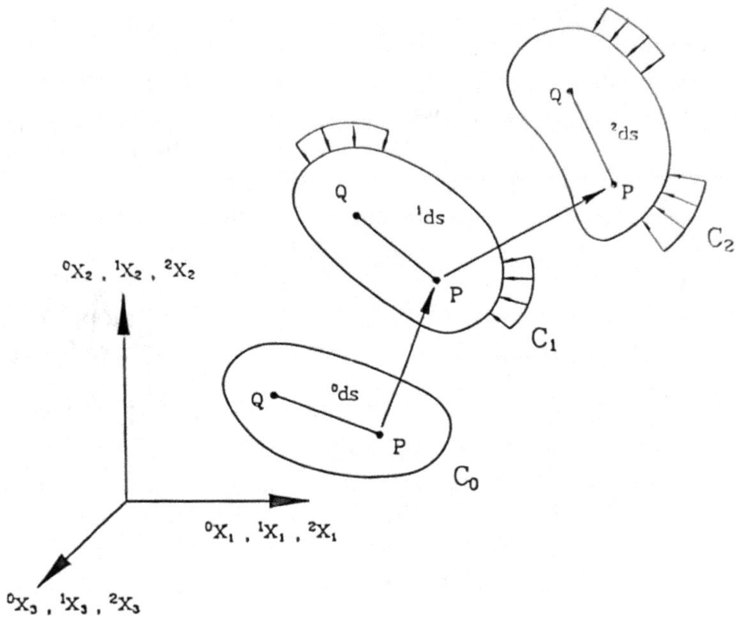

Fig. 1.1. Motion of solid in three-dimensional space.

in which attention was focused on the motion of the fluid through a specific volume. In this book, we shall restrict our discussion to formulations of the Lagrangian type.

The formulation of incremental theories for nonlinear analysis begins by dividing the loading path of the solid into a number of equilibrium configurations. As shown in Figure 1.1, three configurations of the solid can be conceived in terms of a stationary Cartesian coordinate system: the initial undeformed configuration C_0, the last *known* configuration C_1, and the current *unknown* configuration C_2. It is assumed that all the state variables such as stresses, strains, and displacements, together with the loading history, are known up to the C_1 configuration. The problem is then to formulate an incremental theory for determining all the state variables of the solid in the current deformed C_2 configuration, assuming that the external loadings acting on the solid at C_1 have been increased by a small increment to those at C_2. The step characterizing the deformation process of the solid from the C_1 to the C_2 configuration will be referred to as an *incremental step*. While the deformations of the solid within the incremental step from C_1 to C_2 are assumed to be generally small, the accumulated deformations of the solid from C_0 to C_2 can be arbitrarily large.

Depending on which known configuration is selected as the reference state for establishing the governing equilibrium equations of the solid at the current configuration C_2, two types of Lagrangian formulation can be distinguished. In the *UL formulation*, the last calculated configuration C_1 is selected as the reference state, whereas in the *TL formulation*, the initial undeformed configuration C_0 is used for the same purpose. In this book, we shall adopt primarily the UL formulation for most applications, for its higher efficiency in dealing with the beam-type structures in terms of the computational effort required, in comparison with the TL formulation (Bathe and Bolourchi, 1979).

The tensor notation and nomenclature used by Bathe *et al.* (1975) will be adopted with slight modifications, which have been shown to be effective in the formulation of incremental theories for nonlinear problems. In addition, Einstein's summation will be assumed for each tensorial term with a repeated dummy index. Namely, whenever the same index appears twice in a mathematical term, the index is to be given all possible values and the results added together.

The description of the motion of a solid is based on the three configurations C_0, C_1, and C_2. Both left superscript and subscript of a symbol are used to denote these configurations. A left superscript denotes the configuration in which the quantity occurs. The absence of such a superscript indicates that the quantity is an increment between C_1 and C_2. A left subscript denotes the configuration with respect to which the quantity is measured. However, if the quantity under consideration occurs in the same configuration in which it is measured, the left subscript may be omitted, for instance, the Cauchy stresses at C_2 may be denoted by ${}^2\tau_{ij}$ or ${}^2_2\tau_{ij}$, the surface tractions at C_2 may be denoted by ${}^2t_{ij}$ or ${}^2_2t_{ij}$, and so on.

When a solid changes its configuration from C_0 through C_1 to C_2 under some physical actions (Figure 1.1), we assume that the change is continuous, namely, the neighborhoods are changed into neighborhoods. We shall denote the volume of the solid in the three configurations by ${}^0V$, ${}^1V$, ${}^2V$, the surface by ${}^0A$, ${}^1A$, ${}^2A$, and the specific mass by ${}^0\rho$, ${}^1\rho$, ${}^2\rho$. Further, we shall denote the coordinates of a generic material point P within the solid in the three configurations C_0, C_1, and C_2 by $({}^0x_1, {}^0x_2, {}^0x_3)$, $({}^1x_1, {}^1x_2, {}^1x_3)$, and $({}^2x_1, {}^2x_2, {}^2x_3)$, respectively, where the right subscripts refer to the coordinate axes in Figure 1.1. The notation for the displacements of point P can be defined in a similar way. For instance, we shall use the notation $({}^1u_1, {}^1u_2, {}^1u_3)$ and $({}^2u_1, {}^2u_2, {}^2u_3)$ to denote the total displacements of point P at C_1 and C_2, respectively. Accordingly, the coordinates of point P at C_1 and C_2 can be written as follows:

$$ {}^1x_i = {}^0x_i + {}^1u_i \quad (i = 1, 2, 3) \tag{1.1} $$

$$ {}^2x_i = {}^0x_i + {}^2u_i \quad (i = 1, 2, 3) \tag{1.2} $$

and the displacement increments of point P from C_1 to C_2 are

$$ u_i = {}^2u_i - {}^1u_i \quad (i = 1, 2, 3) \tag{1.3} $$

With the present notation for the coordinates and displacements, we are ready to summarize some of the most useful strain and stress tensors for the incremental nonlinear analysis in the following chapters. Throughout the text, new symbols will be defined where they first appear.

1.3 Definition of Strains

In principle, various combinations of strains and stresses can be employed in the analysis of geometric nonlinear problems for solids or structures. However, if the objective is to establish an effective procedure for general finite element analysis, then only a few combinations of strains and stresses can be considered. To this end, we shall restrict our discussion to the Green–Lagrange strain tensors and infinitesimal strain tensors in this section, and the corresponding second Piola-Kirchhoff stress tensors and Cauchy stress tensors in the section to follow, as they are the most convenient strain and stress combinations to formulations of the Lagrangian type. For the case where the use of "Green–Lagrange" may result in the composition of too many words in a single technical term, "Green" will be used as a shorthand.

1.3.1 Green–Lagrange strain tensor

With reference to Figure 1.1, a line element PQ is of length ${}^0ds$ in the C_0 configuration, of length ${}^1ds$ in the C_1 configuration, and of length ${}^2ds$ in the C_2 configuration. The *Green–Lagrange strain tensors* ${}^2_0\epsilon_{ij}$ and ${}^1_0\epsilon_{ij}$ at the C_2 and C_1 configurations, respectively, with respect to the C_0 configuration can be defined as follows:

$$2{}^2_0\epsilon_{ij}d\,{}^0x_id\,{}^0x_j = ({}^2ds)^2 - ({}^0ds)^2 \tag{1.4}$$

$$2{}^1_0\epsilon_{ij}d\,{}^0x_id\,{}^0x_j = ({}^1ds)^2 - ({}^0ds)^2 \tag{1.5}$$

in which

$$({}^2ds)^2 = d\,{}^2x_id\,{}^2x_i \tag{1.6}$$

$$({}^1ds)^2 = d\,{}^1x_id\,{}^1x_i \tag{1.7}$$

$$({}^0ds)^2 = d\,{}^0x_id\,{}^0x_i \tag{1.8}$$

Since

$$d\,{}^2x_i = \frac{\partial\,{}^2x_i}{\partial\,{}^0x_j}d\,{}^0x_j \tag{1.9}$$

$$d\,{}^1x_i = \frac{\partial\,{}^1x_i}{\partial\,{}^0x_j}d\,{}^0x_j \tag{1.10}$$

we can write

$$({}^2ds)^2 = \frac{\partial\,{}^2x_k}{\partial\,{}^0x_i}\frac{\partial\,{}^2x_k}{\partial\,{}^0x_j}d\,{}^0x_id\,{}^0x_j \tag{1.11}$$

$$({}^1ds)^2 = \frac{\partial\,{}^1x_k}{\partial\,{}^0x_i}\frac{\partial\,{}^1x_k}{\partial\,{}^0x_j}d\,{}^0x_id\,{}^0x_j \tag{1.12}$$

Making use of equations (1.8), (1.11),and (1.12), we can derive from equations (1.4) and (1.5) the following expressions for the Green–Lagrange strain tensors ${}^2_0\epsilon_{ij}$ and ${}^1_0\epsilon_{ij}$:

$${}^2_0\epsilon_{ij} = \frac{1}{2}\left(\frac{\partial\,{}^2x_k}{\partial\,{}^0x_i}\frac{\partial\,{}^2x_k}{\partial\,{}^0x_j} - \delta_{ij}\right) \tag{1.13}$$

$${}^1_0\epsilon_{ij} = \frac{1}{2}\left(\frac{\partial\,{}^1x_k}{\partial\,{}^0x_i}\frac{\partial\,{}^1x_k}{\partial\,{}^0x_j} - \delta_{ij}\right) \tag{1.14}$$

where δ_{ij} denotes the Kronecker delta. For $i = j$, $\delta_{ij} = 1$ and for $i \neq j$, $\delta_{ij} = 0$. By the use of equations (1.1) and (1.2), we can further express the strains in terms of the displacement components as follows:

$$ {}_0^2\epsilon_{ij} = \frac{1}{2}\left(\frac{\partial\,{}^2u_i}{\partial\,{}^0x_j} + \frac{\partial\,{}^2u_j}{\partial\,{}^0x_i} + \frac{\partial\,{}^2u_k}{\partial\,{}^0x_i}\frac{\partial\,{}^2u_k}{\partial\,{}^0x_j}\right) \tag{1.15} $$

$$ {}_0^1\epsilon_{ij} = \frac{1}{2}\left(\frac{\partial\,{}^1u_i}{\partial\,{}^0x_j} + \frac{\partial\,{}^1u_j}{\partial\,{}^0x_i} + \frac{\partial\,{}^1u_k}{\partial\,{}^0x_i}\frac{\partial\,{}^1u_k}{\partial\,{}^0x_j}\right) \tag{1.16} $$

where it is noted that the symmetry of the strain tensor holds, i.e., ${}_0^2\epsilon_{ij} = {}_0^2\epsilon_{ji}$ and ${}_0^1\epsilon_{ij} = {}_0^1\epsilon_{ji}$.

Using unbridged notation, we shall replace the C_2 displacements $({}^2u_1,\ {}^2u_2,\ {}^2u_3)$ by $({}^2u,\ {}^2v,\ {}^2w)$, the C_2 coordinates $({}^2x_1,\ {}^2x_2,\ {}^2x_3)$ by $({}^2x,\ {}^2y,\ {}^2z)$, the C_1 displacements $({}^1u_1,\ {}^1u_2,\ {}^1u_3)$ by $({}^1u,\ {}^1v,\ {}^1w)$, the C_1 coordinates $({}^1x_1,\ {}^1x_2,\ {}^1x_3)$ by $({}^1x,\ {}^1y,\ {}^1z)$, and the C_0 coordinates $({}^0x_1,\ {}^0x_2,\ {}^0x_3)$ by $({}^0x,\ {}^0y,\ {}^0z)$. Therefore, the components of the Green–Lagrange strain tensor ${}_0^2\epsilon_{ij}$ at C_2 can be written explicitly as follows:

$$ {}_0^2\epsilon_{xx} = \frac{\partial\,{}^2u}{\partial\,{}^0x} + \frac{1}{2}\left[\left(\frac{\partial\,{}^2u}{\partial\,{}^0x}\right)^2 + \left(\frac{\partial\,{}^2v}{\partial\,{}^0x}\right)^2 + \left(\frac{\partial\,{}^2w}{\partial\,{}^0x}\right)^2\right] \tag{1.17a} $$

$$ {}_0^2\epsilon_{yy} = \frac{\partial\,{}^2v}{\partial\,{}^0y} + \frac{1}{2}\left[\left(\frac{\partial\,{}^2u}{\partial\,{}^0y}\right)^2 + \left(\frac{\partial\,{}^2v}{\partial\,{}^0y}\right)^2 + \left(\frac{\partial\,{}^2w}{\partial\,{}^0y}\right)^2\right] \tag{1.17b} $$

$$ {}_0^2\epsilon_{zz} = \frac{\partial\,{}^2w}{\partial\,{}^0z} + \frac{1}{2}\left[\left(\frac{\partial\,{}^2u}{\partial\,{}^0z}\right)^2 + \left(\frac{\partial\,{}^2v}{\partial\,{}^0z}\right)^2 + \left(\frac{\partial\,{}^2w}{\partial\,{}^0z}\right)^2\right] \tag{1.17c} $$

$$ 2{}_0^2\epsilon_{xy} = \frac{\partial\,{}^2u}{\partial\,{}^0y} + \frac{\partial\,{}^2v}{\partial\,{}^0x} + \frac{\partial\,{}^2u}{\partial\,{}^0x}\frac{\partial\,{}^2u}{\partial\,{}^0y} + \frac{\partial\,{}^2v}{\partial\,{}^0x}\frac{\partial\,{}^2v}{\partial\,{}^0y} + \frac{\partial\,{}^2w}{\partial\,{}^0x}\frac{\partial\,{}^2w}{\partial\,{}^0y} \tag{1.17d} $$

$$ 2{}_0^2\epsilon_{yz} = \frac{\partial\,{}^2v}{\partial\,{}^0z} + \frac{\partial\,{}^2w}{\partial\,{}^0y} + \frac{\partial\,{}^2u}{\partial\,{}^0y}\frac{\partial\,{}^2u}{\partial\,{}^0z} + \frac{\partial\,{}^2v}{\partial\,{}^0y}\frac{\partial\,{}^2v}{\partial\,{}^0z} + \frac{\partial\,{}^2w}{\partial\,{}^0y}\frac{\partial\,{}^2w}{\partial\,{}^0z} \tag{1.17e} $$

$$ 2{}_0^2\epsilon_{zx} = \frac{\partial\,{}^2w}{\partial\,{}^0x} + \frac{\partial\,{}^2u}{\partial\,{}^0z} + \frac{\partial\,{}^2u}{\partial\,{}^0z}\frac{\partial\,{}^2u}{\partial\,{}^0x} + \frac{\partial\,{}^2v}{\partial\,{}^0z}\frac{\partial\,{}^2v}{\partial\,{}^0x} + \frac{\partial\,{}^2w}{\partial\,{}^0z}\frac{\partial\,{}^2w}{\partial\,{}^0x} \tag{1.17f} $$

$$ {}_0^2\epsilon_{yx} = {}_0^2\epsilon_{xy},\ {}_0^2\epsilon_{zy} = {}_0^2\epsilon_{yz},\ {}_0^2\epsilon_{xz} = {}_0^2\epsilon_{zx} \tag{1.17g} $$

and those of the tensor ${}^1_0\epsilon_{ij}$ at C_1 can be obtained by switching the left superscript in each symbol of equation (1.17) from "2" to "1".

1.3.2 *Green strain increment tensor*

In an incremental formulation, the difference between the two strains ${}^2_0\epsilon_{ij}$ and ${}^1_0\epsilon_{ij}$ is defined as the *Green strain increment tensor* ${}_0\epsilon_{ij}$:

$$ {}_0\epsilon_{ij} = {}^2_0\epsilon_{ij} - {}^1_0\epsilon_{ij} \tag{1.18} $$

which is equivalent to

$$ 2{}_0\epsilon_{ij} d\,{}^0x_i d\,{}^0x_j = ({}^2ds)^2 - ({}^1ds)^2 \tag{1.19} $$

By the substitution of equations (1.15) and (1.16) into (1.18), we can obtain

$$ {}_0\epsilon_{ij} = {}_0e_{ij} + {}_0\eta_{ij} \tag{1.20} $$

where the linear components ${}_0e_{ij}$ are

$$ {}_0e_{ij} = \frac{1}{2}\left(\frac{\partial u_i}{\partial\,{}^0x_j} + \frac{\partial u_j}{\partial\,{}^0x_i} + \frac{\partial\,{}^1u_k}{\partial\,{}^0x_i}\frac{\partial u_k}{\partial\,{}^0x_j} + \frac{\partial u_k}{\partial\,{}^0x_i}\frac{\partial\,{}^1u_k}{\partial\,{}^0x_j}\right) \tag{1.21} $$

$$ {}_0\eta_{ij} = \frac{1}{2}\frac{\partial u_k}{\partial\,{}^0x_i}\frac{\partial u_k}{\partial\,{}^0x_j} \tag{1.22} $$

It should be noted that in the preceding two equations, the quantity ${}^1u_i$ denotes the total displacements of the solid from C_0 to C_1 and the quantity u_i the displacement increments from C_1 to C_2.

For the special case of linear analysis, there will be no initial displacements ${}^1u_i$, implying that the two configurations C_1 and C_0 are identical, i.e., $C_1 \equiv C_0$, and that the Green–Lagrange strain tensor at C_1 vanishes, i.e., ${}^1_0\epsilon_{ij} = 0$. Further, it is assumed that the displacement increments u_i are so small that the squares and products of their first derivatives can be neglected. For such problems, the distinction between the two configurations C_1 and C_2 disappears, since it is immaterial whether the derivatives are calculated at the position of a point before or after deformation. By neglecting the nonlinear components and the initial displacements, the Green strain increment tensor ${}_0\epsilon_{ij}$ reduces to the *infinitesimal strain tensor* ${}_0e_{ij}$:

$$ {}_0e_{ij} = \frac{1}{2}\left(\frac{\partial u_i}{\partial x_j} + \frac{\partial u_j}{\partial x_i}\right) \tag{1.23} $$

where all the left superscripts for the displacements and coordinates have been dropped. In unabridged form, we have

$$ {}_0e_{xx} = \frac{\partial u}{\partial x}, \quad {}_0e_{yy} = \frac{\partial v}{\partial y}, \quad {}_0e_{zz} = \frac{\partial w}{\partial z} \tag{1.24a–c} $$

$$ {}_0e_{xy} = \frac{1}{2}\left(\frac{\partial u}{\partial y} + \frac{\partial v}{\partial x}\right) = {}_0e_{yx} \tag{1.24d} $$

$$ {}_0e_{yz} = \frac{1}{2}\left(\frac{\partial v}{\partial z} + \frac{\partial w}{\partial y}\right) = {}_0e_{zy} \tag{1.24e} $$

$$ {}_0e_{zx} = \frac{1}{2}\left(\frac{\partial w}{\partial x} + \frac{\partial u}{\partial z}\right) = {}_0e_{xz} \tag{1.24f} $$

The above are also known as the elasticity shear strain. In practical usage, the elasticity shear strain ${}_0e_{ij}$ $(i \neq j)$ doubled, i.e., $2\,{}_0e_{ij}$, are called the *engineering shear strains* γ_{ij}, which represent the decrease in the right angle of a square of the material initially formed by the sides parallel to the x_i- and x_j-axes.

However, the concept of infinitesimal strain tensor is not restricted to linear analysis as described above. Let us consider the general case of a solid moving from C_0 to C_1 by a number of incremental steps and then from C_1 to C_2 by an incremental step. Though the accumulated displacements ${}^1u_i$ of the solid from C_0 to C_1 can be arbitrarily large, the incremental displacements u_i within the incremental step from C_1 to C_2 are small by definition. As a result, if we refer the strains to the C_2 configuration, we can define the *Euler strain tensor* ${}_2\epsilon_{ij}$ for the solid at C_2 with reference to the axes at the same C_2 as follows:

$$ 2\,{}_2\epsilon_{ij} d\,{}^2x_i d\,{}^2x_j = ({}^2ds)^2 - ({}^1ds)^2 \tag{1.25} $$

Substituting equations (1.6) and (1.7) into equation (1.25) and noting that

$$ d\,{}^1x_i = \frac{\partial\,{}^1x_i}{\partial\,{}^2x_j} d\,{}^2x_j \tag{1.26} $$

we obtain

$$ {}_2\epsilon_{ij} = \frac{1}{2}\left(\delta_{ij} - \frac{\partial\, {}^1x_k}{\partial\, {}^2x_i}\frac{\partial\, {}^1x_k}{\partial\, {}^2x_j}\right) \tag{1.27} $$

From equations (1.1)–(1.3), we have

$$ {}^1x_i = {}^2x_i - u_i \quad (i = 1, 2, 3) \tag{1.28} $$

which can be substituted into equation (1.27) to yield

$$ {}_2\epsilon_{ij} = \frac{1}{2}\left(\frac{\partial u_i}{\partial\, {}^2x_j} + \frac{\partial u_j}{\partial\, {}^2x_i} + \frac{\partial u_k}{\partial\, {}^2x_i}\frac{\partial u_k}{\partial\, {}^2x_j}\right) \tag{1.29} $$

For the present case, the linear part of the Euler strain tensor ${}_2\epsilon_{ij}$ is called the infinitesimal strain tensor ${}_2e_{ij}$:

$$ {}_2e_{ij} = \frac{1}{2}\left(\frac{\partial u_i}{\partial\, {}^2x_j} + \frac{\partial u_j}{\partial\, {}^2x_i}\right) \tag{1.30} $$

which is identical in form to the infinitesimal strain tensor ${}_0e_{ij}$ given in equation (1.23), except that the reference configuration is changed to C_2.

1.3.3 *Updated Green strain increment tensor*

The Green–Lagrange strain tensor ${}^2_0\epsilon_{ij}$ introduced above for the solid at C_2 has been referred to as the initial configuration C_0, which is useful in the TL formulation. When using the UL formulation, we need to refer to the strain tensor of the solid at C_2 to the C_1 configuration instead. Such a strain tensor, denoted as ${}_1\epsilon_{ij}$ or ${}^2_1\epsilon_{ij}$, is called the *updated Green strain increment tensor* (Washizu, 1982), which can be defined as follows:

$$ 2\;{}_1\epsilon_{ij} d\,{}^1x_i d\,{}^1x_j = ({}^2ds)^2 - ({}^1ds)^2 \tag{1.31} $$

Following the procedure from equations (1.25) to (1.29), it is easy to derive the strain increments ${}_1\epsilon_{ij}$ as

$$ {}_1\epsilon_{ij} = \frac{1}{2}\left(\frac{\partial u_i}{\partial\, {}^1x_j} + \frac{\partial u_j}{\partial\, {}^1x_i} + \frac{\partial u_k}{\partial\, {}^1x_i}\frac{\partial u_k}{\partial\, {}^1x_j}\right) \tag{1.32} $$

Similarly, the strain increments ${}_1\epsilon_{ij}$ can be decomposed into two parts as

$$ {}_1\epsilon_{ij} = {}_1e_{ij} + {}_1\eta_{ij} \tag{1.33} $$

where the linear components are

$$ {}_1e_{ij} = \frac{1}{2}\left(\frac{\partial u_i}{\partial\, {}^1x_j} + \frac{\partial u_j}{\partial\, {}^1x_i}\right) \tag{1.34} $$

and

$$ {}_1\eta_{ij} = \frac{1}{2}\frac{\partial u_k}{\partial\, {}^1x_i}\frac{\partial u_k}{\partial\, {}^1x_j} \tag{1.35} $$

As a matter of fact, the possibility exists for the reference configuration of the strain tensors to change from one to the other. For instance, in certain situations, we may need to transform the updated Green strain increment tensor ${}_1\epsilon_{ij}$, which is measured at C_1, to the Green strain increment tensor ${}_0\epsilon_{ij}$, which is measured at C_0, and vice versa. To derive the transformation for the two strain tensors ${}_1\epsilon_{ij}$ and ${}_0\epsilon_{ij}$, let us consider the definitions as given in equations (1.31) and (1.19). By the substitution of equation (1.10), we can obtain the following transformation for the strains ${}_1\epsilon_{ij}$ and ${}_0\epsilon_{ij}$ from these definitions:

$$ {}_0\epsilon_{ij} = \frac{\partial\, {}^1x_p}{\partial\, {}^0x_i}\frac{\partial\, {}^1x_q}{\partial\, {}^0x_j}\, {}_1\epsilon_{pq} \tag{1.36} $$

$$ {}_1\epsilon_{ij} = \frac{\partial\, {}^0x_p}{\partial\, {}^1x_i}\frac{\partial\, {}^0x_q}{\partial\, {}^1x_j}\, {}_0\epsilon_{pq} \tag{1.37} $$

These relations represent the transformations for tensors of the second order, such as strains and stresses, from one reference configuration to the other. Similar transformations can be established for the other strain tensors for switching the reference configuration. In the analysis of large strain problems by the UL formulation, where all the quantities have been referred to the moving configuration C_1, we need to refer all the strains and stresses to a common fixed configuration, such as the C_0 configuration, for the prescription of the constitutive laws, and then to derive the constitutive relations for each incremental step from the ones established for the common fixed configuration. Such an effect with regard to the prescription of constitutive laws has been demonstrated to be significant for large-strain problems (Yang and Leu, 1990). Relations such as those given in equations (1.36) and (1.37) provide the basis for performing the necessary transformation between different reference configurations.

1.4 Definition of Stresses

Although various kinds of stress measures have been proposed by elasticians and mathematicians in the past, only the stress measures that are conjugate in terms of energy to the strain measures presented in the preceding section will be summarized herein. Such stress and strain measures are essential to the Lagrangian formulations to be presented in later chapters. In this section, we shall discuss first the *second Piola–Kirchhoff stress tensors*. For cases where the use of "second Piola–Kirchhoff" may result in technical terms that are too cumbersome to be adopted in an efficient communication, "Kirchhoff" will be used as a shorthand.

1.4.1 *Second Piola–Kirchhoff stress tensor*

To define the Kirchhoff stress tensor, consider a point P enclosed by an infinitesimal rectangular parallelepiped at the initial configuration C_0 by the following six surfaces:

$$ {}^0x_i = \text{const.}, \quad {}^0x_i + d\,{}^0x_i = \text{const.}, \quad (i = 1, 2, 3) \tag{1.38} $$

As the solid containing the point P moves from C_0 to C_1 and then to C_2, we observe that this infinitesimal rectangular parallelepiped will be deformed into a strained infinitesimal parallelepiped (no longer rectangular) at the C_1 and C_2 configurations (Figure 1.2). The Kirchhoff stresses are defined as the internal forces per unit area acting along the normal and two tangential directions of each of the side surfaces of the parallelepiped at the deformed configurations (Washizu, 1982). Only the stresses acting on one side of the parallelepiped are shown in Figure 1.2. The Kirchhoff stress tensors at the C_2 and C_1

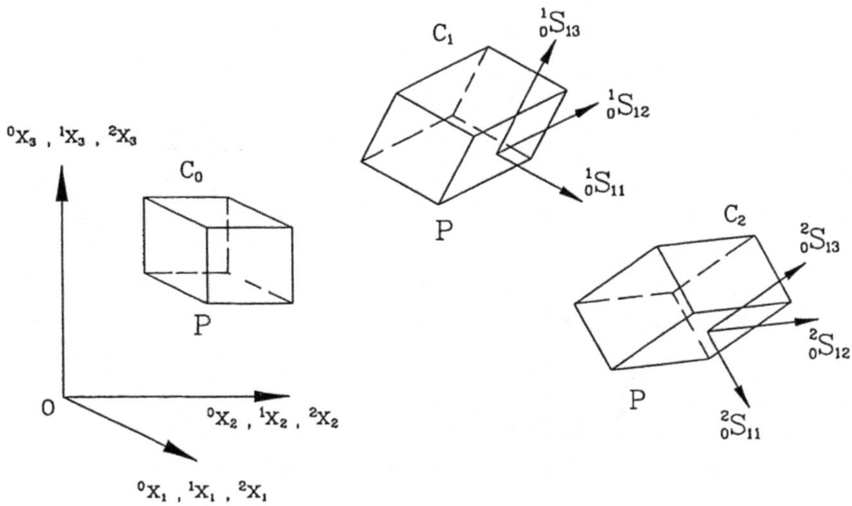

Fig. 1.2. Definition of second Piola–Kirchhoff stress tensor.

configurations are denoted by ${}^2_0S_{ij}$ and ${}^1_0S_{ij}$, respectively. In an incremental analysis, the second Piola–Kirchhoff stress tensor ${}^2_0S_{ij}$ at C_2 can be decomposed as follows:

$$ {}^2_0S_{ij} = {}^1_0S_{ij} + {}_0S_{ij} \tag{1.39}$$

where ${}_0S_{ij}$ is referred to as the *Kirchhoff stress increment tensor.*

1.4.2 *Cauchy stress tensor*

The *Cauchy stress tensor* is characterized by the fact that it is always expressed with respect to the configuration in which the stresses occur. It is also known as the *Euler stress tensor.* The physical meaning for the Cauchy stresses can be easily appreciated from the graphical illustration shown in Figure1.3. Consider again the motion of a generic point P, which is enclosed at C_1 by an infinitesimal rectangular parallelepiped with the following six surfaces:

$$ {}^1x_i = \text{const.}, \quad {}^1x_i + d\,{}^1x_i = \text{const.} \quad (i = 1, 2, 3) \tag{1.40}$$

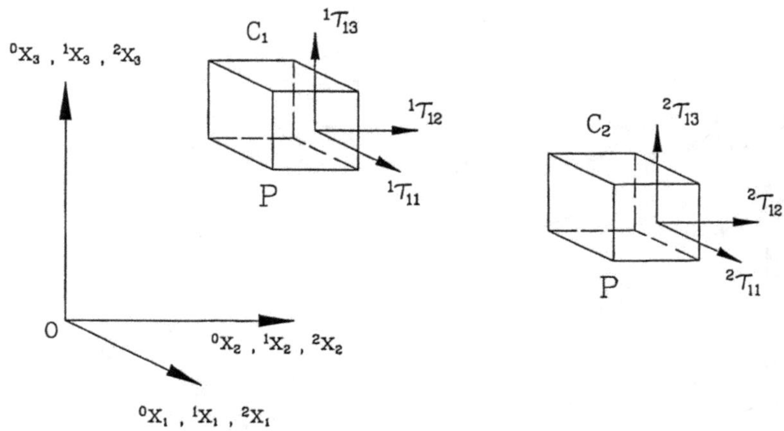

Fig. 1.3. Definition of Cauchy stress tensor.

and at C_2 by another infinitesimal rectangular parallelepiped with the following six surfaces:

$$^2x_i = \text{const.}, \quad ^2x_i + d\,^2x_i = \text{const.} \quad (i = 1, 2, 3) \tag{1.41}$$

The Cauchy stresses $^1\tau_{ij}$ and $^2\tau_{ij}$ are defined as the internal forces per unit area acting along the normal and two tangential directions of each of the side surfaces of the parallelepipeds defined by equations (1.40) and (1.41) for the C_1 and C_2 configurations, respectively.

The Cauchy stress tensor is a natural physical concept but is inconvenient for nonlinear analysis, in which we must relate stresses to strains. If the strains were referred to the initial configuration of a continuum, it would be necessary to define the stresses with respect to the same configuration. In this regard, the Cauchy stress tensor $^2\tau_{ij}$ can be related to the second Piola–Kirchhoff stress tensor $^2_0S_{ij}$ as follows (Fung, 1965):

$$^2_0S_{ij} = \frac{^0\rho}{^2\rho}\frac{\partial\,^0x_i}{\partial\,^2x_p}\frac{\partial\,^0x_j}{\partial\,^2x_q}\,^2\tau_{pq} \tag{1.42}$$

$$^2\tau_{ij} = \frac{^2\rho}{^0\rho}\frac{\partial\,^2x_i}{\partial\,^0x_p}\frac{\partial\,^2x_j}{\partial\,^0x_q}\,^2_0S_{pq} \tag{1.43}$$

where $^0\rho$ and $^2\rho$ represent the mass densities of the material at C_0 and C_2, respectively.

1.4.3 *Updated Kirchhoff stress tensor*

It is possible to define a third kind of stress tensor called the *updated Kirchhoff stress tensor*. Consider the infinitesimal rectangular parallelepiped containing the point P at C_1 enclosed by the six surfaces given by equation (1.40) (see also Figure 1.4). The Cauchy stresses acting at this parallelepiped are denoted by ${}^1\tau_{ij}$. As long as the solid containing point P moves from C_1 to C_2, the rectangular parallelepiped will be deformed into an infinitesimal parallelepiped that is no longer rectangular at C_2. In this case, the coordinates ${}^1x_i$ of the solid at C_1 are employed as the material coordinate system for the solid at C_2. The updated Kirchhoff stresses ${}^2_1S_{ij}$ are defined as the internal forces per unit area acting along the normal and two tangential directions of each of the side surfaces of the parallelepiped at C_2 (Washizu, 1982). In an incremental analysis, the updated Kirchhoff stresses ${}^2_1S_{ij}$ can be decomposed as follows:

$$ {}^2_1S_{ij} = {}^1\tau_{ij} + {}_1S_{ij} \tag{1.44} $$

where ${}^1\tau_{ij}$ should be recognized as the Kirchhoff stresses acting at and referred to as the C_1 configuration, i.e., ${}^1\tau_{ij} = {}^1_1S_{ij}$, and ${}_1S_{ij}$ will be called the *updated Kirchhoff stress increment tensor* (Washizu, 1982).

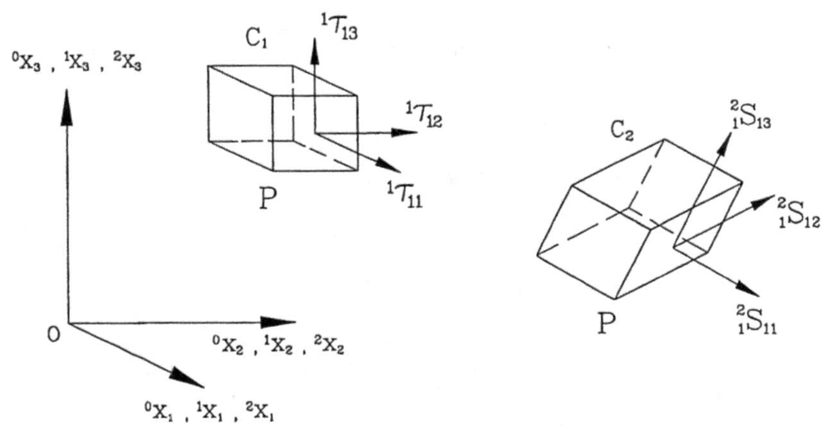

Fig. 1.4. Definition of updated Kirchhoff stress tensor.

1.4.4 Transformation rules

The Cauchy stress tensor ${}^{2}\tau_{ij}$ at C_2 can be related to the updated Kirchhoff stress tensor ${}^{2}_{1}S_{ij}$ by the following formulas:

$$
{}^{2}_{1}S_{ij} = \frac{{}^{1}\rho}{{}^{2}\rho} \frac{\partial\, {}^{1}x_i}{\partial\, {}^{2}x_p} \frac{\partial\, {}^{1}x_j}{\partial\, {}^{2}x_q} {}^{2}\tau_{pq} \tag{1.45}
$$

$$
{}^{2}\tau_{ij} = \frac{{}^{2}\rho}{{}^{1}\rho} \frac{\partial\, {}^{2}x_i}{\partial\, {}^{1}x_p} \frac{\partial\, {}^{2}x_j}{\partial\, {}^{1}x_q} {}^{2}_{1}S_{pq} \tag{1.46}
$$

where ${}^{1}\rho$ represents the mass density of the material at C_1. Since the Cauchy stress tensor ${}^{2}\tau_{ij}$ is symmetric, equations (1.42) and (1.45) indicate that both the second Piola–Kirchhoff stress tensor ${}^{2}_{0}S_{ij}$ and the updated Kirchhoff stress tensor ${}^{2}_{1}S_{ij}$ are also symmetric.

The transformation between the second Piola–Kirchhoff stress tensors with different reference configurations is also possible. For instance, from equations (1.43) and (1.46), one can derive the following relation:

$$
{}^{2}_{0}S_{ij} = \frac{{}^{0}\rho}{{}^{1}\rho} \frac{\partial\, {}^{0}x_i}{\partial\, {}^{1}x_p} \frac{\partial\, {}^{0}x_j}{\partial\, {}^{1}x_q} {}^{2}_{1}S_{pq} \tag{1.47}
$$

Similarly, the following relation exists for the solid at C_1:

$$
{}^{1}_{0}S_{ij} = \frac{{}^{0}\rho}{{}^{1}\rho} \frac{\partial\, {}^{0}x_i}{\partial\, {}^{1}x_p} \frac{\partial\, {}^{0}x_j}{\partial\, {}^{1}x_q} {}^{1}\tau_{pq} \tag{1.48}
$$

where it should be recognized that ${}^{1}\tau_{ij} = {}^{1}_{1}S_{ij}$. Subtracting equation (1.48) from equation (1.47) yields the relation for the incremental stresses ${}_{0}S_{ij}$ and ${}_{1}S_{pq}$:

$$
{}_{0}S_{ij} = \frac{{}^{0}\rho}{{}^{1}\rho} \frac{\partial\, {}^{0}x_i}{\partial\, {}^{1}x_p} \frac{\partial\, {}^{0}x_j}{\partial\, {}^{1}x_q} {}_{1}S_{pq} \tag{1.49}
$$

or inversely,

$$
{}_{1}S_{ij} = \frac{{}^{1}\rho}{{}^{0}\rho} \frac{\partial\, {}^{1}x_i}{\partial\, {}^{0}x_p} \frac{\partial\, {}^{1}x_j}{\partial\, {}^{0}x_q} {}_{0}S_{pq} \tag{1.50}
$$

The preceding two relations are particularly useful in the calculation of the material coefficients for large strain problems in incremental nonlinear analyses based on the UL formulation (Yang and Leu, 1990).

Before concluding this section, let us turn to the relations among the mass densities ${}^0\rho$, ${}^1\rho$, and ${}^2\rho$ of the material at different configurations. For materials that obey the law of conversation of mass, which is always the case within the context of Newtonian mechanics, we can write

$$\int_{{}^2V} {}^2\rho d\,{}^2x_1 d\,{}^2x_2 d\,{}^2x_3 = \int_{{}^0V} {}^0\rho d\,{}^0x_1 d\,{}^0x_2 d\,{}^0x_3 \tag{1.51}$$

where ${}^2V$ is the volume occupied at C_2 by the material that occupied the volume ${}^0V$ at C_0. By a change of the variables of integration from ${}^2x_i$ to ${}^0x_i$, we have

$$\int_{{}^2V} {}^2\rho d\,{}^2x_1 d\,{}^2x_2 d\,{}^2x_3 = \int_{{}^0V} {}^2\rho {}_0^2J d\,{}^0x_1 d\,{}^0x_2 d\,{}^0x_3 \tag{1.52}$$

where the *Jacobian determinant* ${}_0^2J$ is defined as follows:

$${}_0^2J = \left|\frac{\partial\,{}^2x_i}{\partial\,{}^0x_j}\right| = \begin{vmatrix} \frac{\partial\,{}^2x_1}{\partial\,{}^0x_1} & \frac{\partial\,{}^2x_1}{\partial\,{}^0x_2} & \frac{\partial\,{}^2x_1}{\partial\,{}^0x_3} \\ \frac{\partial\,{}^2x_2}{\partial\,{}^0x_1} & \frac{\partial\,{}^2x_2}{\partial\,{}^0x_2} & \frac{\partial\,{}^2x_2}{\partial\,{}^0x_3} \\ \frac{\partial\,{}^2x_3}{\partial\,{}^0x_1} & \frac{\partial\,{}^2x_3}{\partial\,{}^0x_2} & \frac{\partial\,{}^2x_3}{\partial\,{}^0x_3} \end{vmatrix} \tag{1.53}$$

By the use of equations (1.51) and (1.52), we can show that the following relation holds for any material:

$${}^0\rho = {}^2\rho {}_0^2J \tag{1.54}$$

Consequently, equations (1.42) and (1.43) can be rewritten,

$${}_0^2S_{ij} = {}_0^2J \frac{\partial\,{}^0x_i}{\partial\,{}^2x_p} \frac{\partial\,{}^0x_j}{\partial\,{}^2x_q} {}^2\tau_{pq} \tag{1.55}$$

$${}^2\tau_{ij} = \frac{1}{{}_0^2J} \frac{\partial\,{}^2x_i}{\partial\,{}^0x_p} \frac{\partial\,{}^2x_j}{\partial\,{}^0x_q} {}_0^2S_{pq} \tag{1.56}$$

Similarly, by defining the Jacobian determinant ${}_1^2J$ as

$${}_1^2J = \left|\frac{\partial\,{}^2x_i}{\partial\,{}^1x_j}\right| = \begin{vmatrix} \frac{\partial\,{}^2x_1}{\partial\,{}^1x_1} & \frac{\partial\,{}^2x_1}{\partial\,{}^1x_2} & \frac{\partial\,{}^2x_1}{\partial\,{}^1x_3} \\ \frac{\partial\,{}^2x_2}{\partial\,{}^1x_1} & \frac{\partial\,{}^2x_2}{\partial\,{}^1x_2} & \frac{\partial\,{}^2x_2}{\partial\,{}^1x_3} \\ \frac{\partial\,{}^2x_3}{\partial\,{}^1x_1} & \frac{\partial\,{}^2x_3}{\partial\,{}^1x_2} & \frac{\partial\,{}^2x_3}{\partial\,{}^1x_3} \end{vmatrix} \tag{1.57}$$

the following relation remains valid:

$$ {}^{1}\rho = {}^{2}\rho^{2}_{1}J \tag{1.58}$$

It follows that equations (1.45) and (1.46) can be rewritten as

$$ {}^{2}_{1}S_{ij} = {}^{2}_{1}J \frac{\partial\, {}^{1}x_i}{\partial\, {}^{2}x_p} \frac{\partial\, {}^{1}x_j}{\partial\, {}^{2}x_q} {}^{2}\tau_{pq} \tag{1.59}$$

$$ {}^{2}\tau_{ij} = \frac{1}{{}^{2}_{1}J} \frac{\partial\, {}^{2}x_i}{\partial\, {}^{1}x_p} \frac{\partial\, {}^{2}x_j}{\partial\, {}^{1}x_q} {}^{2}_{1}S_{pq} \tag{1.60}$$

In the formulation of incremental theories for nonlinear problems, we need to select appropriately *conjugate* stress and strain measures. By "conjugate", we mean that the stress and strain measures selected with respect to a specified reference configuration can consistently represent the energy measure of the solid at the current configuration. Any expressions with stress and strain measures that are not appropriately conjugate are by themselves lacking a valid physical basis. One sure way of selecting appropriately conjugate stress and strain is to carry out the derivation of the principle of virtual displacements for the problem in question (Malvern, 1969). If the current deformed configuration C_2 is selected as the reference, it will be shown in Section 1.6 that the Cauchy stress tensor ${}^{2}\tau_{ij}$ and infinitesimal strain tensor ${}_{2}e_{ij}$ form a conjugate pair. By application of the transformation rules, we shall show in Section 1.7 that for the TL formulation, for which the reference configuration is the C_0 configuration, we may select the second Kirchhoff stress tensor ${}^{2}_{0}S_{ij}$ and Green–Lagrange strain tensor ${}^{2}_{0}\epsilon_{ij}$ as the conjugate pair, and that for the UL formulation, which is referred to as the C_1 configuration, we may select the updated Kirchhoff stress tensor ${}^{2}_{1}S_{ij}$ and updated Green strain increment tensor ${}_{1}\epsilon_{ij}$ as the possible conjugate pair. The selection of appropriate conjugate stress and strain measures is also important in the derivation of the material coefficients for incremental constitute laws (Yang and Leu, 1991).

1.5 Incremental Constitutive Laws

Constitutive laws, equilibrium equations, and compatibility conditions are the three most essential constituents that must be considered in analysis of the boundary-value problems. In the literature, a great number of constitutive laws have been proposed, which cover a wide spectrum of materials from elastic, viscoelastic, to plastic behaviors, and so on (Fung, 1965; Malvern, 1969). Our objective in this section is not to make a complete survey of all the constitutive laws available in the literature. Instead, we shall focus on the derivation of effective procedures for incorporation in the incremental nonlinear analysis of elastic structures based on the Lagrangian formulations.

In the derivation of finite element matrices for incremental nonlinear analysis, it is necessary to specify the stress–strain relations in an incremental form. For the TL formulation, the constitutive law can be expressed in terms of the Kirchhoff stress increment tensor ${}_0S_{ij}$ and Green strain increment tensor ${}_0\epsilon_{ij}$:

$$ {}_0S_{ij} = {}_0C_{ijkl}\ {}_0\epsilon_{kl} \tag{1.61}$$

where ${}_0C_{ijkl}$ denotes the incremental constitutive tensor with respect to C_0. In parallel, for the UL formulation, the constitutive law can be expressed in terms of the updated Kirchhoff stress increment tensor ${}_1S_{ij}$ and updated Green strain increment tensor ${}_1\epsilon_{ij}$:

$$ {}_1S_{ij} = {}_1C_{ijkl}\ {}_1\epsilon_{kl} \tag{1.62}$$

where ${}_1C_{ijkl}$ denotes the incremental constitutive tensor with respect to C_1.

Conventionally, no distinction has been made between the constitutive coefficients ${}_0C_{ijkl}$ and ${}_1C_{ijkl}$ used in equations (1.61) and (1.62), respectively, for the total and UL formulations. This implied that ${}_0C_{ijkl} = {}_1C_{ijkl}$, regardless of the difference in reference configurations, which is good only for *small-strain* problems, for which the strains of each member of the structure are deemed to be small throughout all the incremental steps. For *large-strain* problems, the load-deflection curves solved for the problem considered can be quite different, especially in the postbuckling response. To resolve the difference in the reference configurations for the total and UL formulations, it is required that the fourth-order transformation to be presented below be performed for converting the material coefficients used in one formulation to the other (Yang and Leu, 1990, 1991).

By substituting equations (1.61) and (1.62) into equation (1.49), along with the use of equation (1.36), we obtain the following fourth-order transformation for the constitutive tensors:

$$ {}_0C_{ijkl} = \frac{{}^0\rho}{{}^1\rho} \frac{\partial\, {}^0x_i}{\partial\, {}^1x_p} \frac{\partial\, {}^0x_j}{\partial\, {}^1x_q} \frac{\partial\, {}^0x_k}{\partial\, {}^1x_r} \frac{\partial\, {}^0x_l}{\partial\, {}^1x_s}\, {}_1C_{pqrs} \tag{1.63} $$

or inversely,

$$ {}_1C_{ijkl} = \frac{{}^1\rho}{{}^0\rho} \frac{\partial\, {}^1x_i}{\partial\, {}^0x_p} \frac{\partial\, {}^1x_j}{\partial\, {}^0x_q} \frac{\partial\, {}^1x_k}{\partial\, {}^0x_r} \frac{\partial\, {}^1x_l}{\partial\, {}^0x_s}\, {}_0C_{pqrs} \tag{1.64} $$

With the preceding two relations, only one set of coefficients, i.e., ${}_0C_{ijkl}$ or ${}_1C_{ijkl}$, needs to be prescribed. The other set of coefficients can be obtained simply by transformation. In doing so, we can therefore ensure that the material properties implied by the total and UL formulations are physically identical to each other, provided that the same step sizes are used in both formulations.

Another question in incremental nonlinear formulations pertains to the way the incremental material coefficients ${}_0C_{ijkl}$ and ${}_1C_{ijkl}$ are specified. Among the great number of alternatives existing in the literature, the simplest way is to define the material law in terms of the second Piola–Kirchhoff stresses ${}_0^2S_{ij}$ and Green–Lagrange strains ${}_0^2\epsilon_{ij}$, which are energetically conjugate, based on experimental data or by postulation,

$$ {}_0^2S_{ij} = f({}_0^2\epsilon_{ij}) \tag{1.65} $$

where f is a single-valued function. A schematic drawing of equation (1.65) is given in Figure 1.5. By dividing the motion of the solid from C_0 to C_2 into an infinite number of infinitesimal steps, we can derive from equation (1.65) the following differential law for the material at $C_1$for the infinitesimal step from C_1 to C_2 as follows:

$$ d{}_0^2S_{ij} = \; f'\left({}_0^1\epsilon_{ij}\right) d{}_0^2\epsilon_{ij} \tag{1.66} $$

where $f'({}_0^1\epsilon_{ij})$ denotes the tangent modulus of the stress–strain curve evaluated at C_1. If the step from C_1 to C_2 is not infinitesimal but of finite size, we may replace the differential quantities $d{}_0^2S_{ij}$ and $d{}_0^2\epsilon_{ij}$

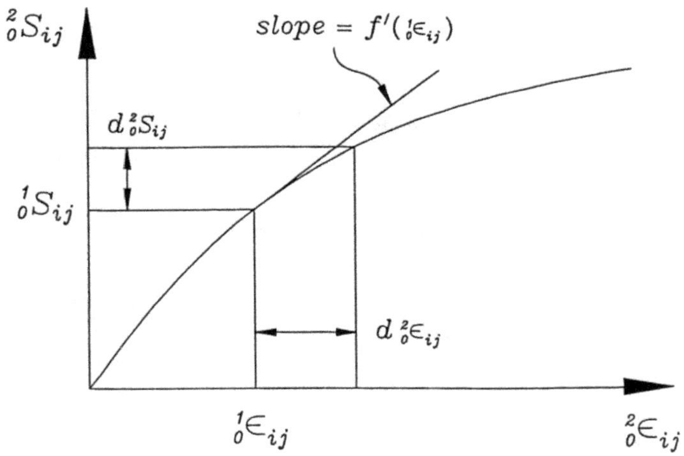

Fig. 1.5. Typical stress–strain curve.

by the stress and strain increments ${}_0S_{ij}$ and ${}_0\epsilon_{ij}$, respectively, and rewrite equation (1.66) as follows:

$$ {}_0S_{ij} = f'({}^1_0\epsilon_{ij})\,{}_0\epsilon_{ij} \tag{1.67} $$

A comparison of equation (1.67) with (1.61) reveals that the material coefficients ${}_0C_{ijkl}$ can be approximately taken as the tangent modulus $f'({}^1_0\epsilon_{ij})$ of the stress–strain curve. The errors introduced by this linearized approximation can be removed through a continuous updating of the material coefficients in an iterative process. Once the set of coefficients ${}_0C_{ijkl}$ is determined, the other set of coefficients ${}_1C_{ijkl}$ can be obtained by the fourth-order transformation given in equation (1.64)

1.6 Principle of Virtual Displacements

For a solid in equilibrium at the deformer configuration C_2, the internal (Cauchy) stresses ${}^2\tau_{ij}$ at any point of the solid should satisfy the equations of equilibrium at that particular configuration C_2 along the three axes using the Cartesian coordinates (Fung, 1965):

$$ \frac{\partial\, {}^2\tau_{ji}}{\partial\, {}^2x_j} + {}^2f_i = 0 \tag{1.68} $$

and

$$^2\tau_{ij} = {}^2\tau_{ji} \tag{1.69}$$

where ${}^2f_i$ denotes the components of body forces per unit volume referred to as the C_2 configuration. In unbridged notation, we have

$$\frac{\partial\,{}^2\tau_{xx}}{\partial\,{}^2x} + \frac{\partial\,{}^2\tau_{yx}}{\partial\,{}^2y} + \frac{\partial\,{}^2\tau_{zx}}{\partial\,{}^2z} + {}^2f_x = 0 \tag{1.70}$$

$$\frac{\partial\,{}^2\tau_{xy}}{\partial\,{}^2x} + \frac{\partial\,{}^2\tau_{yy}}{\partial\,{}^2y} + \frac{\partial\,{}^2\tau_{zy}}{\partial\,{}^2z} + {}^2f_y = 0 \tag{1.71}$$

$$\frac{\partial\,{}^2\tau_{xz}}{\partial\,{}^2x} + \frac{\partial\,{}^2\tau_{yz}}{\partial\,{}^2y} + \frac{\partial\,{}^2\tau_{zz}}{\partial\,{}^2z} + {}^2f_z = 0 \tag{1.72}$$

and

$$^2\tau_{yz} = {}^2\tau_{zy}, \quad {}^2\tau_{zx} = {}^2\tau_{xz}, \quad {}^2\tau_{xy} = {}^2\tau_{yx} \tag{1.73a–c}$$

The equations of equilibrium must be satisfied at all points throughout the volume of the solid considered.

In general, the surface area ${}^2S$ of the solid can be divided into two parts: the part ${}^2S_\tau$ over which the external surface tractions are prescribed and the part ${}^2S_u$ over which the displacements are prescribed. The boundary conditions with prescribed tractions are called the *natural* or *mechanical boundary conditions*, and those with prescribed displacements are called the *geometric* or *rigid boundary conditions*. In an exact analysis, it is required that both types of boundary conditions are satisfied by the solid at C_2.

The natural boundary conditions can be given as follows:

$$^2\tau_{ji}n_j = {}^2t_i \quad \text{on} \quad {}^2S_\tau \tag{1.74}$$

where ${}^2t_i$ denotes the prescribed surface tractions per unit area and n_j the components of the direction cosines of the unit normal ν to the surface of the solid at the point under consideration. In unbridged notation, equation (1.74) can be written as

$$\begin{aligned} {}^2\tau_{xx}l + {}^2\tau_{yx}m + {}^2\tau_{zx}n &= {}^2t_x \\ {}^2\tau_{xy}l + {}^2\tau_{yy}m + {}^2\tau_{zy}n &= {}^2t_y \\ {}^2\tau_{xz}l + {}^2\tau_{yz}m + {}^2\tau_{zz}n &= {}^2t_z \end{aligned} \tag{1.75a–c}$$

on ${}^2S_\tau$, where the direction cosines l, m, and n are defined as

$$l = \cos({}^2x, \nu), \quad m = \cos({}^2y, \nu), \quad n = \cos({}^2z, \nu) \qquad (1.76\text{a–c})$$

A stress field that satisfies the equations of equilibrium (1.68) in the interior of the solid and the boundary conditions (1.74) for which the surface tractions are prescribed is called a *statically admissible* stress field.

On the other hand, when a displacement field satisfies all the boundary conditions with prescribed displacements and possesses continuous first derivatives in the interior of the solid, it is called a *kinematically admissible* displacement field. Since the virtual displacements to be considered are *additional* displacements from the equilibrium configuration, they should be zero on the boundary wherever the actual displacements are prescribed. By letting δu_i denote the infinitesimal virtual displacements from the equilibrium configuration C_2, the rigid boundary conditions can be expressed as follows:

$$\delta u_i = 0 \quad \text{on } {}^2S_u \qquad (1.77)$$

or, in unabridged notation,

$$\delta u_x = 0, \quad \delta u_y = 0, \quad \delta u_z = 0, \quad \text{on } {}^2S_u \qquad (1.78)$$

If the loadings are assumed to remain unchanged during the virtual displacements δu_i, the *external virtual work* ${}^2R$ done by the surface tractions ${}^2t_i$ and body force ${}^2f_i$ can be given by

$${}^2R = \int_{{}^2S} {}^2t_i \delta u_i {}^2dS + \int_{{}^2V} {}^2f_i \delta u_i {}^2dV \qquad (1.79)$$

in which the surface area is ${}^2S = {}^2S_\tau + {}^2S_u$. Based on the fact that ${}^2t_i = {}^2\tau_{ji} n_j$ for the part ${}^2S_\tau$ with prescribed tractions and that $\delta u_i = 0$ for the part ${}^2S_u$ with prescribed displacements, we can rewrite the preceding equation as follows:

$${}^2R = \int_{{}^2S} {}^2\tau_{ji} n_j \, \delta u_i {}^2dS + \int_{{}^2V} {}^2f_i \delta u_i {}^2dV \qquad (1.80)$$

By the divergence theorem, we can transform the first (surface) integral into a volume integral, namely,

$$ {}^2R = \int_{{}^2V} \left[{}^2\tau_{ji} \frac{\partial(\delta u_i)}{\partial\, {}^2x_j} + \delta u_i \left(\frac{\partial\, {}^2\tau_{ji}}{\partial\, {}^2x_j} + {}^2f_i \right) \right] {}^2dV \tag{1.81} $$

The terms in the parentheses vanish by the equations of equilibrium (1.68). By decomposing the derivatives of the virtual displacements δu_i into two parts as the strains $\delta\ {}_2e_{ij}$ and $\delta\ {}_2\omega_{ij}$,

$$ \frac{\partial(\delta u_i)}{\partial\, {}^2x_j} = \delta\ {}_2e_{ij} + \delta\ {}_2\omega_{ij} \tag{1.82} $$

where

$$ \delta\ {}_2e_{ij} = \frac{1}{2}\left(\frac{\partial \delta u_i}{\partial\, {}^2x_j} + \frac{\partial \delta u_j}{\partial\, {}^2x_i} \right) \tag{1.83} $$

$$ \delta\ {}_2\omega_{ij} = \frac{1}{2}\left(\frac{\partial \delta u_i}{\partial\, {}^2x_j} - \frac{\partial \delta u_j}{\partial\, {}^2x_i} \right) \tag{1.84} $$

we can write

$$ {}^2\tau_{ij} \frac{\partial(\delta u_i)}{\partial\, {}^2x_j} = {}^2\tau_{ij}\delta\ {}_2e_{ij} + {}^2\tau_{ij}\delta\ {}_2\omega_{ij} \tag{1.85} $$

Moreover, it can be shown that ${}^2\tau_{ij}\delta\ {}_2\omega_{ij} = 0$, because ${}^2\tau_{ij}$ is symmetric and $\delta\ {}_2\omega_{ij}$ is skew-symmetric. Thus, equation (1.81) takes the following form:

$$ {}^2R = \int_{{}^2V} {}^2\tau_{ij}\ \delta\ {}_2e_{ij} {}^2dV \tag{1.86} $$

Consequently, we have proved that if the stress field is statically admissible, the *principle of virtual displacements* for any admissible virtual displacements δu_i can be written as follows:

$$ \int_{{}^2V} {}^2\tau_{ij}\delta\ {}_2e_{ij} {}^2dV = \int_{{}^2S} {}^2t_i \delta u_i {}^2dS + \int_{{}^2V} {}^2f_i \delta u_i {}^2dV \tag{1.87} $$

The converse proposition states that if the equation of virtual displacements (1.87) is valid for every kinematically admissible virtual displacement field, then the stress field is statically admissible, i.e., it will satisfy the equations of equilibrium inside the solid and the boundary conditions with prescribed surface tractions. A proof for the converse proposition can be found elsewhere (Malvern, 1969).

It is clear from the above statements that the *principle of virtual displacements* as conveyed by equation (1.87) serves as a necessary and sufficient condition for expressing the equilibrium of a solid with prescribed boundary conditions. Another message conveyed by this equation is the conjugateness of the Cauchy stress tensor ${}^{2}\tau_{ij}$ and infinitesimal strain tensor ${}_{2}e_{ij}$ in terms of virtual work. Compared with the principle of minimum potential energy or other energy methods, the principle of virtual displacements has the advantage that it is not restricted to material laws of the elastic type or loadings of the conservative type (i.e., path-independent loadings) because no assumptions have been made in the foregoing derivations either on the material laws or loading functions. Another advantage of the principle of virtual displacements is that the geometric nonlinear effects of the structure can be easily considered through the inclusion of the nonlinear strain components in the incremental Lagrangian formulations, as will be shown in the following section. For these reasons, the principle of virtual displacements will be exclusively adopted in the text as the basis for deriving the theories for various structural elements.

1.7 Incremental Lagrangian Formulations

A fundamental difficulty in the application of the principle of virtual displacements as originally cast in the form of equation (1.87) is that it has been referred to the C_2 configuration of the solid, which is unknown for a nonlinear problem before the equation is solved. This is different from the problems confronting us in linear analysis, in which the displacements are assumed to be so small that the change in geometry of the structure can be neglected and that all quantities can be referred to as the initial configuration C_0, which is known beforehand. To overcome this problem with nonlinear analysis, we can transform the reference configuration from C_2 to any other known configuration. In the TL formulation, the initial configuration C_0 is selected as the reference, whereas in the UL formulation, the last calculated configuration C_1 is used as the reference. In this section, we shall demonstrate how the virtual work equation (1.87) can be transformed to those with reference to the known configurations of C_0 and C_1 for the total and UL formulations, respectively. The resulting equations serve as the basis for the derivation of the finite elements for all the structural members of concern in this text.

1.7.1 *Total Lagrangian formulation*

In the TL formulation, all variables adopted in the virtual work expressions are referred to as the initial configuration C_0. One key step in this formulation is to derive the following relation:

$$\int_{^2V} {}^2\tau_{ij}\delta\, {}_2e_{ij}\, {}^2dV = \int_{^0V} {}^2_0S_{ij}\delta^2_0\epsilon_{ij}\, {}^0dV \tag{1.88}$$

in which ${}^2_0S_{ij}$ denotes the second Piola–Kirchhoff stress and ${}^2_0\epsilon_{ij}$ the associated Green–Lagrange strains. To this end, we have to establish first the relation between ${}^2\tau_{ij}$ and ${}^2_0S_{ij}$ and that between $\delta\, {}_2e_{ij}$ and $\delta^2_0\epsilon_{ij}$. Since the relation between ${}^2\tau_{ij}$ and ${}^2_0S_{ij}$ is available in equations (1.42) and (1.43), in the following we need to concentrate only on the relation between the virtual strains $\delta\, {}_2e_{ij}$ and $\delta^2_0\epsilon_{ij}$.

Noting that $\delta d\, {}^0x_i = 0$, from equations (1.4) and (1.6), we can obtain the following:

$$2\delta^2_0\epsilon_{ij} d\, {}^0x_i d\, {}^0x_j = 2(\delta d\, {}^2x_i) d\, {}^2x_i \tag{1.89}$$

Since $\delta d\, {}^2x_i = d\delta\, {}^2x_i = d\delta u_i$, we can write

$$\delta d\, {}^2x_i = \frac{\partial \delta u_i}{\partial\, {}^2x_i} d\, {}^2x_i \tag{1.90}$$

which can be substituted into equation (1.89) to yield

$$\delta^2_0\epsilon_{ij} d\, {}^0x_i d\, {}^0x_j = \frac{\partial \delta u_i}{\partial\, {}^2x_k} d\, {}^2x_k d\, {}^2x_i \tag{1.91}$$

Further, from equation (1.30),

$$\delta\, {}_2e_{ij} = \frac{1}{2}\left(\frac{\partial \delta u_i}{\partial\, {}^2x_j} + \frac{\partial \delta u_j}{\partial\, {}^2x_i}\right) \tag{1.92}$$

we obtain

$$\delta_0^2\epsilon_{ij} d\,{}^0x_i d\,{}^0x_j = \delta\,{}_2e_{pq} d\,{}^2x_p d\,{}^2x_q \tag{1.93}$$

By the chain rule,

$$\delta_0^2\epsilon_{ij} d\,{}^0x_i d\,{}^0x_j = \delta\,{}_2e_{pq} \frac{\partial\,{}^2x_p}{\partial\,{}^0x_i} \frac{\partial\,{}^2x_q}{\partial\,{}^0x_j} d\,{}^0x_i d\,{}^0x_j \tag{1.94}$$

Hence,

$$\delta_0^2\epsilon_{ij} = \delta\,{}_2e_{pq} \frac{\partial\,{}^2x_p}{\partial\,{}^0x_i} \frac{\partial\,{}^2x_q}{\partial\,{}^0x_j} \tag{1.95}$$

or

$$\delta\,{}_2e_{ij} = \delta_0^2\epsilon_{pq} \frac{\partial\,{}^0x_p}{\partial\,{}^2x_i} \frac{\partial^0x_q}{\partial\,{}^2x_j} \tag{1.96}$$

By the use of the relations in (1.43) and (1.46) and noting that for conserved mass ${}^0\rho^0dV = {}^2\rho^2dV$, we can therefore prove the validity of the relation given in equation (1.88). From this equation, it is evident that the second Piola–Kirchhoff stresses ${}_0^2S_{ij}$ and Green–Lagrange strains ${}_0^2\epsilon_{ij}$ are energetically conjugate.

Further, by defining the surface tractions ${}_0^2t_i$ and body forces ${}_0^2f_i$ with respect to the C_0 configuration as

$$ {}_0^2t_i{}^0dS = {}^2t_i{}^2dS \tag{1.97}$$

$$ {}_0^2f_i{}^0dV = {}^2f_i{}^2dV \tag{1.98}$$

the following relations can be written accordingly

$$\int_{^2S} {}^2t_i\delta u_i{}^2dS = \int_{^0S} {}_0^2t_i\delta u_i{}^0dS \tag{1.99}$$

$$\int_{^2V} {}^2f_i\delta u_i{}^2dV = \int_{^0V} {}_0^2f_i\delta u_i{}^0dV \tag{1.100}$$

With the relations given in (1.88), (1.99), and (1.100), we are able to transform the reference configuration of the virtual work equation (1.87) from the C_2 to the C_0 configuration:

$$\int_{^0V} {}_0^2S_{ij}\delta_0^2\epsilon_{ij}{}^0dV = \int_{^0S} {}_0^2t_i\delta u_i{}^0dS + \int_{^0V} {}_0^2f_i\delta u_i{}^0dV \tag{1.101}$$

of which the external virtual work on the right-hand side will be denoted as ${}_0^2R$ throughout the text, i.e.,

$$ {}_0^2R = \int_{{}^0S} {}_0^2t_i \delta u_i \, {}^0dS + \int_{{}^0V} {}_0^2f_i \delta u_i {}^0dV \tag{1.102} $$

The virtual work equation as given in equation (1.101) is by itself a nonlinear equation of equilibrium for the solid under consideration. Though the reference has been changed from the current configuration C_2 to the initial configuration C_0, the equation remains an exact statement of equilibrium for the structure. Therefore, it can be used as a valid basis for deriving the incremental nonlinear equations in the TL formulation, as will be demonstrated below.

According to equations (1.18) and (1.20), we can write

$$ \delta_0^2\epsilon_{ij} = \delta \, {}_0\epsilon_{ij} = \delta \, {}_0e_{ij} + \delta \, {}_0\eta_{ij} \tag{1.103} $$

where the linear and nonlinear components are obtainable from (1.21) and (1.22), that is,

$$ \delta \, {}_0e_{ij} = \frac{1}{2}\left(\frac{\partial \delta u_i}{\partial \, {}^0x_j} + \frac{\partial \delta u_j}{\partial \, {}^0x_i} + \frac{\partial \, {}^1u_k}{\partial \, {}^0x_i}\frac{\partial \delta u_k}{\partial \, {}^0x_j} + \frac{\partial \delta u_k}{\partial \, {}^0x_i}\frac{\partial \, {}^1u_k}{\partial \, {}^0x_j}\right) \tag{1.104} $$

$$ \delta \, {}_0\eta_{ij} = \frac{1}{2}\left(\frac{\partial \delta u_k}{\partial \, {}^0x_i}\frac{\partial u_k}{\partial \, {}^0x_j} + \frac{\partial u_k}{\partial \, {}^0x_i}\frac{\partial \delta u_k}{\partial \, {}^0x_j}\right) \tag{1.105} $$

In the above, it is noted that $\delta_0^1\epsilon_{ij} = 0$ and $\delta \, {}^1u_k = 0$. By substituting equation (1.39) for the stresses ${}_0^2S_{ij}$ into the virtual work equation (1.101) and making use of (1.103) for the strains $\delta_0^2\epsilon_{ij}$, we obtain

$$ \int_{{}^0V} {}_0S_{ij}\delta \, {}_0\epsilon_{ij} {}^0dV + \int_{{}^0V} {}_0^1S_{ij}\delta \, {}_0\eta_{ij} {}^0dV = {}_0^2R - {}_0^1R \tag{1.106} $$

where ${}_0^1R$ is defined as

$$ {}_0^1R = \int_{{}^0V} {}_0^1S_{ij}\delta \, {}_0e_{ij} {}^0dV \tag{1.107} $$

For a solid in equilibrium at C_1, the following can be written:

$$ {}_0^1R = \int_{{}^0S} {}_0^1t_i \delta u_i \, {}^0dS + \int_{{}^0V} {}_0^1f_i \delta u_i \, {}^0dV \tag{1.108} $$

which can be obtained from the external virtual work in (1.102) by switching all the superscripts from "2" to "1".

A physical interpretation for equation (1.106) can be given as follows: The first term on the left-hand side represents the change in *strain energy* (in variational form) during the incremental step from C_1 to C_2 and the second term is the change in *potential energy* due to the initial stresses ${}^1_0S_{ij}$. The terms ${}^1_0R$ and ${}^2_0R$ on the right-hand side represent the external virtual works done by the surface tractions and body forces acting on the solid at C_1 and C_2, respectively. To put it together, equation (1.106) states that the increase in the virtual works done by the external agencies from C_1 to C_2 is transformed to the increase in the strain energy and potential energy of the solid during the deformation from C_1 to C_2. Since no approximation has been made in the derivation of equation (1.106), this equation remains an exact statement of equilibrium of the solid at C_2, in which all variables of the solid at C_1, including the initial stresses ${}^1_0S_{ij}$, are assumed to have been made available at the beginning of each incremental step in nonlinear analysis.

For the cases where all the stress increments ${}_0S_{ij}$ can be related to the strain increments ${}_0\epsilon_{ij}$ by the incremental constitutive law given in equation (1.61), equation (1.106) can be rewritten as follows:

$$\int_{{}^0V} {}_0C_{ijkl}\, {}_0\epsilon_{kl}\delta\, {}_0\epsilon_{ij}\, {}^0dV + \int_{{}^0V} {}^1_0S_{ij}\delta\, {}_0\eta_{ij}\, {}^0dV = {}^2_0R - {}^1_0R \quad (1.109)$$

where ${}_0C_{ijkl}$ represents the incremental constitutive coefficients. Either equation (1.106) or (1.109) cannot be solved directly, since they are nonlinear in the displacement increments u_i. For problems of which the strain increments can be considered small within each incremental step, approximate solutions can be obtained by making the following assumptions for the *constitutive law*:

$${}_0S_{ij} \cong {}_0C_{ijkl}\, {}_0e_{kl} \quad (1.110)$$

$$\delta\, {}_0\epsilon_{ij} \cong \delta\, {}_0e_{ij} \quad (1.111)$$

Accordingly, equation (1.109) reduces to

$$\int_{{}^0V} {}_0C_{ijkl}\, {}_0e_{kl}\delta\, {}_0e_{ij}\, {}^0dV + \int_{{}^0V} {}^1_0S_{ij}\delta\, {}_0\eta_{ij}\, {}^0dV = {}^2_0R - {}^1_0R \quad (1.112)$$

This is the *linearized*version of the incremental equation of equilibrium for describing the motion of the solid from C_1 to C_2 based on the TL formulation, where all quantities have been expressed with reference to the coordinates at the initial undeformed configuration C_1.

1.7.2 *Updated Lagrangian formulation*

In the UL formulation, all physical quantities will be referred to the last calculated configuration C_1 instead. The first concern herein is to prove the validity of the following relation:

$$\int_{^2V} {}^2\tau_{ij}\delta\ {}_2e_{ij}{}^2dV = \int_{^1V} {}^2_1S_{ij}\delta\ {}_1\epsilon_{ij}{}^1dV \tag{1.113}$$

where it should be noted that the two symbols ${}_1\epsilon_{ij}$ and ${}^2_1\epsilon_{ij}$ are identical, i.e., ${}_1\epsilon_{ij} \equiv {}^2_1\epsilon_{ij}$. Since the relation between the Cauchy stresses ${}^2\tau_{ij}$ and the second Piola–Kirchhofff stresses ${}^2_1S_{ij}$ is already given in equations (1.45) and (1.46), only the relation between the (virtual) infinitesimal strains $\delta\ {}_2e_{ij}$ and Green–Lagranage strains $\delta\ {}_1\epsilon_{ij}$ has to be derived.

Noting that $\delta d\ {}^1x_i = 0$, from equations (1.31) and (1.6),

$$2\ {}_1\epsilon_{ij}\ d\ {}^1x_i d\ {}^1x_j = 2(\delta d\ {}^2x_i)d\ {}^2x_i \tag{1.114}$$

Using equations (1.90) and (1.92), we have

$$\delta\ {}_1\epsilon_{ij}\ d\ {}^1x_i d\ {}^1x_j = \frac{\partial \delta u_i}{\partial\ {}^2x_k} d\ {}^2x_k d\ {}^2x_i = \delta\ {}_2e_{pq} d\ {}^2x_p d\ {}^2x_q \tag{1.115}$$

By the chain rule,

$$\delta\ {}_1\epsilon_{ij}\ d\ {}^1x_i d\ {}^1x_j = \delta\ {}_2e_{pq} \frac{\partial\ {}^2x_p}{\partial\ {}^1x_i} \frac{\partial\ {}^2x_q}{\partial\ {}^1x_j} d\ {}^1x_i d\ {}^1x_j \tag{1.116}$$

Hence,

$$\delta\ {}_1\epsilon_{ij} = \delta\ {}_2e_{pq} \frac{\partial\ {}^2x_p}{\partial\ {}^1x_i} \frac{\partial\ {}^2x_q}{\partial\ {}^1x_j} \tag{1.117}$$

or

$$\delta\ {}_2e_{ij} = \delta\ {}_1\epsilon_{pq} \frac{\partial\ {}^1x_p}{\partial\ {}^2x_i} \frac{\partial\ {}^1x_q}{\partial\ {}^2x_j} \tag{1.118}$$

By the use of equations (1.46) and (1.118) and noting that for conserved mass ${}^1\rho{}^1dV = {}^2\rho{}^2dV$, we can therefore prove the validity of the relation in (1.113). This equation implies that the second Piola–Kirchhoff stresses ${}^2_1S_{ij}$ and Green–Lagrange strains ${}_1\epsilon_{ij}$ are energetically conjugate.

Similarly, by defining the surface tractions ${}^2_1t_i$ and body forces ${}^2_1f_i$ with respect to the C_1 configuration as

$$ {}^2_1t_i\ {}^1dS = {}^2t_i\ {}^2dS \tag{1.119} $$

$$ {}^2_1f_i\ {}^1dV = {}^2f_i\ {}^2dV \tag{1.120} $$

the following relation can be written

$$ \int_{{}^2S} {}^2t_i\delta u_i\ {}^2dS = \int_{{}^1S} {}^2_1t_i\delta u_i\ {}^1dS \tag{1.121} $$

$$ \int_{{}^2V} {}^2f_i\delta u_i\ {}^2dV = \int_{{}^1V} {}^2_1f_i\delta u_i\ {}^1dV \tag{1.122} $$

With the relations given in (1.113), (1.121), and (1.122), we are able to transform the reference configuration of the virtual work equation (1.87) from C_2 to C_1:

$$ \int_{{}^1V} {}^2_1S_{ij}\delta{}^2_1\epsilon_{ij}\,{}^1dV = \int_{{}^1S} {}^2_1t_i\delta u_i{}^1dS + \int_{{}^1V} {}^2_1f_i\delta u_i{}^1dV \tag{1.123} $$

of which the external virtual work on the right-hand side will be denoted as ${}^2_1R$ throughout the text, i.e.,

$$ {}^2_1R = \int_{{}^1S} {}^2_1t_i\delta u_i{}^1dS + \int_{{}^1V} {}^2_1f_i\delta u_i{}^1dV \tag{1.124} $$

The virtual work equation as given in (1.123) is exact and nonlinear in the incremental displacements u_i. This equation can be used as a valid basis for deriving the incremental stiffness equations for various finite elements using the UL formulation.

According to equations (1.33), we can write

$$ \delta\ {}_1\epsilon_{ij} = \delta\ {}_1e_{ij} + \delta\ {}_1\eta_{ij} \tag{1.125} $$

where the linear and nonlinear components can be obtained from (1.34) and (1.35) as follows:

$$\delta\ {}_1e_{ij} = \frac{1}{2}\left(\frac{\partial \delta u_i}{\partial\ {}^1x_j} + \frac{\partial \delta u_j}{\partial\ {}^1x_i}\right) \tag{1.126}$$

$$\delta\ {}_1\eta_{ij} = \frac{1}{2}\left(\frac{\partial \delta u_k}{\partial\ {}^1x_i}\frac{\partial u_k}{\partial\ {}^1x_j} + \frac{\partial u_k}{\partial\ {}^1x_i}\frac{\partial \delta u_k}{\partial\ {}^1x_j}\right) \tag{1.127}$$

By substituting equation (1.44) for the stresses ${}^2_1S_{ij}$ into the virtual work equation (1.123) and making use of (1.125) for the strains $\delta\ {}_1\epsilon_{ij}$, we obtain

$$\int_{{}^1V} {}_1S_{ij}\delta\ {}_1\epsilon_{ij}\ {}^1dV + \int_{{}^1V} {}^1\tau_{ij}\delta\ {}_1\eta_{ij}\,{}^1dV = {}^2_1R - {}^1_1R \tag{1.128}$$

where ${}^1_1R$ is defined as

$${}^1_1R = \int_{{}^1V} {}^1\tau_{ij}\delta\ {}_1e_{ij}\,{}^1dV \tag{1.129}$$

For a solid in equilibrium at C_1, we can write

$${}^1_1R = \int_{{}^1S} {}^1_1t_i\delta u_i\ {}^1dS + \int_{{}^1V} {}^1_1f_i\delta u_i\ {}^1dV \tag{1.130}$$

which can be obtained as an analogy of the external virtual work given in (1.124) by switching all the superscripts from “2” to “1”.

A physical interpretation for equation (1.128) can be given as follows: The first term on the left-hand side represents the change in *strain energy* (in variational form) during the incremental step from C_1 to C_2 and the second term the change in *potential energy* due to the initial stresses ${}^1\tau_{ij}$. The terms ${}^1_1R$ and ${}^2_1R$ on the right-hand side represent the external virtual works done by the surface tractions and body forces acting on the solid at C_1 and C_2, respectively. When all the terms are put together, equation (1.128) states that the difference of the virtual works done by the external agencies at C_2 and C_1 will be transformed to the increase in the strain energy and potential energy of the solid during the incremental step from C_1 to C_2. Since no approximation has been made in the derivation, equation (1.128) remains an exact statement of equilibrium of the solid at C_2 with reference to the C_1 configuration. Moreover, this equation has been presented in a form particularly suitable for the incremental nonlinear analysis of structures.

For the cases where the stress increments ${}_1S_{ij}$ can be related to the strain increments ${}_1\epsilon_{ij}$ by the incremental constitutive law given in equation (1.62), it is convenient to rewrite equation (1.128) in the following form:

$$\int_{{}^1V} {}_1C_{ijkl}\, {}_1\epsilon_{kl}\delta\, {}_1\epsilon_{ij}\, {}^1dV + \int_{{}^1V} {}^1\tau_{ij}\delta\, {}_1\eta_{ij}\, {}^1dV = {}^2_1R - {}^1_1R \qquad (1.131)$$

where ${}_1C_{ijkl}$ represents the incremental constitutive coefficients. It should be emphasized that this equation remains valid only for elastic solids for which all the stress components ${}_1S_{ij}$ can be related to the strain components ${}_1\epsilon_{ij}$ by the incremental constitutive law (1.62). There are cases where not all the stress components can be related to the strain components in the same way, for which case modifications should be made to account for such an effect. One typical example of this is on the study of two- and three-dimensional beams employing the Bernoulli–Euler hypothesis of plane sections for describing the cross-sectional displacements. For problems of this sort, the shear stresses cannot be related to the strains by the constitutive laws, but must be determined from the conditions of equilibrium, because shear deformations are completely restrained for each cross-section of the beam. Such a problem encountered in nonlinear analysis can be skillfully circumvented by invoking the universally valid *rigid body rule*, that forms the focus of the remaining chapters.

The other comment is that the material coefficients ${}_1C_{ijkl}$ involved in equation (1.131) have been given in the *linearized* or incremental form. This is good for a large class of problems for which the strains accumulated from all the incremental steps are not large, such as the beam-type structures. However, for problems with large strains, such as for some trusses in the postbuckling stage, the use of constant coefficients for all incremental steps based on the UL formulation may lead to incorrect results (Yang and Leu, 1990).

Either equation (1.128) or (1.131) cannot be solved directly, since they are nonlinear in the displacement increments u_i. Approximate solutions can be obtained by making the following assumptions for the *constitutive law*:

$$ {}_1S_{ij} \cong {}_1C_{ijkl}\, {}_1e_{kl} \tag{1.132} $$

$$ \delta\, {}_1\epsilon_{ij} \cong \delta\, {}_1e_{ij} \tag{1.133} $$

Accordingly, equation (1.131) reduces to

$$ \int_{{}^1V} {}_1C_{ijkl}\, {}_1e_{kl} \delta\, {}_1e_{ij}\, {}^1dV + \int_{{}^1V} {}^1\tau_{ij} \delta\, {}_1\eta_{ij}\, {}^1dV = {}^2_1R - {}^1_1R \tag{1.134} $$

This is the *linearized* version of the incremental equation of equilibrium for describing the motion of the solid from C_1 to C_2 based on the UL formulation, where all physical quantities have been expressed with reference to the last calculated configuration C_1.

References

Bathe, K.J. and Bolourchi, S. (1979). Large displacement analysis of three-dimensional beam structures, *Int. J. Numer. Meth. Eng.*, 14, 961–986.

Bathe, K.J., Ramm, E., and Wilson, E.L. (1975). Finite element formulation for large deformation analysis, *Int. J. Numer. Meth. Eng.*, 9, 353–386.

Fung, Y.C. (1965). *Foundations of Solid Mechanics*, Prentice-Hall, Englewood Cliffs, New Jersey.

Malvern, L.E. (1969). *Introduction to the Mechanics of a Continuous Medium*, Prentice-Hall, Englewood Cliffs, New Jersey.

Washizu, K. (1982). *Variational Methods in Elasticity and Plasticity*, 3rd edn, Pergamon Press, Oxford, England.

Yang, Y.B. and Kuo, S.R. (1994). *Theory and Analysis of Nonlinear Framed Structures*, Prentice-Hall, Englewood Cliffs, New Jersey.

Yang, Y.B. and Leu, L.J. (1990). Postbuckling analysis of trusses with various Lagrangian formulations, *AIAA J.*, 28(5), 946–948.

Yang, Y.B. and Leu, L.J. (1991). Constitutive laws and force recovery procedures in nonlinear analysis of trusses, *Comp. Meth. Appl. Mech. Eng.*, 92, 121–131.

Chapter 2

Linear Analysis as a Prerequisite

2.1 General

Before proceeding to the following chapters for nonlinear analysis of various structures, we shall summarize some key features of linear analysis, such that the readers will be well equipped with the necessary skills to go for nonlinear analysis. The following are the contents to be covered in this chapter for linear analysis: derivation of element stiffness matrices, assembly of structural stiffness matrices, solution of the structural equations, and element quality tests.

The structures that are considered in the book are mainly *framed structures*, except for the minor adventure to plates and shells. A framed structure is composed of members that are long in comparison with their cross-sectional dimensions, such as width and depth. Members of this sort are called *slender members* and are conventionally represented by one-dimensional line members. Even a line member is a continuum that consists of an infinite number of degrees of freedom, of which the behavior is governed by *differential equations of equilibrium*. For given loading, boundary conditions, and material constitutive law, the solution for the differential equations of the line member is a typical *boundary-value problem*. Though in theory, a direct solution of the boundary-value problem by the classical mathematical approach can yield the exact results, its application in practice is extremely limited and difficult because of the tremendous effort required in formulating and finding a solution for the problem of a large structure that may consist of a large number of degrees of freedom. It is only for very simple frames or individual members with simple loading and boundary conditions that such a classical approach is possible.

The *finite element method* is similar to other numerical methods, such as the finite difference method and boundary element method, in that it approximates the original structural system that has an infinite number of degrees of freedom by a simplified mathematical model that has only a finite number of degrees of freedom. By replacing the original differential equations and continuity and boundary conditions with the finite element matrix equations, which can be solved in a systematic way, we not only reduce the number of degrees of freedom for the problem considered but also circumvent the insurmountable difficulties involved in finding the solution for the classical boundary-value problem.

The finite element method for analyzing structures of the framed type based on the linear theory can be summarized. The design of a structure starts with the selection of cross-sections and dimensions for each member of the structure based on previous experience. As the first step of finite element analysis, the entire framed structure is broken down into a number of line elements that are connected to each other at the nodal points. For the purpose of referencing, each element is assigned an element number and each node a node number. Next, the stiffness equations are formulated for each element in terms of the nodal degrees of freedom. This step ensures the equilibrium of individual elements to be satisfied in a weak or average sense. By transforming the element stiffness equations from the local coordinates for each element to the global coordinates for the structure, and by maintaining the nodal compatibility and equilibrium between connected elements, we can assemble all the element equations to yield the global stiffness equations for the structure. Further, by imposing appropriate geometric boundary conditions, the structure will achieve its kinematical stability, in the sense that rigid body motions are removed, as indicated by the positive definiteness of the global stiffness matrix. For a given set of applied loads, the nodal displacements can then be solved from the global stiffness equations. The final step in finite element analysis is the reverse processing of element displacements to obtain the local quantities such as element forces and stresses. With these, we can check the appropriateness of the cross-section selected for each member of the structure.

As was stated above, the discretization or idealization of the structure constitutes the first step in the finite element analysis procedure. In this step, assumptions have to be made regarding the geometry, cross-sectional properties, connections, materials, boundary conditions, and loading conditions. It is with such assumptions that a structure originally appearing as a continuum with complex loading and boundary conditions can be approximated by a mathematical model with a finite number of degrees of freedom to such an extent that accurate analysis can be carried out. All the information pertaining to the structure model should be prepared as the input data before a finite element analysis program can be executed.

Conventionally, the data preparation stage of a finite element analysis was referred to as the *preprocessing* phase, in contrast with the *processing* phase for execution of the analysis program, and the *postprocessing* phase for display of the analysis result. Although various preprocessors have been developed in the past to enhance the efficiency of data preparation, for instance, using the interactive computer graphics (Pesquera *et al.*, 1983), data preparation remains the most critical part of finite element analysis, in terms of the time and effort spent by engineers in design offices. One reason for this is that the procedure for selecting and checking the geometrical layout of the structure of concern is iterative in nature, and that for large, complex structures, human errors are likely to occur.

In this text, it is assumed that readers are already acquainted with the procedure for linear finite element analysis and the preparation of structural data required by such a procedure. Therefore, no attempt will be made to dig into details of the elementary programming aspects of finite element analysis. Throughout the text, structural members that can carry only axial forces, such as the bracing members commonly used in steel frameworks, will be referred to as the *bar* or *truss elements*. Meanwhile, structural members that can resist axial, flexural, and even torsional actions, such as beams and columns, will be referred to as the *beam* or *frame elements*. In the section to follow, the stiffness matrices for various frame and truss elements based on the linear elastic assumption will be derived first. They are considered as the degenerated cases for the corresponding nonlinear structural members based on the updated Lagrangian formulation. To make the readers easily familiar with the symbols and notation used, the derivation of the equations for the linear problems will be strictly presented as a reduced form of those for the more complicated nonlinear problems to follow in later chapters.

The basic concepts and procedures provided in this chapter for the derivation of the simple beam and truss elements are believed to be useful for readers to proceed with the following chapters on nonlinear analysis. Throughout this book, the symbols $\{\bullet\}$ will be used for a column vector, $\langle\bullet\rangle$ a row vector, $[\bullet]$ a matrix, and the superscript T denotes the transpose of a vector or matrix.

2.2 Derivation of Element Stiffness Matrix

A linear elastic analysis distinguishes from the nonlinear analysis in that the equations of equilibrium for the solid or structure under consideration are established at the initial undeformed configuration C_0, with the effect of change in geometry of the structure assumed to be negligibly small. For problems of this type, there will be no initial displacements or loadings, i.e., ${}^1u_i = 0$ and ${}^1_0f_i = {}^1_1f_i = {}^1_0t_i = {}^1_1t_i = 0$. This means that configurations C_0 and C_1 are identical, i.e., $C_0 \equiv C_1$, and that the external virtual works ${}^1_0R$ and ${}^1_1R$ as defined in equations (1.108) and (1.130) for the initial configuration C_1 can be taken as zero. Moreover, the displacement increments u_i from C_1 to C_2 are so small that the squares and products of their derivatives can be neglected. Thus, no nonlinear components of the strains need to be considered. Both the total and updated Green-Lagrange strain tensors ${}_0\epsilon_{ij}$ and ${}_1\epsilon_{ij}$ reduce to the same infinitesimal strain tensor, i.e., ${}_0\epsilon_{ij} \equiv {}_1\epsilon_{ij} \equiv {}_0e_{ij} \equiv {}_1e_{ij} \equiv e_{ij}$ in linear analysis.

With the foregoing assumptions, linear analysis can be regarded as a special case of nonlinear analysis that requires only a *single increment of loading.* Since no difference can be made between the total and updated Lagrangian formulations, both equations (1.109) and (1.131) reduce to the same following form:

$$\int_V C_{ijkl}\, e_{kl}\delta e_{ij} dV = R \tag{2.1}$$

where the external work R is defined as follows:

$$R = \int_S t_i \delta u_i dS + \int_V f_i \delta u_i dV \tag{2.2}$$

with t_i denoting the surface tractions and f_i the body forces. In equations (2.1) and (2.2), each variable has been written with its left subscript and superscript dropped, since it is known that all quantities in these two equations occur in the C_0 configuration and are referred to the same configuration. In the following, the stiffness matrices will be derived for various frame and truss elements based on equation (2.1). For the purpose of illustration, the procedure employed in deriving the plane frame element will be presented with sufficient details.

2.2.1 *Plane frame element*

Let $(x,\ y)$ be a set of orthogonal Cartesian coordinates with the x-axis denoting the centroidal axis of the plane frame element in Figure 2.1(a). For this element, only the axial strain e_{xx} is significant; all the other strain components can be taken as zero. Consequently,

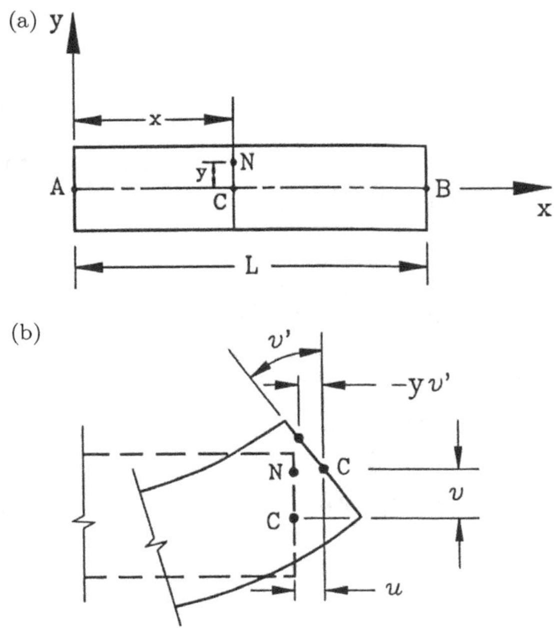

Fig. 2.1. Plane frame element: (a) coordinates; (b) deformed shape.

equation (2.1) reduces to

$$\int_V E e_{xx} \delta e_{xx} dV = R \tag{2.3}$$

where E denotes Young's modulus of the material and the integration is taken over the entire volume of the beam.

From equation (1.24a), the axial strain e_{xx} at a generic point N of the beam can be related to the axial displacement u_x at the same point as

$$e_{xx} = \frac{\partial u_x}{\partial x} \tag{2.4}$$

Based on the *Bernoulli–Euler hypothesis*, a cross-section that is plane and normal to the longitudinal axis of the beam remains plane and normal to the same axis after deformation. As a consequence, the axial and transverse displacements u_x and u_y of the generic point N can be related to the displacements u and v of the centroid C of the beam at the same cross-section as in 2.1(b):

$$u_x = u - yv' \tag{2.5}$$

$$u_y = v \tag{2.6}$$

in which a prime denotes differentiation with respect to the coordinate x. Substituting equation (2.5) for u_x into equation (2.4) yields

$$e_{xx} = u' - yv'' \tag{2.7}$$

Further, by substituting the preceding expression for the strain e_{xx} into equation (2.3), we can express the virtual work equation as

$$\int_0^L (EA\, u'\delta u' + EI_z v''\delta v'')dx = R \tag{2.8}$$

where L denotes the length of the beam, the cross-sectional area is assumed to be constant throughout the length of the element, and I_z denotes the moment of inertia of the cross-section about the z-axis,

$$I_z = \int_A y^2 dA \tag{2.9}$$

In arriving at equation (2.8), the following orthogonality condition for the principal centroidal coordinates has been adopted:

$$\int_A y dA = 0 \tag{2.10}$$

At this point, we are ready to derive the external virtual work term R. By neglecting the body forces f_i and assuming the surface tractions t_i to be concentrated only at the two ends A and B of the element, the external virtual work R_b done by the surface tractions t_i at end B can be integrated over the cross-sectional area A of the end:

$$R_b = \int_S t_i \delta u_i ds = \int_A (t_x \delta u_x + t_y \delta u_y) dA \tag{2.11}$$

where the surface tractions t_x and t_x are taken positive when directed along the positive x- and y-axes, respectively.

Based on the conditions of equilibrium, the axial force F_{xb}, shear force F_{yb}, and bending moment M_{zb} at section B can be related to the surface tractions t_i as

$$F_{xb} = \int_A t_x dA \tag{2.12}$$

$$F_{yb} = \int_A t_y dA \tag{2.13}$$

$$M_{zb} = -\int_A t_x y dA \tag{2.14}$$

Substituting equations (2.5) and (2.6) for the displacements u_x and u_y into equation (2.11), and making use of the preceding expressions for the nodal actions, we can calculate the external virtual work R_b as follows:

$$R_b = \{\delta u_b\}^T \{f_b\} \tag{2.15}$$

where superscript T denotes the transpose of a vector, $\{\bullet\}$ a column vector, $\{u_b\}$ the displacement vector associated with end B of the beam, and $\{f_b\}$ the corresponding force vector:

$$\{u_b\}^T = \langle u_b \quad v_b \quad \theta_b \rangle \tag{2.16}$$

$$\{f_b\}^T = \langle F_{xb} \quad F_{yb} \quad M_{zb} \rangle \tag{2.17}$$

where $\langle \bullet \rangle$ denotes a row vector and $\theta_b = v_b'$ for small rotations.

In a similar manner, the external virtual work R_a done by the surface tractions t_i at end A of the beam can be written as

$$R_a = \int_A (t_x \delta u_x + t_y \delta u_y) dA \tag{2.18}$$

where the integration is performed for the cross-sectional area of end A. Following the same procedure, we can express the external virtual work R_a in terms of the nodal quantities at end A of the element as

$$R_a = \{\delta u_a\}^T \{f_a\} \tag{2.19}$$

where $\{u_a\}$ is the displacement vector at end A and $\{f_a\}$ the corresponding force vector,

$$\{u_a\}^T = \langle u_a \quad v_a \quad \theta_a \rangle \tag{2.20}$$

$$\{f_a\}^T = \langle F_{xa} \quad F_{ya} \quad M_{za} \rangle \tag{2.21}$$

with $\theta_a = v'_a$ for small rotations.

Consequently, the total external virtual work R of the beam can be obtained as the sum of R_b in (2.15) and R_a in (2.19),

$$R = R_a + R_b = \{\delta u\}^T \{f\} \tag{2.22}$$

where $\{u\}$ denotes the element displacement vector and $\{f\}$ the associated force vector (see Figure 2.2),

$$\{u\}^T = \langle u_a \quad v_a \quad \theta_a u_b \quad v_b \quad \theta_b \rangle \tag{2.23}$$

$$\{f\}^T = \langle F_{xa} \quad F_{ya} \quad M_{za} F_{xb} \quad F_{yb} \quad M_{zb} \rangle \tag{2.24}$$

Thus, we have shown that the virtual work equation of equilibrium (2.3) for the plane frame element can be written as follows:

$$\int_0^L (EA\, u' \delta u' + EI_z v'' \delta v'') dx = \{\delta u\}^T \{f\} \tag{2.25}$$

It should be noted that with present*finite-element sign convention* for the nodal forces in Figure 2.2(b), a direct force is taken positive when acting rightwards or upwards, and a bending moment is positive when counterclockwise. Such a sign convention is different

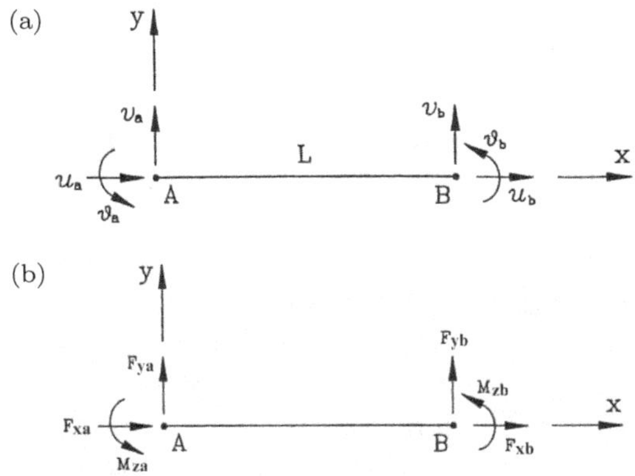

Fig. 2.2. Plane frame element: (a) nodal degrees of freedom; (b) nodal forces and moments.

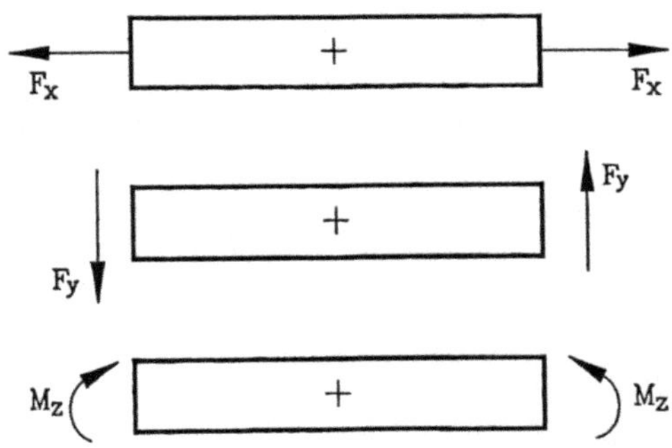

Fig. 2.3. Positive forces based on deformation sign convention.

from the *deformation sign convention* commonly used in the study of mechanics problems (Gere and Timoshenko, 1997), in which an axial force is taken positive when it tends to elongate the member on which it acts, a shear force is positive when it tends to rotate the member in the counterclockwise sense on the x–y plane, and a bending moment is positive when it tends to deform the member in the concave upward manner, as shown in Figure 2.3.

For the two sign conventions, it is easy to see that the forces acting on the right end B (for $x = L$) of the element are identical, but there exists a sign difference for the forces acting on the left end A (for $x = 0$). For this reason, when talking about the external virtual works or others, *we prefer to use the right end* B (for $x = L$)

as the end of illustration for the derivation. This is what has been adopted in the above, which will also be adopted in the following chapters.

As a side remark, we like to mention that a finite element formulation based on the preceding virtual work formulation is just as rational as that based on the *governing differential equations* and *boundary conditions* for the problem considered. Though such a finite element formulation based on the virtual work principle can be carried out basically without knowledge of the associated governing differential equations, as will be presented later on in this section, a preview of the related governing differential equations provides valuable clues for the selection of the interpolation functions and nodal degrees of freedom in the finite element formulation. It is noted that when using the variational procedures, the governing differential equations can also be derived as the *Euler–Lagrange equations* of the functional of the problem considered.

To derive the finite element matrices, let us integrate equation (2.25) by parts to obtain the virtual quantities δu and δv,

$$\begin{aligned}&\int_0^L (-EAu''\delta u + EI_z v''''\delta v)dx + EAu'\delta u|_0^L\\&\quad + EI_z v''\delta v'|_0^L - EI_z v'''\delta v|_0^L = \{\delta u\}^T\{f\}\end{aligned} \tag{2.26}$$

Rewriting,

$$\begin{aligned}&\int_0^L (-EAu''\delta u + EI_z v''''\delta v)dx - (F_x - EAu')\delta u|_0^L\\&\quad - (M_z - EI_z v'')\delta v'|_0^L - (F_y + EI_z v''')\delta v|_0^L = 0\end{aligned} \tag{2.27}$$

where use has been made of the following for the nodal displacements $\{u\}$,

$$u = u_a, \quad v = v_a, \quad v' = \theta_a \quad \text{for } x = 0 \tag{2.28}$$

$$u = u_b, \quad v = v_b, \quad v' = \theta_b \quad \text{for } x = L \tag{2.29}$$

and of the following for the nodal forces $\{f\}$ at the two ends in (2.24),

$$F_{xa} = -F_x, \quad F_{ya} = -F_y, \quad M_{za} = -M_z \quad \text{for } x = 0 \tag{2.30}$$

$$F_{xb} = F_x, \quad F_{yb} = F_y, \quad M_{zb} = M_z \quad \text{for } x = L \tag{2.31}$$

Admitting the arbitrary nature of the virtual displacements δu and δv, we can obtain from (2.27) the differential equations of equilibrium for the plane element as the Euler–Lagrange equations of the functional,

$$EAu'' = 0 \tag{2.32}$$

$$EI_z v'''' = 0 \tag{2.33}$$

along with the boundary conditions for the two ends of the beam,

$$\delta u = 0 \quad \text{or} \quad F_x = EAu' \tag{2.34}$$

$$\delta v' = 0 \quad \text{or} \quad M_z = EI_z v'' \tag{2.35}$$

$$\delta v = 0 \quad \text{or} \quad F_y = -EI_z v''' \tag{2.36}$$

where the first set of boundary conditions, i.e., $\delta u = 0$, $\delta v' = 0$, and $\delta v = 0$, is known as the *geometric boundary conditions*, and the second set as the *natural boundary conditions*. The differential equations of equilibrium as derived in (2.32) and (2.33) are exactly valid for plane frame elements free of any distributed loads. For this problem, the *exact solution* for the axial displacement u can be represented by a function linear in x and that for the transverse displacement v by a cubic polynomial function.

In the above derivation, we have shown that the virtual work equation as presented in (2.25) is equivalent to the governing differential equations and boundary conditions commonly used in classical analysis. Thus, equation (2.25) remains itself a valid statement of equilibrium for the plane frame element under the action of axial forces, shear forces, and bending moments. It can therefore be used as a valid basis for deriving the plane frame element, as will be presented below.

In the displacement-based finite element formulation, the displacements at any point within a finite element can be related to those at the nodal points of the finite element by properly chosen *interpolation* or *shape functions*. Errors may arise if the interpolation functions do not satisfy exactly the governing differential equations of the subject element under consideration. For the plane frame element, we can interpolate the axial displacement u and transverse displacement v by linear and cubic polynomial functions, respectively,

$$u = \{n_1\}^T\{\overline{u}\} \tag{2.37}$$

$$v = \{n_3\}^T\{\overline{v}\} \tag{2.38}$$

where, by letting $i = x/L$, the interpolation functions $\{n_1\}$ and $\{n_3\}$ can be written as follows:

$$\{n_1\}^T = \langle(1-i) \quad i\rangle \tag{2.39}$$

$$\{n_3\}^T = \langle(1-3i^2+2i^3) \quad (i-2i^2+i^3) \quad (3i^2-2i^3) \quad (i^3-i^2)\rangle \tag{2.40}$$

and the displacement vectors $\{\overline{u}\}$ and $\{\overline{v}\}$ are defined as

$$\{\overline{u}\}^T = \langle u_a \quad u_b\rangle \tag{2.41}$$

$$\{\overline{v}\}^T = \langle v_a \quad L\theta_a \quad v_b \quad L\theta_b\rangle \tag{2.42}$$

As was stated previously, the above interpolation functions are exact for the plane frame element that are free of distributed loads. By substituting equations (2.37) and (2.38) for the displacements u and v into equation (2.25), we obtain

$$\begin{aligned}&\{\delta\overline{u}\}^T\int_0^1 \frac{EA}{L}\{n_1'\}\{n_1'\}^T di\{\overline{u}\}\\&+\{\delta\overline{v}\}^T\int_0^1 \frac{EI_z}{L^3}\{n_3''\}\{n_3''\}^T di\{\overline{v}\} = \{\delta u\}^T\{f\}\end{aligned} \tag{2.43}$$

where $di = dx/L$ and a prime denotes derivative with respect to the dimensionless coordinate i.

To facilitate the present derivation, it is convenient to introduce the following notation for integrals involving various shape functions (Rajasekaran, 1977):

$$\left[K_{gh}^{stv}\right] = \int_0^1 i^v \left\{n_g^s\right\}\left\{n_h^t\right\}^T di \tag{2.44}$$

where the subscripts g and h denote the degrees of the interpolation functions $\{n\}$, the superscripts s and t the order of differentiation, and "v" the exponent of the multiplying factor i. Various integral matrices based on this definition that are useful to derivation of the stiffness matrices in this book have been given in Appendix A.

Using the above notation for the integrals, equation (2.43) can be rewritten in a more compact form as follows:

$$\{\delta\overline{u}\}^T\frac{EA}{L}\left[K_{11}^{110}\right]\{\overline{u}\}+\{\delta\overline{v}\}^T\frac{EI_z}{L^3}\left[K_{33}^{220}\right]\{\overline{v}\}=\{\delta u\}^T\{f\} \quad (2.45)$$

Noting that the virtual displacements $\{\delta\overline{u}\}$ and $\{\delta\overline{v}\}$ are correlated to $\{\delta u\}$ [see equations (2.41), (2.42), and (2.23)], by taking into account the arbitrary nature of virtual displacements, we can derive the following equations of equilibrium for the element from equation (2.43):

$$\frac{EA}{L}\left[K_{11}^{110}\right]\{\overline{u}\}=\{\overline{f}_x\} \quad (2.46)$$

$$\frac{EI_z}{L^3}\left[K_{33}^{220}\right]\{\overline{u}\}=\{\overline{f}_y\} \quad (2.47)$$

where the nodal forces vectors $\{\overline{f}_x\}$ and $\{\overline{f}_y\}$ correspond to the nodal displacement vectors $\{\overline{u}\}$ and $\{\overline{v}\}$, respectively,

$$\{\overline{f}_x\}=\langle F_{xa} \quad F_{xb}\rangle \quad (2.48)$$

$$\{\overline{f}_y\}=\left\langle F_{ya} \quad \frac{M_{za}}{L} \quad F_{yb} \quad \frac{M_zb}{L}\right\rangle \quad (2.49)$$

With the submatrices $\left[K_{11}^{110}\right]$ and $\left[K_{33}^{220}\right]$ given in Appendix A, the equations of equilibrium as given in (2.46) and (2.47) for the axial and flexural actions can be combined in a matrix form,

$$[k]\{u\}=\{f\} \quad (2.50)$$

where $[k]$ represents the *elastic stiffness matrix*, $\{u\}$ the displacement vector, and $\{f\}$ the force vector of the plane frame element. For this element, both the $\{u\}$ and $\{f\}$ have been given in equations (2.23) and (2.24), respectively. The stiffness matrix $[k]$ can be given

as follows:

$$[k] = \begin{bmatrix} \frac{EA}{L} & 0 & 0 & -\frac{EA}{L} & 0 & 0 \\ & \frac{12EI_z}{L^3} & \frac{6EI_z}{L^2} & 0 & -\frac{12EI_z}{L^3} & \frac{6EI_z}{L^2} \\ & & \frac{4EI_z}{L} & 0 & -\frac{6EI_z}{L^2} & \frac{2EI_z}{L} \\ & & & \frac{EA}{L} & 0 & 0 \\ & Symm. & & & \frac{12EI_z}{L^3} & -\frac{6EI_z}{L^2} \\ & & & & & \frac{4EI_z}{L} \end{bmatrix} \tag{2.51}$$

From this equation, it can be seen that the axial and bending actions of the plane frame element are not coupled. The elastic stiffness matrix $[k]$ will also be referred to as the *linear stiffness matrix* $[k_e]$ in this book, to distinguish it from the *geometric stiffness matrix* $[k_g]$ to be presented for nonlinear finite elements in later chapters.

2.2.2 *Space frame element*

Extension of the procedure in Section 2.2.1 to the derivation of the stiffness matrix for the space frame element is straightforward. Only those equations that are different or not covered by the plane case will be presented herein. In this regard, we note that in addition to the axial strain e_{xx} caused by axial and bending actions, the shear strain e_{xy} and e_{xz} caused by the torsional action of the member should also be included in the energy expression of equation (2.1), i.e.,

$$\int_V (Ee_{xx}\delta e_{xx} + 4Ge_{xy}\delta e_{xy} + 4Ge_{xz}\delta e_{xz})\, dV = R \tag{2.52}$$

where G is the shear modulus. The factor "4" has been included in equation (2.52) to account for the difference between the elasticity shear strains (e_{xy}, e_{xz}) and engineering shear strains $(\gamma_{xy}, \gamma_{xz})$. They have a difference of two times, i.e., $e_{xy} = \gamma_{xy}/2$ and $e_{xz} = \gamma_{xz}/2$.

The displacements of a generic point of the space frame element now contain three components u_x, u_y, and u_z. To describe the three-dimensional beam, we shall adopt *the principal centroidal coordinate system* (x, y, z), with x denoting the longitudinal axis, and y and z the major and minor principal axes, respectively, of the cross-section. Based on the *Bernoulli–Euler hypothesis* for bending and *St. Venant's theory* of torsion, the following displacement field can be written for a space frame element,

$$u_x = u - yv' - zw' \tag{2.53}$$

$$u_y = v - z\theta_x \tag{2.54}$$

$$u_z = w + y\theta_x \tag{2.55}$$

where (u, v, w) denote the displacements of the centroid C along the three axes and θ_x the angle of twist of the member. Accordingly, the three strain components are

$$e_{xx} = u' - yv'' - zw'' \tag{2.56}$$

$$e_{xy} = -z\theta_x' \tag{2.57}$$

$$e_{xz} = y\theta_x' \tag{2.58}$$

where it can be seen that the shear strains e_{xy} and e_{xz} are caused by the torsional action alone.

Substituting the preceding expressions for the strains into equation (2.52) yields the following virtual work equation:

$$\int_0^L (EAu'\delta u' + EI_z v''\delta v'' + EI_y w''\delta w'' + GJ\theta_x'\delta\theta_x')dx = R \tag{2.59}$$

where J is the torsional constant, and I_y and I_z are the moments of inertia of the cross-section about the y- and z-axis, respectively,

$$J = \int_A (y^2 + z^2)dA \tag{2.60}$$

$$I_y = \int_A z^2 dA \tag{2.61}$$

In addition to equation (2.10), the following orthogonality conditions for the principal centroidal coordinates have been adopted in the derivation of equation (2.59),

$$\int_A z\,dA = 0 \tag{2.62}$$

$$\int_A yz\,dA = 0 \tag{2.63}$$

With regard to the external virtual work R, we first note that the expression for R_b in (2.11) should be modified to include the contribution from the third direction z,

$$R_b = \int_A (t_x \delta u_x + t_y \delta u_y + t_z \delta u_z) dA \tag{2.64}$$

where t_z denotes the traction on the end section B of the element along the z-axis. The same modification should also be made for the virtual work R_a in (2.18). Based on the conditions of equilibrium, three more components of stress resultants can be defined for the space frame element, in addition to those of (2.12)–(2.14). Namely,

$$F_{zb} = \int_A t_z dA \tag{2.65}$$

$$M_{yb} = \int_A t_x z dA \tag{2.66}$$

$$M_{xb} = \int_A (t_z y - t_y z) dA \tag{2.67}$$

By substituting equations (2.53)–(2.55) for the displacements into equation (2.64), and utilizing the stress resultant definitions in (2.12)–(2.14) and (2.65)–(2.67), an expression identical in form to that of equation (2.15) can be obtained for the virtual work R_b. However, the displacement and force vectors in this expression should be augmented for end B of the beam as

$$\{u_b\}^T = \langle u_b \quad v_b \quad w_b \quad \theta_{xb} \quad \theta_{yb} \quad \theta_{zb} \rangle \tag{2.68}$$

$$\{f_b\}^T = \langle F_{xb} \quad F_{yb} \quad F_{zb} \quad M_{xb} \quad M_{yb} \quad M_{zb} \rangle \tag{2.69}$$

where $\theta_{yb} = -w_b'$ for the case of small rotations. Thus, we have shown that the use of six degrees of freedom, i.e., three translations and three rotations, can properly account for the external virtual work

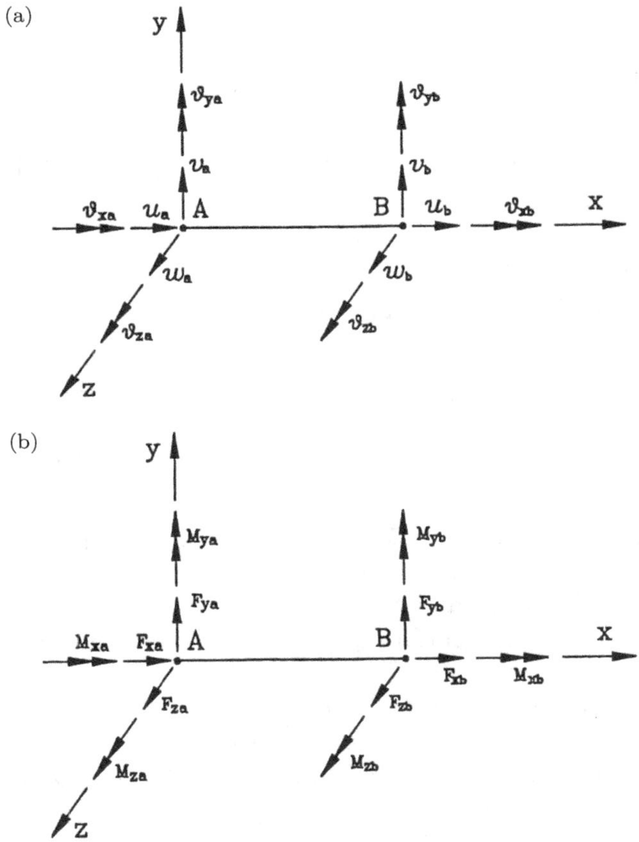

Fig. 2.4. Space frame element: (a) nodal degrees of freedom; (b) nodal forces and moments.

associated with each end of the space frame element. Corresponding to these nodal degrees of freedom, there are six stress resultants, i.e., three forces and moments. A graphical representation of the nodal quantities for the space frame element is shown in Figure 2.4.

Similarly, we may show that the relation (2.19) remains valid for R_a for end A of the space frame element, except that the displacement and force vectors should be augmented as follows:

$$\{u_a\}^T = \langle u_a \quad v_a \quad w_a \quad \theta_{xa} \quad \theta_{ya} \quad \theta_{za} \rangle \tag{2.70}$$

$$\{f_a\}^T = \langle F_{xa} \quad F_{ya} \quad F_{za} \quad M_{xa} \quad M_{ya} \quad M_{za} \rangle \tag{2.71}$$

where $\theta_{ya} = -w'_a$ is assumed. Realizing that $R = R_a + R_b$, we can proceed to show that the virtual work equation for the space frame element is

$$\int_0^L (EAu'\delta u' + EI_z v''\delta v'' + EI_y w''\delta w'' + GJ\theta'_x\delta\theta'_x)dx = \{\delta u\}^T\{f\} \tag{2.72}$$

in which

$$\{u\}^T = \langle u_a \quad v_a \quad w_a \quad \theta_{xa} \quad \theta_{ya} \quad \theta_{za} \quad u_b \quad v_b \quad w_b \quad \theta_{xb} \quad \theta_{yb} \quad \theta_{zb}\rangle \tag{2.73}$$

$$\{f\}^T = \langle F_{xa}\ F_{ya}\ F_{za}\ M_{xa}\ M_{ya}\ M_{za}\ F_{xb}\ F_{yb}\ F_{zb}\ M_{xb}\ M_{yb}\ M_{zb}\rangle \tag{2.74}$$

As was stated previously, a virtual work equation such as the one given in (2.72) serves as a valid basis for derivation of the space frame element. However, to provide judgments regarding the selection of interpolation functions and element degrees of freedom, an overview of the associated differential equations and boundary conditions is still helpful.

Following the same lines of derivation, the differential equations and boundary conditions that are associated with the space from element can be derived from (2.72). For the sake of brevity, only the differential equations that are not covered by the plane case are listed as follows:

$$EI_y w'''' = 0 \tag{2.75}$$

$$GJ\theta''_x = 0 \tag{2.76}$$

Correspondingly, the geometric and natural boundary conditions are

$$\delta w' = 0 \quad \text{or} \quad M_y = -EI_y w'' \tag{2.77}$$

$$\delta w = 0 \quad \text{or} \quad F_z = EI_y w''' \tag{2.78}$$

$$\delta\theta'_x = 0 \quad \text{or} \quad M_x = GJ\theta'_x \tag{2.79}$$

As can be seen from equations (2.75) and (2.76), the exact solutions for the transverse displacement w and angle of twist θ_x can be represented by cubic and linear polynomial functions, respectively. Consequently, they can be interpolated as follows:

$$w = \{n_3\}^T\{\overline{w}\} \tag{2.80}$$

$$\theta_x = \{n_1\}^T\{\overline{\theta}_x\} \tag{2.81}$$

in which

$$\{\overline{w}\}^T = \langle w_a \quad -L\theta_{ya} \quad w_b \quad -L\theta_{yb}\rangle \tag{2.82}$$

$$\{\overline{\theta}_x\}^T = \langle\theta_{xa} \quad \theta_{xb}\rangle \tag{2.83}$$

By interpolating the displacements $u, v, w,$ and θ_x by their nodal degrees of freedom, i.e., through substitution of (2.37), (2.38), (2.80), and (2.81), we can obtain from equation (2.72) the following:

$$\begin{aligned}&\{\delta\overline{u}\}^T\frac{EA}{L}\left[K_{11}^{110}\right]\{\overline{u}\} + \{\delta\overline{v}\}^T\frac{EI_z}{L^3}\left[K_{33}^{220}\right]\{\overline{v}\}\\&+\{\delta\overline{w}\}^T\frac{EI_y}{L^3}\left[K_{33}^{220}\right]\{\overline{w}\}\\&+\{\delta\overline{\theta}_x\}^T\frac{GJ}{L}\left[K_{11}^{110}\right]\{\overline{\theta}_x\} = \{\delta u\}^T\{f\}\end{aligned} \tag{2.84}$$

Since the virtual displacements $\{\delta\overline{u}\}$, $\{\delta\overline{v}\}$, $\{\delta\overline{w}\}$, and $\{\delta\overline{\theta}_x\}$ are correlated to $\{\delta u\}$ [see (2.41), (2.42), (2.82), (2.83), and (2.73)], we can take advantage of their arbitrary nature to derive from (2.84) the following matrix equations:

$$\frac{EI_y}{L^3}\left[K_{33}^{220}\right]\{\overline{w}\} = \{\overline{f}_z\} \tag{2.85}$$

$$\frac{GJ}{L}\left[K_{11}^{110}\right]\{\overline{\theta}_x\} = \{\overline{f}_\theta\} \tag{2.86}$$

in addition to those presented in equations (2.46) and (2.47). The corresponding force vectors $\{\overline{f}_z\}$ and $\{\overline{f}_\theta\}$ are defined as follows:

$$\{\overline{f}_z\} = \left\langle F_{za} \quad -\frac{M_{ya}}{L} \quad F_{zb} \quad -\frac{M_{yb}}{L}\right\rangle \tag{2.87}$$

$$\{\overline{f}_\theta\} = \langle M_{xa} \quad M_{xb}\rangle \tag{2.88}$$

Combining equations (2.85) and (2.86) with equations (2.46) and (2.47) yields the stiffness equations for the space frame element, $[k]\{u\} = \{f\}$, in which the element displacement vector $\{u\}$ and force vector $\{f\}$ are exactly those defined in equations (2.73) and (2.74), respectively. The stiffness matrix $[k]$ for the space frame element, which has a dimension of 12×12, can be given as follows:

$$[k] = \begin{bmatrix} [k_1] & [k_2] \\ [k_2]^T & [k_3] \end{bmatrix} \tag{2.89}$$

where the submatrices are

$$[k_1] = \begin{bmatrix} \frac{EA}{L} & 0 & 0 & 0 & 0 & 0 \\ & \frac{12EI_z}{L^3} & 0 & 0 & 0 & \frac{6EI_z}{L^2} \\ & & \frac{12EI_y}{L^3} & 0 & -\frac{12EI_y}{L^2} & 0 \\ & & & \frac{GJ}{L} & 0 & 0 \\ & Symm. & & & \frac{4EI_y}{L} & 0 \\ & & & & & \frac{4EI_z}{L} \end{bmatrix} \tag{2.90a}$$

$$[k_2] = \begin{bmatrix} -\frac{EA}{L} & 0 & 0 & 0 & 0 & 0 \\ 0 & -\frac{12EI_z}{L^3} & 0 & 0 & 0 & \frac{6EI_z}{L^2} \\ 0 & 0 & -\frac{12EI_y}{L^3} & 0 & -\frac{6EI_y}{L^2} & 0 \\ 0 & 0 & 0 & -\frac{GJ}{L} & 0 & 0 \\ 0 & 0 & \frac{6EI_y}{L^2} & 0 & \frac{2EI_y}{L} & 0 \\ 0 & -\frac{6EI_z}{L^2} & 0 & 0 & 0 & \frac{2EI_z}{L} \end{bmatrix} \tag{2.90b}$$

$$[k_3] = \begin{bmatrix} \frac{EA}{L} & 0 & 0 & 0 & 0 & 0 \\ & \frac{12EI_z}{L^3} & 0 & 0 & 0 & -\frac{6EI_z}{L^2} \\ & & \frac{12EI_y}{L^3} & 0 & \frac{6EI_y}{L^2} & 0 \\ & & & \frac{GJ}{L} & 0 & 0 \\ & Symm. & & & \frac{4EI_y}{L} & 0 \\ & & & & & \frac{4EI_z}{L} \end{bmatrix} \tag{2.90c}$$

2.2.3 *One-dimensional truss element*

A truss element distinguishes itself from a frame element in that it cannot resist bending, shearing or torsional actions. As both ends of the truss element are taken to be hinged, it can be treated as the special case of the plane frame element with zero moment of inertia I_z subjected to zero bending moments. The following is a summary of the procedure for derivation of the one-dimensional truss element.

For a truss element of constant cross-section, the axial displacement u_x can be interpreted as the displacement of the centroid of the cross-section, i.e., $u_x \equiv u$. Substituting equation (2.4) for the axial strain e_{xx} into equation (2.3), along with the use of equation (2.22) for the external work, yields

$$\int_0^L EAu'\delta u' dx = \{\delta u\}^T\{f\} \tag{2.91}$$

For a one-dimensional truss element, the following definitions can be adopted for the nodal displacement and force vectors (Figure 2.5):

$$\{u\}^T = \langle u_a \quad u_b \rangle \tag{2.92}$$

$$\{f\}^T = \langle F_{xa} \quad F_{xb} \rangle \tag{2.93}$$

since only axial deformations and actions are involved.

Fig. 2.5. One-dimensional truss element: (a) nodal degrees of freedom; (b) nodal forces.

Adopting the same linear interpolation as in equation (2.37) for the axial displacement u, we can write

$$u = \{n_1\}^T \{u\} \tag{2.94}$$

where the displacement vector $\{u\}$ has been defined in equation (2.92). Substituting the preceding equation for the axial displacement u into equation (2.91), and using the notation of (2.44), we can obtain

$$\{\delta u\}^T \frac{EA}{L} \left[K_{11}^{110}\right] \{u\} = \{\delta u\}^T \{f\} \tag{2.95}$$

By the arbitrary nature of the virtual displacement $\{\delta u\}$, it is possible to derive

$$\frac{EA}{L} \left[K_{11}^{110}\right] \{u\} = \{f\} \tag{2.96}$$

or in symbolic form as $[k]\{u\} = \{f\}$. For the one-dimensional truss element, the stiffness matrix $[k]$ is simply

$$[k] = \begin{bmatrix} \frac{EA}{L} & -\frac{EA}{L} \\ -\frac{EA}{L} & \frac{EA}{L} \end{bmatrix} \tag{2.97}$$

which is the simplest stiffness matrix that can be found for a single element in structural mechanics. Again, it should be added that in the nonlinear analysis of trusses in later chapters, the elastic stiffness matrix $[k]$ will also be referred to as the $[k_e]$ matrix.

2.2.4 *Plane and space truss elements*

One basic characteristic of the truss or bar element is that it can resist only axial forces, but not shear forces, bending moments, or torsional actions. Such a characteristic remains true regardless of whether the element is one-, two-, or three-dimensional. To achieve a systematic treatment for the assembly of element stiffness matrices in the analysis of plane and space trusses, it is helpful to have the element stiffness equations augmented in such a manner as to include the effects of forces and displacements in the transverse directions. For instance, the displacement vector $\{u\}$ and force vector $\{f\}$ can be augmented for the *plane truss element* as follows:

$$\{u\}^T = \langle u_a \quad v_a \quad u_b \quad v_b \rangle \tag{2.98}$$

$$\{f\}^T = \langle F_{xa} \quad F_{ya} \quad F_{xb} \quad F_{yb} \rangle \tag{2.99}$$

as was depicted in Figure 2.6. Correspondingly, the stiffness matrix $[k]$ for the two-dimensional truss can be augmented by adding zero entries to the rows and columns related to the transverse degrees of freedom as

$$[k] = \begin{bmatrix} \frac{EA}{L} & 0 & -\frac{EA}{L} & 0 \\ 0 & 0 & 0 & 0 \\ -\frac{EA}{L} & 0 & \frac{EA}{L} & 0 \\ 0 & 0 & 0 & 0 \end{bmatrix} \tag{2.100}$$

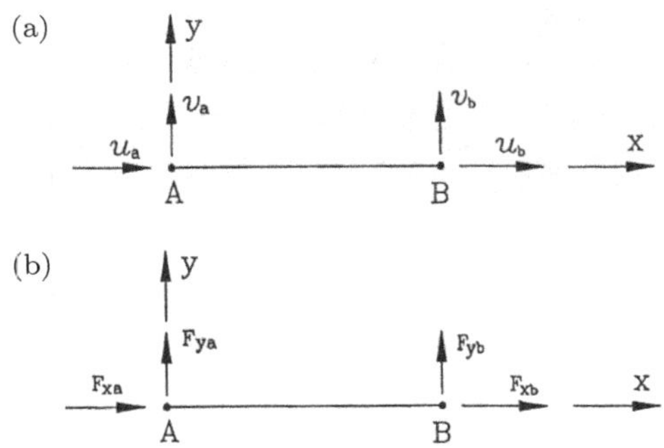

Fig. 2.6. Plane truss element: (a) nodal degrees of freedom; (b) nodal forces.

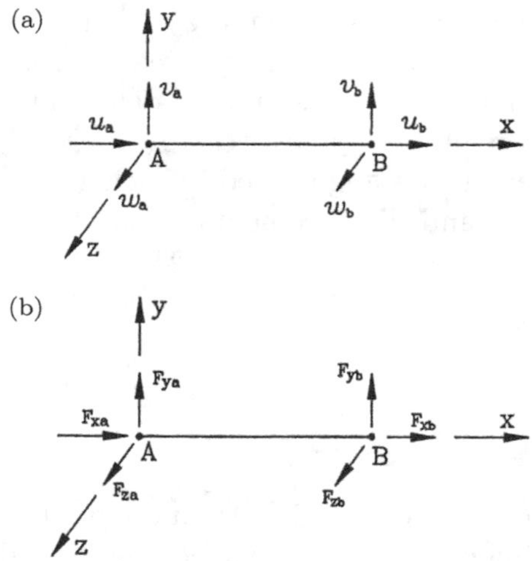

Fig. 2.7. Space truss element: (a) nodal degrees of freedom; (b) nodal forces.

On the other hand, for a *space truss element*, the displacement vector $\{u\}$ and force vector $\{f\}$ can be augmented to consist of three degrees of freedom at each node of the element:

$$\{u\}^T = \langle u_a \quad v_a \quad w_a \quad u_b \quad v_b \quad w_b \rangle \tag{2.101}$$

$$\{f\}^T = \langle F_{xa} \quad F_{ya} \quad F_{za} \quad F_{xb} \quad F_{yb} \quad F_{zb} \rangle \tag{2.102}$$

as plotted in Figure 2.7. Correspondingly, the augmented stiffness matrix $[k]$ is

$$[k] = \begin{bmatrix} \frac{EA}{L} & 0 & 0 & -\frac{EA}{L} & 0 & 0 \\ 0 & 0 & 0 & 0 & 0 & 0 \\ 0 & 0 & 0 & 0 & 0 & 0 \\ -\frac{EA}{L} & 0 & 0 & \frac{EA}{L} & 0 & 0 \\ 0 & 0 & 0 & 0 & 0 & 0 \\ 0 & 0 & 0 & 0 & 0 & 0 \end{bmatrix} \tag{2.103}$$

From the above expressions of the stiffness matrix $[k]$, it is obvious that a truss element can resist only the axial actions, but not the other actions.

Before closing this section, a comment should be given for the element loads. In the foregoing derivations, we have assumed that all the applied loads are concentrated at the nodal points simply for the sake of convenience. Such an assumption should not be regarded as a restriction of the theory, if one realizes that all loads not directly applied at the nodal points of an element can be easily transformed to those at the nodal points based on the concept of lumped loads or equivalent nodal loads (McGuire *et al.*, 2000).

2.3 Formation of Structure Equations

A fundamental issue in matrix structural analysis is to treat the structure as an assemblage of discrete finite elements. In linear analysis by the stiffness method, the formation of the structure equations is generally separated into two parts pertaining to the individual elements and the assemblage. Each of the two parts is formed in terms of the geometry, statics, and constitutive relations. In the preceding section, methods have been presented for derivation of the element stiffness equations:

$$[k]\{u\} = \{f\} \tag{2.104}$$

based on the principle of virtual displacements. In this section, we shall proceed to demonstrate how the stiffness equations for each element of the structure can be assembled to form the structure stiffness equations:

$$[K]\{U\} = \{P\} \tag{2.105}$$

following the procedure of the direct stiffness method. Here $[K]$ denotes the stiffness matrix, $\{U\}$ the displacement vector, and $\{P\}$ the load vector of the structure. It is not the purpose herein to make a comprehensive presentation on all variations of the procedure. Textbooks on matrix structural analysis or finite element methods, such as McGuire *et al.* (2000), Bathe and Wilson (1976), Rao (1982), and Cook *et al.* (1989), may be consulted for more details. In this section, only the most fundamental features of the procedure leading to the structure equations will be summarized.

First of all, we note that all the element equations derived in the preceding section have been referred to their *element* or *local*

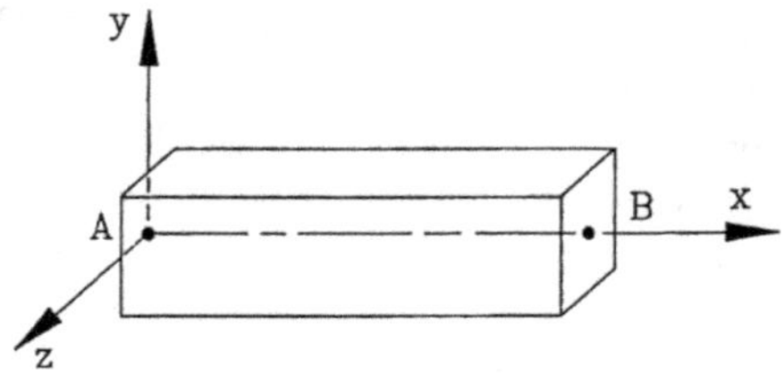

Fig. 2.8. Element coordinates.

coordinates. Before these equations can be assembled to form the structure equations, they must be transformed from their individual local coordinates to the *structure* or *global coordinates*. To this end, the orientation of each finite element must be specified with respect to the common global coordinate system.

To define the orientation of a space frame element with respect to the global coordinate axes X, Y, and Z, we normally attach to each element an element coordinate system x, y, and z. The x-axis, conventionally taken to be the centroidal axis of the element, is uniquely determined by the direction vector that extends from node A to node B of the element. As shown in Figure 2.8, the y-axis can be conveniently taken as the minor principal axis of the cross-section of the element, of which the direction can be determined either through the specification of a third node on the axis or a direction vector with reference to the global coordinates. The z-axis, on the other hand, can be computed as the cross product of the two vectors that represent the x- and y-axes. With the local axes x, y, and z specified for each element, the orientation of each element of the structure with respect to the global axes X, Y, and Z can be completely determined. Since a truss element has neither a major nor minor axis, the y- and z-axes of the truss element can be arbitrarily taken to be any pair of perpendicular axes lying on the plane at node A normal to the x-axis of the element.

Figure 2.9 shows a force vector $\overrightarrow{P}$ with its components expressed both in the local coordinate system (x, y, z) and in the global coordinate system (X, Y, Z). Using, respectively, the symbols λ, μ and ν, with appropriate subscripts, to designate the *direction cosines* of the angle between the subscripted local axis and each of the global axes

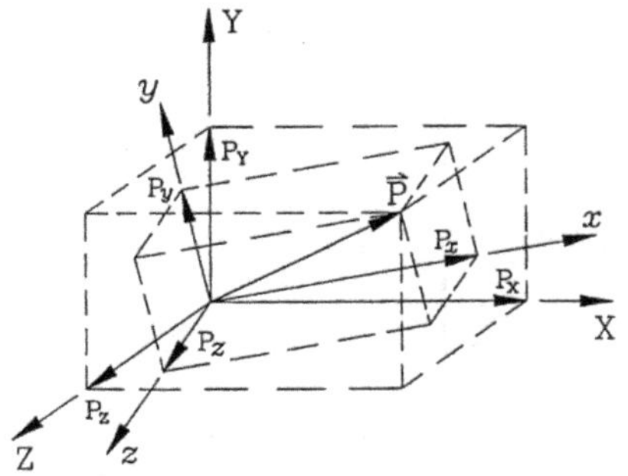

Fig. 2.9. Force in local and global coordinates.

(X, Y, Z), we can relate the components of the force vector $\overrightarrow{P}$ in the local axes to those in the global axes as follows:

$$\begin{Bmatrix} P_x \\ P_y \\ P_z \end{Bmatrix} = \begin{bmatrix} \lambda_x & \mu_x & \nu_x \\ \lambda_y & \mu_y & \nu_y \\ \lambda_z & \mu_z & \nu_z \end{bmatrix} \begin{Bmatrix} P_X \\ P_Y \\ P_Z \end{Bmatrix} \tag{2.106}$$

or in abridged form,

$$\{p\} = [\gamma]\{\hat{p}\} \tag{2.107}$$

in which $\{p\}$ and $\{\hat{p}\}$ denote the components of the force vector $\overrightarrow{P}$ in the local and global coordinate systems, respectively. Using the present notation, an element matrix or vector with reference to the global coordinates is designated by a hat "^". The *rotation matrix* $[\gamma]$ is known to be an *orthogonal matrix*, because its inverse is equal to its transpose, i.e., $[\gamma]^{-1} = [\gamma]^T$ (McGuire *et al.*, 2000). Recognizing that the same rule of transformation applies to the direct forces, moments, translations, and rotations, we can directly use equation (2.107) in the transformation of the displacement and force vectors for the space frame element:

$$\{u\} = [\Gamma]\{\hat{u}\} \tag{2.108}$$

$$\{f\} = [\Gamma]\{\hat{f}\} \tag{2.109}$$

where the *transformation matrix* $[\Gamma]$ for the space frame element with 12 degrees of freedom is simply

$$[\Gamma] = \begin{bmatrix} [\gamma] & & & \\ & [\gamma] & & \\ & & [\gamma] & \\ & & & [\gamma] \end{bmatrix} \tag{2.110}$$

In equations (2.108) and (2.109), $\{u\}$ and $\{f\}$ denote the element displacement and force vectors for the space frame element, which have been defined in equations (2.73) and (2.74), respectively, and $\{\hat{u}\}$ and $\{\hat{f}\}$ denote the corresponding quantities in the global coordinate system. Clearly, the transformation matrix $[\Gamma]$ is orthogonal, as the submatrices $[\gamma]$ contained in it are orthogonal.

Now, by substituting the relations (2.108) and (2.109) into the element stiffness equation (2.104), we get

$$[k][\Gamma]\{\hat{u}\} = [\Gamma]\{\hat{f}\} \tag{2.111}$$

Using the property of orthogonality for the $[\Gamma]$ matrix, i.e., $[\Gamma]^T[\Gamma] = [\Gamma]^{-1}[\Gamma] = [I]$, where $[I]$ is a unity matrix containing only unity in the diagonal entries, and zero for the remaining,

$$[\Gamma]^T[k][\Gamma]\{\hat{u}\} = \{\hat{f}\} \tag{2.112}$$

Thus, we have shown that the element stiffness equations can be transformed to the global coordinates as

$$[\hat{k}]\{\hat{u}\} = \{\hat{f}\} \tag{2.113}$$

where the matrix $[\hat{k}]$ in the global coordinates is related to the matrix $[k]$ in the local coordinates as

$$[\hat{k}] = [\Gamma]^T[k][\Gamma] \tag{2.114}$$

and the load vector $\{\hat{f}\}$ as

$$\{\hat{f}\} = [\Gamma]^T\{f\} \tag{2.115}$$

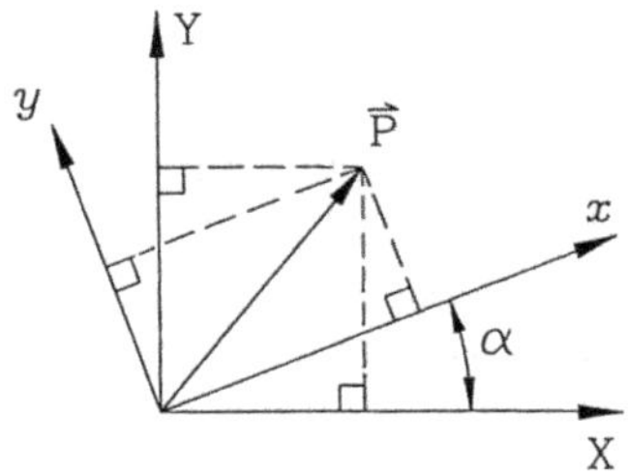

Fig. 2.10. Force in element and global coordinates – plane case.

The preceding two equations define what is called the *congruent transformation*. The transformed global matrix $[\hat{k}]$ is symmetric, so long as the original matrix $[k]$ is symmetric.

The transformation equations for other types of elements, such as the plane frame element, and plane and space truss elements, can be considered as special cases of the equations derived above for the space frame element. For example, for a plane frame element lying on the X–Y plane, as in Figure 2.10, there will be only three degrees of freedom, i.e., two translations and one rotation, existing at each end of the element. For this case, the *transformation matrix* $[\Gamma]$ reduces to

$$[\Gamma] = \begin{bmatrix} [\gamma] & \\ & [\gamma] \end{bmatrix} \tag{2.116}$$

and the *rotation matrix* $[\gamma]$ is simply

$$[\gamma] = \begin{bmatrix} \lambda_x & \mu_x & 0 \\ \lambda_y & \mu_y & 0 \\ 0 & 0 & 1 \end{bmatrix} \tag{2.117}$$

where $\lambda_x = \mu_y = \cos\alpha$ and $\mu_x = -\lambda_y = \sin\alpha$, with α denoting the angle between the local x and global coordinate axes (see Figure 2.10).

Once the element matrices and vectors are transformed to the global coordinate system, the next step is to construct the *master* or *global stiffness equations* for the entire structure based on the conditions of *compatibility* and *equilibrium* at the nodal points of the structure. By compatibility, we mean that all the elements meeting at a common node have the same displacement for each degree of freedom at the common node. By equilibrium, we mean that the nodal forces exerted by all the elements meeting at the same node will add up to counterbalance the external loads applied on the related degrees of freedom, as represented by their respective element stiffness equations. Based on the above considerations, the stiffness coefficients of all the elements meeting at the common node can be summed up for the related degrees of freedom to form the global stiffnesses. Such a process has been referred to as the *stiffness assembly* process. In the meantime, the load vector can be formed simply according to where (i.e., the degrees of freedom) the loads are applied.

The effect of *boundary conditions*, i.e., restraints, can be directly considered in the assembly of the structure stiffness equations, through the use of an identification vector for distinguishing the *active* from the *nonactive* (restrained) degrees of freedom. For instance, if some degrees of freedom are zero (restrained). Then the "columns" corresponding to these degrees of freedom in the global stiffness matrix can just be deleted. Meanwhile, the "rows" in the global stiffness matrix relate to the reactions to be induced for these degrees of freedom. They are nonactive in the solution process and can also be deleted. The final result is a reduced set of stiffness matrix (which is still a square one) with all the restrained degrees of freedom removed.

Let E denote the total number of elements of which the structure is composed, and N the total active number of active degrees of freedom of the structure excluding the restrained degrees of freedom. Then the stiffness matrix $[K]$ and load vector $\{P\}$ of the structure consisting of only the active degrees of freedom will have a dimension of $N \times N$ and $N \times 1$, respectively. Let n denote the number of degrees for each finite element. We have, for instance, $n = 6$ for a plane frame element, $n = 12$ for a space frame element, $n = 4$ for a plane truss element, and $n = 6$ for a space truss element. The element stiffness matrix $[\hat{k}]$ and load vector $\{\hat{f}\}$ (for the loads assigned elementwise) in the global coordinates are of the dimension of $n \times n$ and $n \times 1$, respectively. Conceptually, they can be augmented to the dimension of the global stiffness matrix $[K]$ and load vector $\{P\}$, i.e., $N \times N$ and $N \times 1$, respectively, by including zeros in the dummy entries and by deleting the entries corresponding to the restrained degrees of freedom. Also, the sequential numbers of all active degrees of freedom of each element can be assigned identically to those of the structure by bookkeeping. In so doing, the stiffness matrix $[K]$ and load vector $\{P\}$ of the structure can be established by algebraic addition as

$$[K] = \sum_{i=1}^{E} [\hat{k}^{(i)}] \tag{2.118}$$

and

$$\{P\} = \sum_{i=1}^{E} \{\hat{f}^{(i)}\} \tag{2.119}$$

where $[\hat{k}^{(i)}]$ is the augmented stiffness matrix and $\{\hat{f}^{(i)}\}$ the augmented force vector of element i with the restrained degrees of freedom excluded. In computer programming, all the entries of both $[K]$ and $\{P\}$ are assigned zero values before the summation is executed.

In the above assembly process, only the loads associated with each element has been considered. The implication is that the external loads are not applied at the nodal points, but are applied on the elements and then included in the element force vector $\{\hat{f}^{(i)}\}$ by the concept of *equivalent nodal loads* and the like. For the more general case where the external loads are applied not only on the elements but also at the nodal points of the structure, equation (2.119) should be modified to

$$\{P\} = \sum_{i=1}^{E} \{\hat{f}^{(i)}\} + \{P\}_n \tag{2.120}$$

where the vector $\{P\}_n$ denotes the concentrated loads that are directly applied at the nodal points of the structure.

It should be noted that both equations (2.118) and (2.119) remain valid even if the assemblage contains different types of elements. In this case, the number of element degrees of freedom, n, may change from element to element. In practice, the expansion of the element matrices $[\hat{k}^{(i)}]$ and $\{\hat{f}^{(i)}\}$ from their original dimensions of $n \times n$ and $n \times 1$ to match the dimensions of the structure matrices $[K]$ and $\{P\}$, i.e., $N \times N$ and $N \times 1$, is not necessary. One simple way is to first identify the locations of the entries of the element matrices $[\hat{k}^{(i)}]$ and $\{\hat{f}^{(i)}\}$ in the structure matrices $[K]$ and $\{P\}$, respectively, and then add them to the existing values in those locations as we loop over all the elements, i.e., with the index i changing for the element from 1 to E.

In preparation of the finite element mesh, we attach to each node a node number and each element an element number. The connection of each element to the assemblage and therefore, the locations of $[\hat{k}^{(i)}]$ and $\{\hat{f}^{(i)}\}$ in $[K]$ and $\{P\}$, respectively, can be determined from the degrees of freedom associated with the labels of the nodal points at the two ends of each element.

The procedure presented in this section serves only to illustrate the basic concept involved in the formation of the structure stiffness equations. Variants of the procedure that take into account the efficiency of computer memory and other factors, such as the symmetry, bandwidth, and sparsity of the stiffness matrix $[K]$, are available elsewhere. Readers interested in such topics should refer to finite element books, such as Tong and Rossettos (1977) and Bathe and Wilson (1976), for more information.

2.4 Solution of Simultaneous Equations

Assume that a structure is properly restrained against rigid body motions and that all restrained degrees of freedom of the structure have been removed in the formation of the structure stiffness equations:

$$[K]\{U\} = \{P\} \tag{2.121}$$

where $[K]$ denotes the stiffness matrix, $\{U\}$ the displacement vector to be solved, and $\{P\}$ is the load vector of the structure, assumed to be given. If the structure has a total of N degrees of freedom, then the preceding matrix equation will contain N simultaneous equations. One typical problem in the finite element analysis is to solve for this set of N simultaneous equations. For linear problems, the stiffness matrix $[K]$ is assumed to be constant, whereas for nonlinear problems, the stiffness matrix $[K]$ are functions of the initial displacements or initial element forces. In this section, only linear problems will be dealt with.

Methods for solving a system of equations of high dimensions can be classified either as *direct methods* or *iterative methods*. An iterative method differs from the direct method in that the solution of the algebraic equations is obtained by an iterative process, of which the number of iterations can be large. Iterative methods are not suitable for the solution no of nonlinear systems to be discussed in Chapter 7, which may involve iterations of other sort to account for the variation in stiffness of the structure. Thus, iterative methods are excluded from the discussion in this section.

Among the direct methods, Gaussian elimination has been widely used in the solution of simultaneous equations. Description of the procedure for Gaussian elimination can be found in almost any textbook on matrix structural analysis or finite element methods. In this section, we shall focus on the direct methods that rely on the decomposition schemes, especially on those methods that can be effectively adopted in solving structural systems not only of the stable nature, of which the stiffness matrix is *positive-definite*, but also of the unstable nature, of which the stiffness matrix may be *positive-semidefinite* or *indefinite*. By positive-definite, we mean that all the eigenvalues of a matrix are positive. By positive-semidefinite, we mean that all the eigenvalues are greater than or equal to zero. And by indefinite, we mean that the eigenvalues of a matrix are either positive, zero, or negative. The consideration of *definiteness* of a matrix is extremely important in the solution of critical loads for structures by the eigenvalue approach or in the solution of the postbuckling behavior for structures using the incremental-iterative approaches.

In this book, we shall deal mainly with *conservative structures*, in the sense that the work done by the loadings applied to the structure depends solely on the initial and final configurations of the structure, but not the path of deformation. For a conservative structure, the stiffness matrix $[K]$ will always be *symmetric*. In this case, the stiffness matrix $[K]$ can be factored as follows:

$$[K] = [L][D][L]^T \tag{2.122}$$

where $[L]$ is a lower triangular matrix and $[D]$ is a diagonal matrix,

$$[L] = \begin{bmatrix} L_{11} & & & \\ L_{21} & L_{22} & & \\ \vdots & & \ddots & \\ L_{N1} & L_{N2} & \cdots & L_{NN} \end{bmatrix} \quad \text{and} \quad [D] = \begin{bmatrix} D_1 & & & \\ & D_2 & & \\ & & \ddots & \\ & & & D_N \end{bmatrix} \tag{2.123}$$

By expansion,

$$K_{ij} = \sum_{r=1}^{j} L_{ir} D_r L_{jr} \quad (j \leq i;\ i = 1, 2, \ldots, N) \tag{2.124}$$

Since the stiffness matrix $[K]$ is symmetric, i.e., $K_{ij} = K_{ji}$, the preceding expressions contain at most $N(N+1)/2$ independent equations, but there are $N(N+1)/2$ entries of L_{ij} and N entries of D_i to be evaluated. We therefore have at least N free choices in assigning the values of L_{ij} and D_i.

With the *Cholesky method*, all of the N entries D_i of the diagonal matrix $[D]$ are set equal to unity. In accordance, we have the following recurrence formulas for L_{ij}:

$$L_{ii} = \sqrt{K_{ii} - \sum_{r=1}^{i-1} L_{ir}^2} \tag{2.125}$$

$$L_{ij} = \frac{K_{ij} - \sum_{r=1}^{j-1} L_{ir} L_{jr}}{L_{jj}} \quad (i > j) \tag{2.126}$$

which may be applied by columns. One drawback with the Cholesky method is that it works only for positive-definite systems. For systems that are not positive-definite, we may encounter numerical difficulties such as the occurrence of negative numbers within the square root of equation (2.125) in calculating the diagonal entries L_{ii}. Because of this restriction, the method is not recommended for use beyond the solution of positive-definite systems.

Alternatively, we may choose to set the N diagonal entries L_{ii} equal to unity. In this case, the recurrence formulas for L_{ij} and D_i are as follows:

$$D_i = K_{ii} - \sum_{r=1}^{i-1} D_r L_{ir}^2 \tag{2.127}$$

$$L_{ij} = \frac{K_{ij} - \sum_{r=1}^{j-1} L_{ir} D_r L_{jr}}{D_i} \quad (i > j) \tag{2.128}$$

This method has been referred to as the *modified Cholesky method* or *triple-factoring method*. Since the square root is no longer required, the numerical difficulties associated with the original Cholesky method have been circumvented. Accordingly, the modified method can be effectively used in the solution of both positive-definite and indefinite systems.

As a side remark, we like to mention that the modified Cholesky decomposition has other important properties, in addition to the computation advantages mentioned above. One property is that the determinant of the stiffness matrix $[K]$ is equal to that of the diagonal matrix $[D]$,

$$\det[K] = \det\left([L][D][L]^T\right) = \prod_{i=1}^{N} D_{ii} \tag{2.129}$$

based on the fact that all the diagonal entries of the lower triangular matrix $[L]$ are equal to one, thereby resulting in $\det[L] = 1$. The second property has been referred to as the *Sturm sequence property*. It states that the number of negative entries in the diagonal matrix $[D]$ is equal to the number of negative eigenvalues for the stiffness matrix $[K]$. Because of these properties, the modified Cholesky decomposition can be advantageously used in the postbuckling analysis of structures involving the transition from stable to unstable phases, and vice versa.

By replacing the stiffness matrix $[K]$ by $[L][D][L]^T$, the structure stiffness equations become

$$[L][D][L]^T\{U\} = \{P\} \tag{2.130}$$

The displacement vector $\{U\}$ can be solved in three steps. First, by letting $[D][L]^T\{U\} = \{G\}$, equation (2.130) reduces to

$$[L]\{G\} = \{P\} \tag{2.131}$$

Since $[L]$ is a lower triangular matrix, we can determine $\{G\}$ by forward substitution starting from the first equation of the system. This will yield the following for each entry of the $\{G\}$ vector:

$$G_i = \frac{1}{L_{ii}}\left(P_i - \sum_{j=1}^{i-1} L_{ij}G_j\right) \tag{2.132}$$

where for $i = 1$, we have $G_1 = P_1/L_{11}$. The second step is to divide both sides of $[D][L]^T\{U\} = \{G\}$ by $[D]$, leading to

$$[L]^T\{U\} = \{H\} \tag{2.133}$$

where $\{H\} = [D]^{-1}\{G\}$ or, equivalently, $H_i = G_i/D_i$. The final step is to solve the preceding equation for $\{U\}$. This can be accomplished in a simple process by backward substitution from the last equation of the system. The following is the recurrence formula for each entry of $\{U\}$:

$$U_i = \frac{1}{L_{ii}}\left(H_i - \sum_{j=i+1}^{N} L_{ji}U_j\right) \tag{2.134}$$

where for starting, we let $U_N = H_N/L_{NN}$.

Thus, we have shown how the structure stiffness equations can be solved by the modified Cholesky method. After the structure displacements $\{U\}$ are obtained, the displacements at the two ends of each element can be determined from the geometric connectivity of the element to the structure, with those for the restrained degrees of freedom set equal to zero, if there are any. Note that the element displacements directly extracted from $\{U\}$ were expressed in the global coordinates, which can be transformed into the element coordinates to yield the element displacement vector $\{u\}$. In linear analysis, we can multiply the element displacements $\{u\}$ by the element stiffness matrix $[k]$ to obtain the element forces $\{f\}$ in the same coordinates. The procedures in this section and the preceding one are typical for the analysis of structural frames and trusses concerning the linear behaviors.

2.5 Quality Tests for Linear Elements

Before we talk about the quality of finite elements, let us briefly review some basic features of the most notable approximate method, the *Rayleigh–Ritz method*, for solving the boundary-value problems. Using this method, the first step is to approximate the exact displacement field of a structure by functions which contain a finite number of independent coefficients. By substituting the trial displacement field into the potential of the structure, and by requiring the potential to be a minimum, the variational principle can be employed to derive a set of simultaneous algebraic equations, from which the independent coefficients can be determined. One basic requirement with the Rayleigh–Ritz method is that the assumed functions must be *kinematically admissible*, that is, they must satisfy the internal compatibility and geometric boundary conditions. There is, however, no requirement on the satisfaction of the natural boundary conditions. If we can represent the trial displacement field by a series of functions that form a complete set, then by adding more functions to the series, the trial displacement field will converge to the exact solution for the problem (Shames and Dym, 1985).

The finite element method can be regarded as a variant of the Rayleigh-Ritz method. By using the finite element method, a structure is broken down into a number of finite elements that are connected at the nodal points. Note that in addition to the interelement nodal points, internal nodal points may also exist within a finite element. However, the degrees of freedom associated with the internal nodal points can always be eliminated by the static condensation procedure.

The approximate displacement field of the structure is determined from the displacements at the nodal points, including the internal nodal points, if there are any, through the use of the *interpolation functions*, also known as the *shape functions*, for each finite element. The accuracy of the finite element analysis can be improved through refinement of the finite element mesh used to model the structure, i.e., by increasing the number of nodal degrees of freedom. To ensure convergence of the finite element solutions to the exact one, certain criteria have to be satisfied. All these criteria fall into three groups concerning the *compatibility*, *completeness*, and *stability* of the finite elements used. For elements that fulfill all these criteria, the convergence characteristics of the approximate solutions to the exact one can be shown to be monotonic or nearly asymptotic.

2.5.1 *Convergence criteria*

With respect to the *compatibility* of the finite elements, it is required that the displacements within each element and across the element boundaries be continuous. For the frame and truss elements discussed in this book, polynomial functions are used to interpolate the element displacements. There is no problem in maintaining a continuous displacement field within the finite element.

Elements that satisfy both the compatibility and completeness conditions (the latter will be discussed below) are called *conforming elements*, while those that satisfy only the condition of completeness but violate the condition of interelement compatibility are called *nonconforming* or *incompatible elements*. A nonconforming element can still be used in practice to yield accurate results if the element can approach the state of *constant strain* as the finite element mesh is refined. One difference is that the manner of convergence exhibited by the finite element models with nonconforming elements may not be monotonic. It should be mentioned that the elements to be discussed in this book, including the truss element and frame element, either two- or three-dimensional, all fall in the category of conforming elements, as they all satisfy the interelement compatibility conditions.

The requirement of *completeness* implies that the displacement functions of the elements must be able to represent the *rigid body modes* and the *constant strain states*. From the point of finite element formulation, we would expect the assumed displacement field of a finite element to be able to represent the exact displacement field as close as possible. However, this is not always achievable in practice. In the finite element formulation, the ability of a finite element to represent the rigid body modes becomes the most fundamental requirement. When using the polynomial functions as the interpolation functions, the rigid body modes can be easily considered through the inclusion of the constant and linear terms in the polynomial series.

From a physical point of view, the necessity for finite elements to represent the constant strain states can be understood if we try to use more and more elements to model a structure. In the limit when the finite elements get smaller and smaller, the strain in each element should approach a constant value. Such a capability is essential to the modeling of structures showing complex variation in strain. In beams and plates, where bending deformations are involved, the strains should be interpreted as *generalized strains*. Obviously, the state of constant strains incorporates the rigid body modes as a special case with zero strains.

By *stability*, we mean that the solution of the structure stiffness equations (2.121) must remain bounded and unique. This implies that the stiffness matrix $[K]$ must be nonsingular and the structure must be stable, given just enough restraint conditions to prevent rigid body motions. For a stable structure, the displacements $\{U\}$ solved from equations (2.121) contain no zero-energy modes. The stability condition has excluded the possibility of any divergence or numerical overflow in the solution process. It is therefore a sufficient condition of convergence. In contrast, the completeness and compatibility conditions mentioned above represent merely a necessary condition of convergence.

2.5.2 *The patch test*

All the above requirements, that is, the constant strain conditions, rigid body capabilities, interelement compatibility, and stability conditions, can be checked by a simple test originated by Irons, now known as the *patch test* (Bazeley *et al.*, 1965; Taylor *et al.*, 1986). In this test, we first assemble a "patch" of elements in such a way that at least one node is completely surrounded by elements. Boundary nodes are loaded by consistently derived nodal loads corresponding to a state of constant stress. Internal nodes are neither loaded or restrained. To prevent the patch from rigid body motion, only a minimal number of restraints needs to be provided to the boundary nodes. By solving the structure stiffness equations, we obtain the nodal displacements $\{U\}$ and compute the element stresses (or strains) therefrom. If the computed stresses (or strains) within each element do agree with the exact values within certain tolerances, then the patch test is said to be passed.

The above procedure has the advantage that any singularity of the $[K]$ matrix can be immediately detected. This represents a check on the stability of the elements that form the patch. It should be mentioned that in a standard patch test, all possible constant strain states demanded of the element should be examined. Thus, an element is considered to have met all the convergence criteria if it can pass the patch test. When elements of this type are used in modeling the structure, it can be assured that the approximate solutions will converge to the exact one upon mesh refinement.

The test procedure described above serves both as a necessary and sufficient condition for convergence of the finite element solution to the exact one. Theoretically speaking, the patch test has to be passed for elements with infinitesimal size. However, for most elements that adopt the polynomials as the interpolation functions, the patch size did not in fact enter into consideration. And the requirement that the patch test be passed for any element size becomes a standard. Occasionally, an element fails to pass the patch test when it is of finite size may pass the patch test when it becomes infinitesimally small. For such elements, the patch test is said to be passed in the weak sense, and convergence to correct results can still be assured.

Variants of the patch test exist in the literature. For instance, an alternate version of the patch test is to assign a set of displacements $\{U\}$ consistent with the constant strain state to all nodal degrees of freedom, and then compute the corresponding nodal forces $\{P\}$:

$$\{P\} = [K]\{U\} \tag{2.135}$$

If all the nodal forces inside the boundary of the patch can represent the corresponding state of constant stress, then the patch test is said to be passed. Since a test of this form verifies only the satisfaction of the basic differential equations, but not the stability condition and the approximation in boundary conditions, it serves merely as a necessary condition for convergence.

2.5.3 *Eigenvalue test*

The eigenvalue test can detect the instability, lack of invariance, and other defects of a single element, by which the quality of competing elements can be estimated (Cook *et al.*, 1989). It is one of the most widely used procedures for checking the element quality. Consider the following eigenvalue equation for an unrestrained finite element:

$$([k] - \lambda[I])\{u\} = \{0\} \tag{2.136}$$

where $[k]$ denotes the complete element matrix, $[I]$ a unit matrix, $\{0\}$ a zero vector, λ the eigenvalue, and $\{u\}$ the corresponding eigenvector. There are as many eigenvalues λ as the number of degrees of freedom contained in the eigenvector $\{u\}$. Let the eigenvector $\{u\}_i$ be normalized such that

$$\{u\}_i^T \{u\}_i = 1 \tag{2.137}$$

We can premultiply equation (2.136) by $\{u\}_i^T$ to obtain

$$\{u\}_i^T [k]\{u\}_i = \lambda_i \quad or \quad 2U_i = \lambda_i \tag{2.138}$$

where U_i is the strain energy of the element associated with the eigenvector $\{u\}_i$. As can be seen from equation (2.138), a consistent stiffness matrix $[k]$ should yield a *zero eigenvalue*, i.e., $\lambda_i = 0$, if the associated eigenvector $\{u\}_i$ represents a *rigid body mode*. For a two-dimensional finite element, three linearly independent rigid body motions are possible. Therefore, the eigenvalues λ_i obtained from equation (2.136) should contain three zero roots. Similarly, for three-dimensional finite elements, there should be six zero roots λ_i to account for all possible rigid body motions (i.e. three rotations and three displacements in three-dimensional space). In testing a finite element, we first check that the stiffness matrix $[k]$ has as many zero eigenvalues as expected. Too few suggest that the element will be artificially strained when subjected to rigid body motions. Too many suggest that certain kinds of *mechanisms* have been introduced into the element in the formulation stage (due to inconsistencies in theory) or in the programming stage (due to hidden coding errors). A finite element having too many zero eigenvalues may lose its stability under particular mesh patterns or loading conditions.

The basic idea of this section is to shed light on the physical principles underlying the various tests that have been frequently used in the literature for linear elements. Since the linear elements presented in Section 2.2 do not violate any of the compatibility, completeness, and stability requirements, it is confirmed that they can all pass the tests presented in this section. In the following chapter, we shall proceed to demonstrate how the rigid body properties presented herein can be extended to the testing of nonlinear elements in the incremental form.

References

Bathe, K.J. and Wilson, E.L. (1976). *Numerical Methods in Finite Element Analysis*, Prentice Hall, Englewood Cliffs, N.J.

Bazeley, G.P., Cheung, Y.K., Irons, B.M., and Zienkiewicz, O.C. (1965). Triangular elements in plate bending – Conforming and nonconforming elements, in *Proc. Conf. on Matrix Methods in Structural Mechanics*, Wright-Patterson Air Force Base, Ohio, pp. 547–576.

Cook, R.D., Malkus, D.S., and Plesha, M.E., (1989), *Concepts and Applications of Finite Element Analysis*, 3rdedn., John Wiley, New York, N.Y.

Gere, J.M. and Timoshenko, S.P. (1997). *Mechanics of Materials*, 4thedn., PWS, Boston, MA.

McGuire, W., Gallagher, R.H., and Ziemian, R.D. (2000). *Matrix Structural Analysis*, 2bdedn., John Wiley & Sons, New York, N.Y.

Pesquera, C.I., McGuire, W., and Abel, J.F. (1983). Interactive graphical preprocessing of three-dimensional structures, *Comput. Struct.*, 17(1), 1–12.

Rajasekaran, S. (1977). Finite element method for plastic beam-columns, in Chen, W.F. and Atsuta, T.(eds.),*Theory of Beam-Columns, Vol. 2: Space Behavior and Design*, McGraw-Hill, New York, N.Y., 539–608.

Rao, S.S. (1982). *The Finite Element Method in Engineering*, Pergamon Press, Oxford, England.

Shames, I.H. and Dym, C.L. (1985). *Energy and Finite Element Methods in Structural Mechanics*, McGraw-Hill, New York, N.Y.

Taylor, R.L., Simo, J.C., Zienkiewicz, O.C. and Chan, A.C.H. (1986). The patch test – a condition for assessing FEM convergence, *Int. J. Numer. Meth. Eng.*, 22(1), 39–62.

Tong, P. and Rossettos, J.N. (1977). *Finite Element Method – Basic Techniques and Implementation*, MIT Press, Cambridge, MA.

Chapter 3

Rigid Body Rule and Plane Frame Element

3.1 Introduction

While we have talked about the quality test for linear finite elements, there is virtually no test available for ensuring the quality of nonlinear elements to be used in an incremental nonlinear analysis. A nonlinear element to be used in the incremental analysis is different from the linear elements in that it may be stressed under the initial nodal loads. Such an effect of initial nodal loads (or stresses) has to be taken into account in proposing the quality test for nonlinear finite elements in relation to the rigid body motions.

To the knowledge of the authors, the *rigid body rule* proposed by Yang and Chiou (1987) for testing the quality of plane beam elements is the first test existing for nonlinear finite elements. This test requires that the initial forces acting on a finite element rotate and translate with the rigid body motion, while the magnitudes of the initial forces remain unchanged, so as to preserve the equilibrium of the finite element prior to and after the rigid body rotation. Such a test is consistent with the basic physical laws and is, therefore, more complete than the previous tests for linear finite elements on rigid body motions, especially with regard to the existing initial forces on the finite element.

In fact, the rigid body rule is a very useful tool for structural nonlinear analysis. As was stated above, it can be used to test the quality of a nonlinear element. Later, in the incremental-iterative procedure, the rigid body rule can be employed to update the nodal forces existing at an incremental or iterative step (known as the corrector, Chapter 7), or in guiding the direction of trial iterations (in the predictor, Chapter 7). Moreover, the rigid body rule can be adopted as well to derive rigid-qualified geometric stiffness matrices (Chapter 8). For the above reasons, the methods and procedures to be presented in this book are referred to as the *physics-based computational method*. It is shown that only when a consistent finite element is used, along with a physically justified incremental-iterative procedure, convergent and accurate solutions can be ensured for the nonlinear problem of concern.

For a finite element to be legitimate in linear analysis, it should exhibit the state of zero strains and zero stresses (or nodal forces) when subjected to rigid body motions. Such a requirement is implied by the *patch test*, which determines the quality of the finite elements used by checking whether or not a patch of elements surrounding a common node can consistently reproduce the state of zero strains when subjected to rigid body motions. Applications of the patch test have been restricted to linear problems and may be found in textbooks, such as Zienkiewicz (1977), Irons and Ahmad (1980), and Cook *et al.* (1989), among others.

However, comparatively little work has been conducted on the quality test of finite elements to be used for nonlinear problems. To the authors' knowledge, there is no simple test like the patch test that can be used to determine whether a finite element is qualified for nonlinear analysis. In a geometrically nonlinear analysis based on the stiffness formulation, the geometric stiffness matrix $[k_g]$ should be included in addition to the linear (or elastic) stiffness matrix $[k_e]$ in any of the tests. The linear stiffness matrix $[k_e]$, being derived from the strain energy, will generate zero element forces when subjected to a rigid body motion if it is properly derived or if it passes the patch test. The geometric stiffness matrix $[k_g]$, which represents the effect of initial forces on a finite element due to geometry change and is the source of structural instability, will produce non-zero forces when subjected to rigid body motion. Thus, the geometric stiffness matrix contains the most important message that describes the behavior of a prestressed element undergoing a rigid body motion. However, to examine the legitimacy of a nonlinear finite element, we cannot rely solely on the geometric stiffness matrix but must consider the incremental element stiffness equation as a whole, say, based on the updated Lagrangian formulation, to see if the finite element can cope with the basic physical rule on the rigid body motion.

Previously, the geometric stiffness matrix was decomposed by Porter and Powell (1971) into two parts: the *internal* and *external* stiffness matrices. The internal stiffness matrix, as well as the linear stiffness matrix, accounts for the changes in the nodal forces due to natural deformations of the element in an incremental loading step, while the external stiffness matrix accounts for the changes in the nodal forces due to the rigid body movements of the finite element in the same incremental step. It was demonstrated by Jagannathan *et al.* (1975) that neglect of certain higher order terms in the strain-displacement relations may result in fictitious strains for the finite element undergoing a rigid rotation. In that case, inconsistencies in structural behavior may occur.

The concept of internal and external stiffness matrices was exploited further by Gattass (1982) to explain the equilibrium of a two-dimensional beam element undergoing a rigid body rotation based on the updated Lagrangian formulation. Using the same Lagrangian formulation, the stiffness equation was derived by Yang and McGuire (1986a) for a three-dimensional I-shaped beam element, which consists of a total of seven degrees of freedom at each end including the warping degree. In a companion paper by Yang and McGuire (1986b), a more general variational method was devised to establish the external and internal stiffness matrices for the I-shaped beam element. The equilibrium of such an element undergoing various rigid body motions in the three dimensions was extensively discussed. One problem with the aforementioned works is that they focus only on the behavior of the geometric stiffness matrix with regard to rigid body motions. In effect, to judge whether a nonlinear finite element is justified, we should examine the behavior of the entire element stiffness equation, including the geometric stiffness matrix, in the process of rigid body movements.

Although the effect of rigid body motions on equilibrium of the finite element has been considered in previous research, no attention has been paid to the effect of rigid body motions on the magnitudes and directions of the initial forces on the finite element prior to the publication of the paper by Yang and Chiou (1987). This chapter has been revised from the paper by Yang and Chiou (1987), aiming to fill such a gap. For illustration of the basic physical behaviors involved, a two-dimensional beam element will be derived based on the updated Lagrangian formulation. Then, the basic *rigid body rule* will be introduced, which governs the behavior of the initial loads acting on the finite element when subjected to rigid body motions. Consequently, we expect that for a finite element with initial loads to be legitimate in a nonlinear analysis, the finite element must exhibit the same behavior as that prescribed by the rigid body rule for rigid body motions. It is based on such a concept that the *rigid body motion test* is proposed for the finite elements to be used in an incremental nonlinear analysis. In the numerical example, it is demonstrated that very good nonlinear solutions can be obtained if the finite element used in modeling the structure can pass the rigid body motion test and if a consistent procedure is adopted for calculating the element forces in the incremental-iterative process. The basic concept of the rigid body motion test exploited herein for the two-dimensional beam element can be extended to other more complicated solid elements concerning their formulation for nonlinear analysis, for instance, regarding the legitimacy of the space frame element presented in Chapter 6.

3.2 Principle of Virtual Displacements

Conventionally, both the updated Lagrangian (UL) formulation and total Lagrangian (TL) formulation have been used in the incremental nonlinear analysis of structures. When consistently developed, the two formulations should yield identical results. Both the two formulations were reviewed by Bathe *et al.* (1975). In this chapter, the UL formulation is adopted since it was shown to be computationally more efficient for the analysis of beam structures (Bathe and Bolourchi, 1979).

The derivations to be presented in the following follow basically the same lines as those of Yang and McGuire (1986a). Consider a two-dimensional beam element in its path of motion in Figure 3.1. We use configuration C_0 to denote the initial undeformed state of the beam, configuration C_1 the last (known) deformed equilibrium state, and configuration C_2 the current (desired) deformed state. In the UL formulation, the reference state for each element is its last

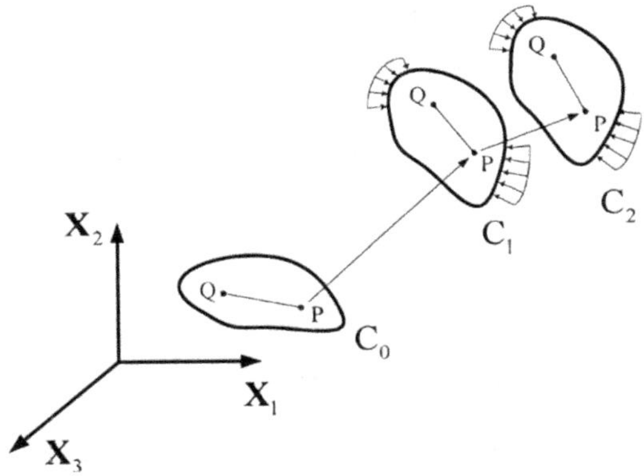

Fig. 3.1. Motion of a solid structure.

calculated configuration C_1, and the incremental deformation of the beam element from C_1 to C_2 is assumed to be small.

The description of the motion of the beam element is based on the three configurations C_0, C_1, and C_2; left subscripts and superscripts on a symbol are used to denote these configurations. Particularly, a left superscript denotes the configuration in which the quantity occurs. The absence of such a superscript indicates that the quantity is an increment between C_1 and C_2. A left subscript denotes the configuration in which the quantity is measured.

The principle of virtual displacements requires that the equilibrium of the solid be represented at the current deformed state C_2:

$$\int_{{}^2V} {}^2\tau_{ij}\delta_2 e_{ij}{}^2dV = {}^2_2R \tag{3.1}$$

in which the external work ${}^2_2R$ is

$${}^2_2R = \int_{{}^2A} {}^2_2t_k \delta u_k{}^2dA \tag{3.2}$$

neglecting the body force. In equations (3.1) and (3.2), ${}^2\tau_{ij}$ is the Cauchy stress tensor, $\delta_2 e_{ij}$ the variation of the linear strain tensor, ${}^2_2t_k$ the surface tractions, and δu_k the variation of the displacement components at C_2.

In the updated Lagrangian formulation, equation (3.1) can be transformed into a linearized incremental form as

$$\int_{^1V} {}_1C_{ijkl}e_{kl}\delta_1 e_{ij}{}^1dV + \int_{^1V} {}^1\tau_{ij}\delta_1\eta_{ij}{}^1dV = {}^2_1R - {}^1_1R \tag{3.3}$$

in which

$${}^2_1R = \int_{^1A} {}^2_1t_k\delta u_k{}^1dA \tag{3.4}$$

$${}^1_1R = \int_{^1V} {}^1\tau_{ij}\delta_1 e_{ij}{}^1dV \tag{3.5}$$

In equations (3.3)–(3.5), ${}_1e_{ij}$ and ${}_1\eta_{ij}$ are the linear and nonlinear components of the Green–Lagrange incremental strain tensor, ${}^1\tau_{ij}$ denotes the Cauchy stress tensor at C_1, ${}_1C_{ijkl}$ the components of the constitutive tensor, and ${}^2_1t_k$ the surface at C_2 but measured with reference to C_1.

It is noted that for a solid in equilibrium at C_1, the internal virtual work should be equal to the external virtual work, i.e.,

$${}^1_1R = \int_{^1V} {}^1\tau_{ij}\delta_1 e_{ij}{}^1dV = \int_{^1A} {}^1_1t_k\delta u_k{}^1dA \tag{3.6}$$

in which ${}^1_1t_k$ denotes the surface tractions at C_1. Note that equation (3.6) may also be obtained from equations (3.1) and (3.2) by switching all the subscripts and superscripts from “2” to “1”.

Figure 3.2 shows a two-dimensional beam at C_1. Let (x, y) be the orthogonal Cartesian coordinate system such that x coincides with the centroidal axis of the beam. For the two-dimensional beam, there are two independent stress components: ${}^1\sigma_{xx}(= {}^1\tau_{xx})$ and ${}^1\tau_{xy}$. Correspondingly, there are two independent strain components. The linear and nonlinear parts of the strain components are denoted by $({}_1e_{xx}, {}_1e_{xy})$ and $({}_1\eta_{xx}, {}_1\eta_{xy})$, respectively. Hence, the virtual work equation (3.3) reduces to the following for the two-dimensional beam:

$$\int_V (Ee_{xx}\delta e_{xx} + 2Ge_{xy}\delta e_{xy})dV$$
$$+ \int_V (\sigma_{xx}\delta\eta_{xx} + 2\tau_{xy}\delta\eta_{xy})dV = {}^2R - {}^1R \tag{3.7}$$

in which E and G are the moduli of elasticity and rigidity, respectively, and the factor of 2 has been added to account for the symmetry of the shear stresses and strains, i.e., $\tau_{xy} = \tau_{yx}$, $e_{xy} = e_{yx}$,

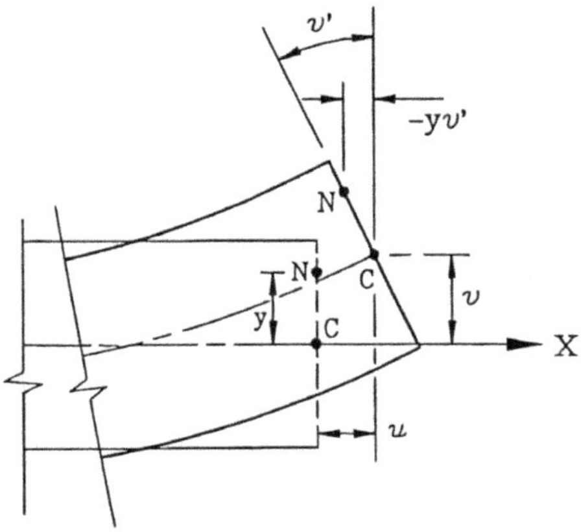

Fig. 3.2. Coordinates of two-dimensional beam.

and $\eta_{xy} = \eta_{yx}$. In equation (3.7), the subscripts and superscripts "1" on the stress and strain components and external virtual works have been dropped for clarity.

The first and second integrals on the left-hand side of equation (3.7) represent the change in *strain energy* and *potential energy*, respectively, caused by the incremental deformation of the beam from C_1 to C_2. The terms on the right-hand side of equation (3.7), i.e., ${}^2R - {}^1R$, represent the increment in external virtual work resulting from the increase of the surface tractions from ${}^1_1t_k$ at C_1 to ${}^2_1t_k$ at C_2.

3.2.1 *Strain–displacement relations*

Let u_x and u_y denote the longitudinal and transverse displacements of a *generic point* N at section x, as shown in Figure 3.2. The strains at point N may be expressed in terms of the displacements at the same point as

$$e_{xx} = u_{x,x} \tag{3.8a}$$

$$e_{xy} = \frac{1}{2}(u_{x,y} + u_{y,x}) \tag{3.8b}$$

$$\eta_{xx} = \frac{1}{2}(u_{x,x}^2 + u_{y,x}^2) \tag{3.8c}$$

$$\eta_{xy} = \frac{1}{2}(u_{x,y}u_{x,x} + u_{y,y}u_{y,x}) \tag{3.8d}$$

where a comma denotes differentiation with respect to the coordinate following. Let u and v denote the displacements of the *centroid* C at section x. Based on the *Bernoulli–Euler hypothesis* of plane sections remaining plane and orthogonal to the centroidal axis after deformation, the displacements of point N at the same section can be given by

$$u_x = u - yv' \tag{3.9a}$$

$$u_y = v \tag{3.9b}$$

where a prime denotes differentiation with respect to coordinate x. Substituting equations (3.9) into equations (3.8) yields

$$e_{xx} = u' - yv'' \tag{3.10a}$$

$$e_{xy} = 0 \tag{3.10b}$$

$$\eta_{xx} = \frac{1}{2}(u'^2 - 2yu'v'' + y^2v''^2 + v'^2) \tag{3.10c}$$

$$\eta_{xy} = \frac{1}{2}(-u'v' + yv'v'') \tag{3.10d}$$

It should be noted that some of the terms in η_{xx} and η_{xy} do not represent exactly the physical reality but are caused by the adoption of the Bernoulli–Euler hypothesis. This hypothesis is known to yield good results for linear analysis. But when it is "over" extended to nonlinear analysis, it will result in some nonlinear terms that cannot be physically justified. For this reason, we shall omit the last term $yv'v''$ in η_{xy} of equation (3.10d) in the following derivation since it cannot be well interpreted physically when judged from the rigid body consideration. With this omission, *the finite element derived is ensured to pass the rigid body motion,* as will be demonstrated later on in the related test. On the contrary, if this term is included in the derivation, some artificial forces will be induced when the element, initially loaded, is subjected to rigid body motions.

3.2.2 *Stress resultants*

The stress resultants for a two-dimensional beam at C_1 include the axial force ${}^1F_x$, transverse shear ${}^1F_y$, and bending moment ${}^1M_z$, which are defined as follows:

$$^1F_x = \int_A {}^1\sigma_{xx} dA \tag{3.11a}$$

$$^1F_y = \int_A {}^1\tau_{xy} dA \tag{3.11b}$$

$$^1M_z = -\int_A {}^1\sigma_{xx} y dA \tag{3.11c}$$

where A is the cross-sectional area of the beam and the stresses inside the integrals are recognized as the Cauchy stresses.

3.2.3 *External virtual works*

In the following, the symbol $\langle\ \rangle$ will be used for a row vector and $\{\ \}$ for a column vector. For the present purposes, we assume that only loadings of the concentrated type are allowed to act at the two ends A and B of the beam. Consequently, the external virtual work ^1R in equation (3.7) done by the external loads on the beam at C_1 can be expressed as the product of the nodal forces $\{^1f\}$ and the virtual displacements $\{\delta u\}$:

$$^1R = \{u\}^T\{^1f\} \tag{3.12}$$

where the displacement vector $\{u\}$, as shown in Figure 3.3, is

$$\{u\}^T = \langle u_a \quad v_a \quad \theta_a \quad u_b \quad v_b \quad \theta_b \rangle \tag{3.13}$$

which subscripts a and b are used for ends A and B, respectively, and θ is the rotation, taken to be the derivative of the displacement v.

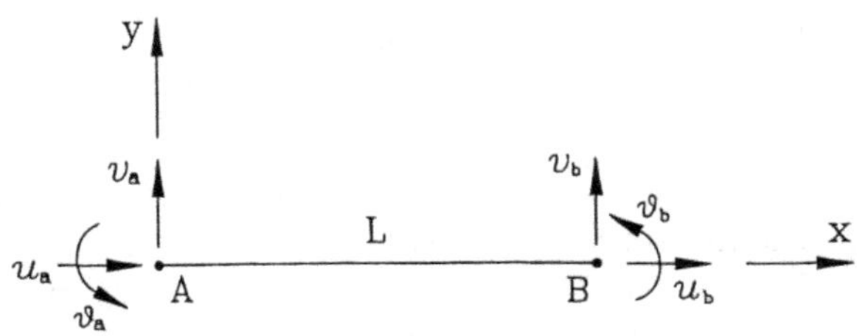

Fig. 3.3. Nodal degrees of freedom of two-dimensional beam element.

Correspondingly, the force vector $\{^1f\}$ acting at C_1 is

$$\{^1f\}^T = \langle ^1F_{xa} \quad ^1F_{ya} \quad ^1M_{za} \quad ^1F_{xb} \quad ^1F_{ya} \quad ^1M_{zb}\rangle \tag{3.14}$$

Similarly, the virtual work ^2R done by the external loads on the beam at C_2 can be expressed as the product of the nodal forces $\{^2f\}$ and the corresponding virtual displacements $\{\delta u\}$:

$$^2R = \{u\}^T\{^2f\} \tag{3.15}$$

where the nodal force vector $\{^2f\}$ can be obtained by switching the superscripts in equation (3.14) from "1" to "2".

3.2.4 *Virtual work equation*

Substituting the expressions for the strains, equations (3.10a)–(3.10d), into equation (3.7), noting that $dV = dAdx$, and using the stress resultant definitions, equations (3.11a)–(3.11c), together with the expressions for the external virtual works in equations (3.12) and (3.15), we can arrive at the following equation for an element of length L:

$$\begin{aligned}
&\int_0^L (EAu'\delta u' + EI_z v''\delta v'')dx \\
&\quad + \frac{1}{2}\int_0^L \left[^1F_x\delta(u'^2 + v'^2) + {^1F_x}\frac{I_z}{A}\delta(v'')^2 \right] dx \\
&\quad + \int_0^L [^1M_z\delta(u'v'') - {^1F_y}\delta(u'v')]dx = \langle \delta u\rangle(\{^2f\} - \{^1f\})
\end{aligned} \tag{3.16}$$

where I_z is the moment of inertia and the displacements (u, v) should be recognized as displacements of the centroid C of the cross-section.

The first integral in equation (3.16) is the *strain energy* of the element in variational form. From this term, the elastic stiffness matrix $[k_e]$ can be derived, which has been presented as equation (2.51) in Chapter 2, and is reproduced later on for easy referencing. The second integral in equation (3.16) is the *potential energy* in variational form, which is a function of the initial forces acting on the element. From this term, the geometric stiffness $[k_g]$ can be derived, as is shown later on.

While there is a general consensus on the composition of the strain energy for the beam and truss element, there is a concern that the potential energy should cover the effects of all the forces acting on the element (i.e., beam or truss element) with regard to structural instability, as is the case presented in equation (3.16). A detailed review of most existing geometric stiffnesses will show that not all the force components have been properly included (Przemieniecki, 1968; Porter and Powell, 1971; Chajes, 1974; Chen and Lui, 1987; Cook *et al.*, 1989). As such, they have not been presented in a unified form.

What we like to emphasize here is that the procedure for deriving the element stiffness matrices is not a foolproof one, using either the virtual work principle or the potential energy approach. There are always decisions that must be made regarding the retaining or omission of higher-order terms. Before we proceed to derive the stiffness matrices for the two-dimensional beam, it is instructive to have scrutiny at the governing differential equations and boundary conditions for describing the behavior of the structural member. This can help unveil the basic mechanical behavior of the beam under rigid body motions. Besides, we can judge if the theory is really qualified for formulating the finite element.

3.3 Governing Differential Equations and Boundary Conditions

The virtual work equation given in equation (3.16) is used to derive the governing differential equations and boundary conditions. Integrating equation (3.16) for the beam by parts to obtain the virtual displacement δu and δv, we can obtain

$$
\begin{aligned}
&\int_0^L [-EAu'' - {}^1F_x u'' - ({}^1M_z v'')' + {}^1F_y v'']\delta u dx \\
&\quad + \int_0^L \left[EI_z v'''' - {}^1F_x v'' + {}^1F_x \frac{I_z}{A} v'''' + ({}^1M_z u')'' + {}^1F_y u''\right] \delta v dx \\
&\quad + (EAu' + {}^1F_x u' + {}^1M_z v'' - {}^1F_y v')\delta u\Big|_0^L
\end{aligned}
$$

$$+\left[-EI_z v''' + {}^1F_x v' - {}^1F_x \frac{I_z}{A} v''' - ({}^1M_z u')' - {}^1F_y u'\right] \delta v\Big|_0^L$$
$$+ \left(EI_z v'' + {}^1F_x v'' + {}^1M_z u'\right) \delta v'\big|_0^L$$
$$= \langle \delta u \rangle \left(\{{}^2 f\} - \{{}^1 f\}\right) \tag{3.17}$$

where it should be noted that for a beam with no distributed loads, both the axial force ${}^1F$ and shear force ${}^1F_y$ are constant throughout its length.

Admitting that the virtual displacement δu, δv, and $\delta v'(= \delta\theta)$ are arbitrary in nature, we can derive from the preceding equation the *buckling differential equations* as the Euler–Lagrange equations of the functional for the beam:

$$EAu'' + {}^1F_x u'' + ({}^1M_z v'')' - {}^1F_y v'' = 0 \tag{3.18}$$

$$EI_z v'''' - {}^1F_x v'' + {}^1F_x \frac{I_z}{A} v'''' + ({}^1M_z u')'' + {}^1F_y u'' = 0 \tag{3.19}$$

Evidently, the effect of all the initial forces acting on the two-dimensional beam at C_1 has been taken into account. As a byproduct of the variational procedure, two sets of boundary conditions can be obtained for the beam. The *natural boundary conditions* at ends A and B are

$$EAu' + {}^1F_x u' + {}^1M_z v'' - {}^1F_y v' = {}^2F_x - {}^1F_x \tag{3.20}$$

$$-EI_z v'''' + {}^1F_x v' - {}^1F_x \frac{I_z}{A} v''' - ({}^1M_z u')' - {}^1F_y u' = {}^2F_y - {}^1F_y \tag{3.21}$$

$$EI_z v'' + {}^1F_x \frac{I_z}{A} v'' + {}^1M_z u' = {}^2M_z - {}^1M_z \tag{3.22}$$

in which the deformation sign convention in Figure 2.3 should be used for the force parameters involved. Correspondingly, the geometric boundary conditions at ends A and B are

$$\delta u = 0 \tag{3.23}$$

$$\delta v = 0 \tag{3.24}$$

$$\delta v' = \delta\theta = 0 \tag{3.25}$$

The buckling differential equations and boundary conditions as derived in equations (3.18)–(3.19) and (3.20)–(3.22) are the equations of the *general theory* for the two-dimensional beam (Yang and Kuo, 1994) since they have been derived from the virtual work equation (3.16) with complete coverage of the nonlinear strain components, except the term $yv'v''$ in η_{xy} of equation (3.10d) deemed to be not rigid-body qualified.

In the literature, the geometric boundary conditions are also known as the *essential boundary conditions.* They correspond to prescribed displacements and rotations of the beam under consideration. Meanwhile, the natural boundary conditions correspond to prescribed boundary forces and moments. In a typical boundary-value problem, either the geometric or the natural boundary conditions, but not both, should be prescribed at each end of the beam.

3.4 Finite Element Formulation

In structural mechanics, there is basically no argument on the elastic stiffness matrices used for the beam and truss elements, as they conform excellently to the basic governing equations based on linear theories. They are not the focus of the discussion of this book. However, the same is not true for the geometric stiffness matrix derived, particularly with regard to its qualification for use in an incremental nonlinear analysis.

For a two-dimensional beam with nodes A and B (see Figure 3.3), the nodal displacement vector $\{u\}$ is given in equation (3.13) and the initial load vector $\{^1f\}$ existing at the beginning of the incremental step is given in equation (3.14). The load vector $\langle^2f\rangle$ existing at the end of the incremental step can be obtained by switching the superscripts in equation (3.14) from “1” to “2”. An important feature of the initial forces $\{^1f\}$ is that they are in equilibrium at C_1.

In the present formulation, a linear displacement field is adopted for the axial displacement u_x, and a cubic displacement field for the transverse displacement u_y, i.e.,

$$u_x = \langle n_1\rangle\{\bar{u}_x\} \tag{3.26a}$$

$$u_y = \langle n_3\rangle\{\bar{u}_y\} \tag{3.26b}$$

in which, with $i(= x/L)$, the interpolation functions can be expressed as follows:

$$\langle n_1 \rangle = \langle (1-i) \quad i \rangle \tag{3.27a}$$

$$\langle n_3 \rangle = \langle (1-3i^2+2i^3) \quad (i-2i^2+i^3) \quad (3i^2-2i^3) \quad (i^3-i^2) \rangle \tag{3.27b}$$

and the axial and transverse displacement vectors are (see Figure 3.3)

$$\langle \bar{u}_x \rangle = \langle u_a \quad u_b \rangle \tag{3.28a}$$

$$\langle \bar{u}_y \rangle = \langle v_a \quad L\theta_a \quad v_b \quad L\theta_b \rangle \tag{3.28b}$$

Based on the conditions of equilibrium of the beam element at C_1, the stress resultants at section x of the beam element in C_1 can be related to those at the two nodes as

$${}^1F_x = {}^1F_{xb} \tag{3.29a}$$

$${}^1F_y = -\frac{{}^1M_{za} + {}^1M_{zb}}{L} \tag{3.29b}$$

$${}^1M_y = -{}^1M_{ya}(1-i) + {}^1M_{yb}i \tag{3.29c}$$

In an incremental nonlinear analysis, the forces acting on each element should be in equilibrium at C_1, as is implied above, if the previous iterations have been properly performed.

Now, by performing the following steps to the virtual work equation in (3.16), (1) substituting the expressions for the incremental displacements in equations (3.26a), (3.26b), (2) substituting the expressions for the stress resultants in equations (3.29a)–(3.29c), and (3) admitting the arbitrary nature of virtual displacements, we can obtain after some operation the *incremental stiffness equation* of the beam element as

$$[k_e]\{u\} + [k_g]\{u\} = \{^2f\} - \{^1f\} \tag{3.30}$$

in which $[k_e]$ and $[k_g]$ are the *linear* (or *elastic*) *stiffness* and *geometric stiffness matrices*, respectively. The linear stiffness matrix $[k_e]$ was denoted as $[k]$ in equation (2.51), which is reproduced as follows:

$$[k_e] = \begin{bmatrix} \frac{EA}{L} & 0 & 0 & -\frac{EA}{L} & 0 & 0 \\ & \frac{12EI_z}{L^3} & \frac{6EI_z}{L^2} & 0 & -\frac{12EI_z}{L^3} & \frac{6EI_z}{L^2} \\ & & \frac{4EI_z}{L} & 0 & -\frac{6EI_z}{L^2} & \frac{2EI_z}{L} \\ & & & \frac{EA}{L} & 0 & 0 \\ & \text{Symm.} & & & \frac{12EI_z}{L^3} & -\frac{6EI_z}{L^2} \\ & & & & & \frac{4EI_z}{L} \end{bmatrix} \tag{3.31}$$

The geometric stiffness matrix $[k_g]$ is a function of the initial nodal forces $\{^1f\}$ acting on the element at C_1:

$$[k_g] = \begin{bmatrix} [A] & [B] \\ [B]^T & [C] \end{bmatrix} \tag{3.32}$$

where

$$[A] = \begin{bmatrix} \frac{^1F_{xb}}{L} & 0 & -\frac{^1M_{za}}{L} \\ 0 & \frac{6\,^1F_{xb}}{5L} + \frac{12I_z\,^1F_{xb}}{AL^3} & \frac{^1F_{xb}}{10} + \frac{6I_z\,^1F_{xb}}{AL^2} \\ -\frac{^1M_{za}}{L} & \frac{^1F_{xb}}{10} + \frac{6I_z\,^1F_{xb}}{AL^2} & \frac{2L\,^1F_{xb}}{15} + \frac{4I_z\,^1F_{xb}}{AL} \end{bmatrix} \tag{3.32a}$$

$$[B] = \begin{bmatrix} -\frac{^1F_{xb}}{L} & 0 & -\frac{^1M_{zb}}{L} \\ 0 & -\frac{6\,^1F_{xb}}{5L} - \frac{12I_z\,^1F_{xb}}{AL^3} & \frac{^1F_{xb}}{10} + \frac{6I_z\,^1F_{xb}}{AL^2} \\ \frac{^1M_{za}}{L} & -\frac{^1F_{xb}}{10} - \frac{6I_z\,^1F_{xb}}{AL^2} & -\frac{L\,^1F_{xb}}{30} + \frac{2I_z\,^1F_{xb}}{AL} \end{bmatrix} \tag{3.32b}$$

$$[C] = \begin{bmatrix} \frac{^1F_{xb}}{L} & 0 & \frac{^1M_{zb}}{L} \\ 0 & \frac{6\,^1F_{xb}}{5L} + \frac{12I_z\,^1F_{xb}}{AL^3} & -\frac{^1F_{xb}}{10} - \frac{6I_z\,^1F_{xb}}{AL^2} \\ \frac{^1M_{zb}}{L} & -\frac{^1F_{xb}}{10} - \frac{6I_z\,^1F_{xb}}{AL^2} & \frac{2L\,^1F_{xb}}{15} + \frac{4I_z\,^1F_{xb}}{AL} \end{bmatrix} \tag{3.32c}$$

The geometric stiffness matrix is the source of structural instability. Our focus in this chapter is to assess the quality of the geometric stiffness matrix $[k_g]$.

With the incremental stiffness equation derived in equation (3.30) for each element, we can now proceed to assemble all the element equations to form the *incremental stiffness equation* for the structure. Such a procedure has been presented in Section 2.3 by considering the conditions of compatibility and equilibrium for all the nodal points of the structure, along with the imposition of boundary condition. Corresponding to the element stiffness equation (3.30), the following is the incremental stiffness equation for the structure:

$$[K_e]\{U\} + [K_g]\{U\} = \{^2P\} - \{^1P\} \tag{3.33}$$

where, corresponding to the element stiffness equation (3.30), $\{U\}$ is the displacement vector, $[K_e]$ the elastic stiffness matrix, and $[K_g]$ the geometric stiffness matrix of the structure. The load vectors $\{^1P\}$ and $\{^2P\}$ denote the loads acting on the structure at C_1 and C_2, respectively. In a typical incremental step, $\{^2P\}$ is also interpreted as the total applied loads and $\{^1P\}$ the total initial loads of the structure at C_1.

3.5 Basic Physical Law for Rigid Motion

Consider a bar sitting on the surface of the earth and subjected to a gravitational force P on the top. For equilibrium, a reaction of the same magnitude P will be induced on the bottom, as shown in Figure 3.4(a). Such a loading configuration can be regarded as the C_1 configuration of the bar. Now, suppose that the earth rotates as a rigid body by angle θ_r. In accordance, the bar will rotate to the C_2 configuration as shown in Figure 3.4(b). From this, we observe that the line of action of force P rotates with the movement, while its magnitude remains unchanged. An overall result is the preservation of equilibrium of the bar in the C_2 configuration.

The kind of rigid behavior described herein is the minimum requirement for a solid when it is initially acted upon by a set of *conservative loads*, e.g., of the type of gravitational force, with its direction of action pointed toward the gravity center. It is true that the solid may be further deformed (aside from rigid rotation) when subjected to additional loads. But any theory or finite elements describing the behavior of the solid should be able to cope with the extreme or minimum case of rigid body rotation. Fortunately, the rigid body behavior of the solid as depicted in Figure 3.4 can be perfectly simulated in the incremental sense for a solid moving from C_1 to C_2 if care is duly exerted.

For instance, if the bar in Figure 3.4 is represented by a finite element that is initially stressed by a set of forces in equilibrium at C_1 in

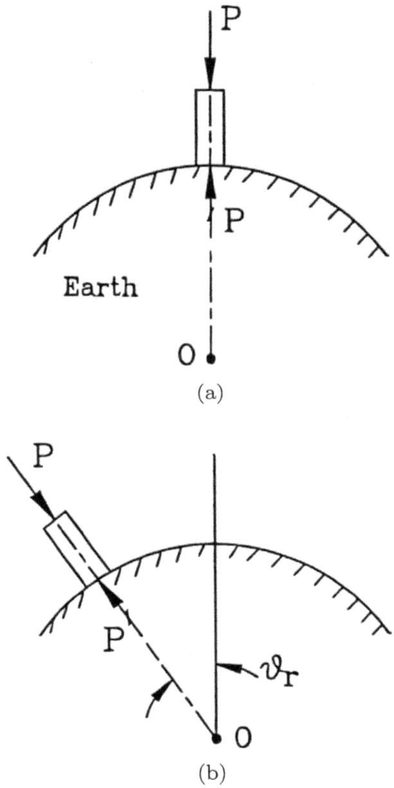

Fig. 3.4. Initially stressed bar: (a) prior to rigid rotation; (b) after rigid rotation.

Figure 3.5, which is the situation encountered at the beginning of an incremental step in nonlinear analysis. It is anticipated that the finite element should be able to cope with the rigid rotation (an extreme case of deformation) from C_1 to C_2, to end up with the situation at C_2 in Figure 3.5. Based on this, we are saying that *for an initially stressed finite element to be subjected to a rigid rotation, the initial forces acting on the element in equilibrium at C_1 must rotate with the rigid rotation, while their magnitudes remain unchanged. The final result is the preservation of equilibrium of the element at C_2, i.e., after the rigid rotation.* This is exactly the philosophy underlying the rigid body motion test proposed by Yang and Chiou (1987) for testing the quality of nonlinear finite elements.

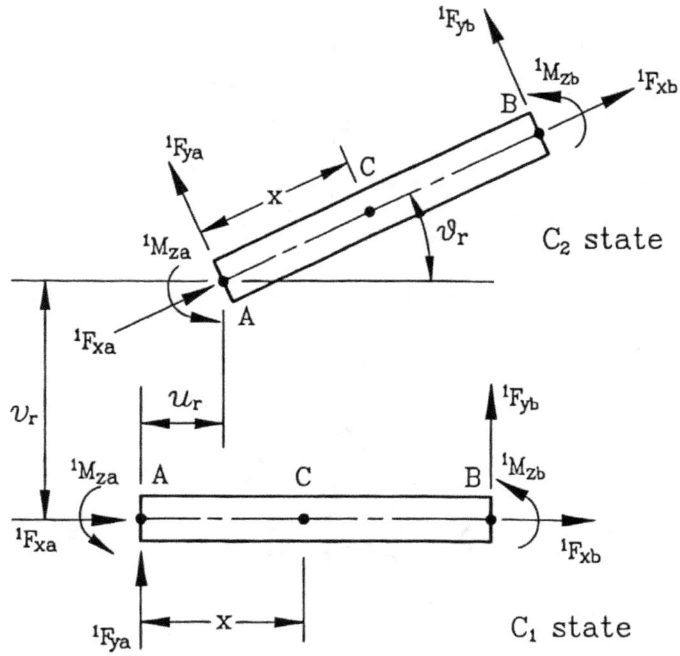

Fig. 3.5. Initially stressed beam prior to and after the rigid body motion.

Unlike the *patch test* that deals with the *non-stressed* finite elements under rigid displacements, the rigid body motion test presented herein is more general in that it includes the effect of initial loads or initial stresses existing on the finite element. Such a test should be conducted in the incremental sense, rather than the linear sense, as is presented in the following.

3.6 Rigid Body Motion Test for Beam Theory

Based on the general theory, the virtual work equation (3.16) has been adopted as the basis for deriving the element stiffness matrices in Section 3.4. The element stiffness equations so derived can be regarded as a numerical representation (or approximation) of the governing differential equations in (3.18)–(3.19) and natural boundary conditions in (3.20)–(3.22), with the understanding that the geometric boundary conditions are normally satisfied in the finite element modeling. Any errors that have been generated during the finite element formulation are called *formulation errors*. Moreover, we know that the natural boundary conditions in (3.20)–(3.22) can in effect be obtained through integration of the governing differential equations in (3.18)–(3.19). As such, the finite element equations derived, which relate the nodal forces to nodal displacements, can also be regarded as a numerical representation of the natural boundary conditions (3.20)–(3.22).

To obtain reliable solutions for nonlinear problems, it is essential that the element stiffness equations employed in analysis pass the rigid body test as a minimum requirement. Based on the fact that the element stiffness equations are a numerical representation of the natural boundary conditions, it is reasonable to expect that the underlying natural boundary conditions should pass the rigid body motion test as a prerequisite. This serves as a minimum criterion for justifying the legitimacy of the underlying nonlinear theory.

In the following, the rigid body motion test will be employed to justify the natural boundary conditions previously derived for the beam in (3.20)–(3.22). With reference to Figure 3.5, we consider that a two-dimensional beam initially equilibrated by a set of nodal forces, i.e., ${}^1F_{xa}$, ${}^1F_{ya}$, ${}^1M_{za}$, ${}^1F_{xb}$, ${}^1F_{yb}$, and ${}^1M_{zb}$, is subjected to a rigid body motion. According to the rigid body rule, the initial forces acting on the beam after the rigid body motion at C_2 should remain unchanged in magnitude but rotate with the rigid rotation. To examine if the natural boundary conditions of the beam can really cope with such a phenomenon, let us assume that the rigid body displacement to be imposed on the beam consists of translations u_r and v_r along the x- and y-axes, and a rotation θ_r about the z-axis. With this, the rigid displacement field for the centroid C of the beam can be written as follows:

$$u = u_r \tag{3.34}$$

$$v = v_r + x\theta_r \tag{3.35}$$

where x denotes the position of the section and the rotation θ_r is small in magnitude.

By substituting the rigid displacements in equations (3.34) and (3.35) into the natural boundary conditions (3.20)–(3.22), the following expressions can be obtained for the forces acting on the two ends of the beam after the rigid body motion at C_2:

$$-{}^1F_y\theta_r = {}^2F_x - {}^1F_x \tag{3.36}$$

$${}^1F_x\theta_r = {}^2F_y - {}^1F_y \tag{3.37}$$

$$0 = {}^2M_z - {}^1M_z \tag{3.38}$$

Consequently, the forces acting at end B of the beam after the rigid body motion at C_2 are

$${}^2F_{xb} = {}^1F_{xb} - {}^1F_{yb}\theta_r \tag{3.39}$$

$${}^2F_{yb} = {}^1F_{yb} + {}^1F_{xb}\theta_r \tag{3.40}$$

$${}^2M_{zb} = {}^1M_{za} \tag{3.41}$$

and those at end A can be expressed in a similar way, i.e., by switching the subscript b in the preceding equations to a. Due to the difference between the finite element sign convention (i.e., rightward, upward, and counterclockwise positive) and the deformation sign convention (see Figure 2.3), the nodal forces can be related to the member forces as follows: ${}^2F_{xa} = -{}^2F_x$, ${}^2F_{ya} = -{}^2F_y, \ldots,$ ${}^1F_{xa} = -{}^1F_x$, ${}^1F_{ya} = -{}^1F_y$ for end A and ${}^2F_{xb} = {}^2F_x$, ${}^2F_{yb} = {}^2F_y, \ldots,$ ${}^1F_{xb} = {}^1F_x$, ${}^1F_{yb} = {}^1F_y$ for end B.

The member forces acting on end B of the beam at C_2 as represented by equations (3.39)–(3.41) (and similarly for end A) have been plotted in Figure 3.5. Obviously, the initial forces acting on the beam at C_1 have rotated following the rigid rotation, while their magnitudes remain unchanged. Thus, the rigid body test has been passed for the theory.

3.7 Rigid Body Motion Test for Plane Frame Element

We now use the incremental stiffness equation (3.30) of the beam element to conduct the rigid body motion test. It should be noted that in equation (3.30), the initial nodal forces appear not only in the geometric stiffness matrix $[k_g]$ but also in the initial force vector $\{{}^1f\}$. Whether the finite element can pass the rigid body motion test depends on how well the incremental stiffness equation (3.30) is satisfied when subjected to the rigid rotation.

With regard to the incremental stiffness equation (3.30) for the element, the elastic stiffness matrix $[k_e]$ and geometric stiffness matrix $[k_g]$ are readily available in equations (3.31) and (3.32), respectively. Particularly, the geometric stiffness matrix $[k_g]$ is a function of the initial force vector $\{^1f\}$ acting on the element at C_1. For the beam element to be in equilibrium at C_1, the following can be written:

$$^1F_{xa} = -^1F_{xb} \tag{3.42a}$$

$$^1F_{yb} = -^1F_{ya} = -\frac{^1M_{za} + {^1M_{zb}}}{L} \tag{3.42b}$$

Consequently, the initial nodal forces $\{^1f\}$ can be written as

$$\{^1f\}^T = \Big\langle -^2F_{xb} \quad \frac{@1M_{za} + {^1M_{zb}}}{L} \quad {^1M_{za}} \quad \overset{F}{xb}$$
$$-\frac{^1M_{za} + \overset{M}{zb}}{L} \quad {^1M_{zb}} \quad {^1M_{zb}} \Big\rangle \tag{3.43}$$

which has been visually shown in Figure 3.6(a).

In theory, the displacement vector $\{u\}$ should be representative of all kinds of displacements of the element encountered during the incremental step from C_1 to C_2, including specifically the rigid displacements. For the present purposes, let us assume that the element is subjected to a rigid rotation θ_r, assumed to be small in magnitude for the convenience of operation. Correspondingly, the displacement vector $\{u\}_r$ can be written as

$$\{u\}_r^T = \langle 0 \quad 0 \quad \theta_r \quad 0 \quad L\theta_r \quad \theta_r \rangle \tag{3.44}$$

To conduct the rigid body motion test, we have to check if all the terms in the incremental element equation (3.30) can work together to represent the rigid body behavior.

First, in the case of conservative forces, the force vector $\{^1f\}$ after the rigid rotation is the one shown in Figure 3.6(b). Next, it is easy to show that the elastic stiffness matrix $[k_e]$ generates zero forces upon the rigid rotation, i.e.,

$$[k_e]\{u\}_r = \{0\} \tag{3.45}$$

where $\{0\}$ is a zero vector. The preceding result is as expected since it is known that the elastic stiffness matrix $[k_e]$ as given in equation

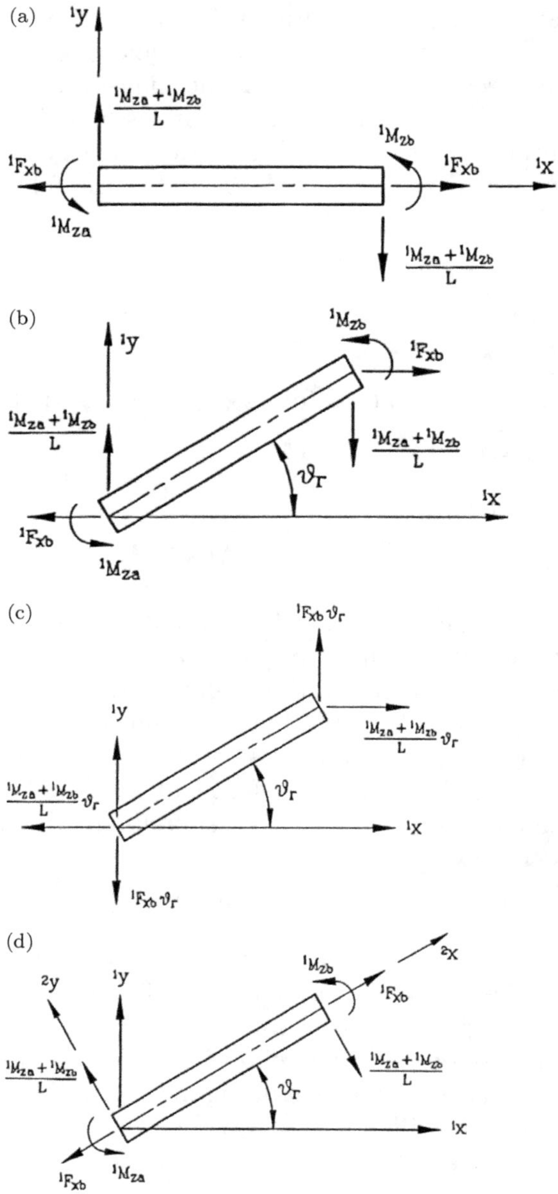

Fig. 3.6. Behavior of beam with rigid rotation: (a) initial forces $\{^1f\}$ at C_1; (b) initial forces $\{^1f\}$ at C_2; (c) forces generated by $[k_g]$; (d) final forces $\{^2f\}$ at C_2.

Note: Left superscript "1" for forces was omitted for clarity.

(3.31) is exact for the beam element based on the Bernoulli–Euler hypothesis.

Then, we find that a rigid body force vector will be generated by geometric stiffness matrix $[k_g]$ upon the rigid rotation:

$$[k_g]\{u\}_r = \left\langle -\frac{{}^1M_{za} + 1M_{zb}}{L}\theta_r \quad -{}^1F_{xa}\theta_r \quad 0 \right.$$
$$\left. \frac{{}^1M_{za} + {}^1M_{zb}}{L}\theta_r \quad {}^1F_{xb}\theta_r \quad 0 \right\rangle^T \tag{3.46}$$

which is also visually shown in Figure 3.6(c). As can be seen, the forces generated by the geometric stiffness matrix $[k_g]$ upon rigid rotation are neither zero nor in equilibrium.

As the final step in the test, we have to substitute all the terms calculated above into the incremental stiffness equation to see if the resulting forces $\{^2f\}$ predicted by the equation for the finite element at C_2 after the rigid rotation θ_r can really conform to the rigid body rule. To this end, we can calculate the resulting forces $\{^2f\}$ according to equation (3.30), i.e., by substituting equations (3.43)–(3.46) into equation (3.30), to obtain

$$\{^2f\}^T = \left\langle -{}^1F_{xb} - \frac{{}^1M_{za} + {}^1M_{zb}}{L}\theta_r, \quad -{}^1F_{xb}\theta_r + \frac{{}^1M_{za} + {}^1M_{zb}}{L}, \right.$$
$${}^1M_{za} \quad {}^1F_{xb} + \frac{{}^1M_{za} + {}^1M_{zb}}{L}\theta_r,$$
$$\left. {}^1F_{xb}\theta_r - \frac{{}^1M_{za} + {}^1M_{zb}}{L}, \quad {}^1M_{zb} \right\rangle \tag{3.47}$$

which has also been plotted in Figure 3.6(d). Alternatively, the forces $\{^2f\}$ may also be obtained as the summation of Figures 3.6(b) and 3.6(c).

From Figure 3.6(d), we can observe that the initial forces $\{^1f\}$ acting on the finite element have been rotated by an amount equal to the rigid rotation θ_r, while their magnitudes remain unchanged. Naturally, the ability of the derived finite element, particularly the geometric stiffness matrix $[k_g]$, to deal with the rigid body rotation has been confirmed. At this point, we like to point out that the above test has been conducted for the case of rigid rotation, but not for the case of rigid translation, as the latter is trivial. Theoretically, the test should be conducted for all possible types of rigid motions for a finite element.

Another observation from the rigid body motion test is that the element forces $\{^2f\}$ acting on the beam element after the rigid rotation, when referred to the C_1 axes, can be related to the initial forces $\{^1f\}$ by the following relation:

$$\{^2_1f\} = [T]\{^1_1f\} \tag{3.48}$$

where the subscript "1" has been added to signify that both force vectors are referred to the C_1 axes. The matrix $[T]$ that accounts for the *transformation* from the C_2 axes to the C_1 axes is

$$[T] = \begin{bmatrix} [R] & \\ & [R] \end{bmatrix} \tag{3.49}$$

For small rigid rotation θ_r, the *rotation matrix* $[R]$ is

$$[R] = \begin{bmatrix} \cos\theta_r & -\sin\theta_r & 0 \\ \sin\theta_r & \cos\theta_r & 0 \\ 0 & 0 & 1 \end{bmatrix} \cong \begin{bmatrix} 1 & -\theta_r & 0 \\ \theta_r & 1 & 0 \\ 0 & 0 & 1 \end{bmatrix} \tag{3.50}$$

But when the resulting forces $\{^2_2f\}$ are referred to the C_2 (rotated or deformed) axes, it can be related to the initial forces $\{^1_1f\}$ referred to the C_1 axes simply as

$$\{^2_2f\} \equiv \{^1_1f\} \tag{3.51}$$

where the equal sign "$\equiv$" is used to mean that both vectors are of the same magnitude. The preceding equation is exactly the relation that can be used in treating the initial forces $\{^1_1f\}$ during the *rigid rotation* from C_1 to C_2 in an incremental-iterative analysis, which is good for any magnitude of rigid rotations.

3.8 Remarks on Rigid Body Motion Test

The following remarks are made on the rigid body motion test for nonlinear finite elements based on the updated Lagrangian formulation, in particular, on the geometric stiffness that accounts for the instability effect of the finite element.

3.8.1 *Remark on magnitude of rigid rotations*

There is virtually no restriction on the magnitude of rigid rotations concerning the validity of the rigid body rule. As revealed by the conceptual model in Figure 3.4, the earth on which the rod is fixed is allowed to rotate for any amount. The same property should be carried over to any nonlinear finite element. By assuming the rigid rotation θ_r to be *small* as in equations (3.35) and (3.44), we are in fact talking about the *minimum* requirement that must be satisfied by a finite element, for which the rigid body motion test can be carried out in a mathematically simple way. On the other hand, we shall see that in the incremental-iterative analysis of structures in later chapters, what is assumed for each incremental step is that the deformations occurring with each incremental step are small, but the rigid rotation involved can be large.

Nevertheless, we should not forget that the relation $\{^2_2f\} \equiv \{^1_1f\}$ derived in equation (3.51) is universally valid. Namely, it is good for any magnitude of rigid rotations. Such a property will be further exploited in the recovery of element forces concerning the rigid rotations at each incremental step of the incremental-iterative nonlinear analysis.

3.8.2 *Remark on incremental equations using* $\{\Delta f\}$

To conduct the rigid body motion test, it is essential that the equation of the finite element be given in the form of equation (3.30). It is with this form that the roles of the initial forces $\{^1f\}$ and total forces $\{^2f\}$ can be fully revealed during the rigid body motion. In many previous studies on nonlinear analysis, the incremental stiffness equation (3.30) and structure equation (3.33), respectively, have been replaced by the following equations:

$$[k_e]\{u\} + [k_g]\{u\} = \{\Delta f\} \tag{3.52}$$

$$[K_e]\{U\} + [K_g]\{U\} = \{\Delta P\} \tag{3.53}$$

where the symbol Δ was used to signify an incremental quantity. It is clear that the rigid body motion test *cannot* be conducted using the form in equation (3.52), since the role of the initial forces $\{^1f\}$ has been excluded, but only partially covered in the geometric stiffness matrix $[k_g]$ (which is a function of the initial forces $\{^1f\}$).

The other drawback with the incremental equation (3.52) can be observed when it is used to recover the element force increment $\{\Delta f\}$ by the following expression:

$$\{\Delta f\} = [k_e]\{u\} + [k_g]\{u\} \tag{3.54}$$

The quantity $\{\Delta f\}$ calculated herein should be regarded as the elastic force increment. Beyond this, no provision has been provided for treating the initial nodal forces $\{^1 f\}$. Obviously, we cannot judge whether the initial force $\{^1 f\}$ acting on the element at C_1 have been properly treated in the face of rigid rotations for each incremental step.

3.8.3 *Remark on kinematic hypothesis adopted*

In the analysis of beam-type structures, various kinematic hypotheses have often been adopted to simplify the analysis of a three-dimensional solid to a one-dimensional structure. In the present formulation, the Bernoulli–Euler hypothesis has been adopted which states that plane sections of the beam remain plane after deformation. Such a hypothesis is good for elastic linear analysis of the beam, and that is what has been adopted all the time. When the Bernoulli–Euler hypothesis was first proposed, researchers did not really know that it would be extended such far from linear to nonlinear analysis.

Here we should understand the linear and nonlinear strain components derived from elasticity are universally valid. They were originally able to accommodate all kinds of deformations including rigid rotations. However, the employment of kinematic hypotheses brings some artificial restraint on the displacement field used to describe the behavior of a solid. For this reason, it may introduce terms in the nonlinear strains that may not be physically interpretable and even violate the rigid body rule. One typical example is the appearance of the last term $yv'v''$ in the nonlinear shear strain η_{xy} in equation (3.10d). It is obvious that if this term were adopted in the foregoing derivation for the element stiffness matrices, we are going to encounter some terms that cannot cope with the rigid body rule. From this point of view, it is justified to neglect such a high-order term in the derivation.

In Chapter 4, we demonstrate that for the truss element, there is no need to employ any kinematic hypothesis in describing the cross-sectional behavior of the truss member. As such, all the high-order components of the nonlinear strains included in the derivation do not introduce any difficulty for the truss element to pass the rigid body motion test.

3.9 Force Recovery Based on Rigid Body Rule

Rigid body rotation is not a rare but a prevailing phenomenon in nonlinear analysis. For instance, let us use the cantilever and portal frame in Figures 3.7(a) and 3.7(b), respectively, as the vehicle for illustrating the ideas involved. First of all, either the cantilever or portal frame is divided into a number of beam elements in nonlinear analysis.

When in buckling, it is easy to see that *all* the elements used to represent the structures are subjected to rigid rotations of certain magnitudes, including the ones with one end fixed at the supports. Such a phenomenon can be better visualized from the typical element exaggerated in the circles in the figure. As a matter of fact, rigid rotation is a *prevailing* phenomenon, but not a rare phenomenon, for all the beam elements used to represent a framed structure in nonlinear or buckling analysis.

In an incremental-iterative nonlinear analysis, it is convenient to physically decompose the displacement increment $\{u\}$ of each element into two parts as the *rigid displacements* $\{u\}_r$ and *natural deformations* $\{u\}_n$, as schematically shown in Figure 3.7 for the cantilever and portal frame. Obviously, the *rigid displacements* $\{u\}_r$ constitute a *large* portion of the buckling displacements of each element of the structure. In comparison, the magnitudes of *natural deformations* $\{u\}_n$ are relatively *small*, as illustrated in Figure 3.7, for the structures represented by a sufficient number of elements. This is consistent with the assumption of small incremental steps adopted in the nonlinear formulation.

The other concern is that the initial forces $\{{}^1_1 f\}$ acting on each element of the structure may be generally *large* since they have been accumulated step-by-step in the nonlinear analysis. Here the subscript "1" has been restored to signify the reference configuration

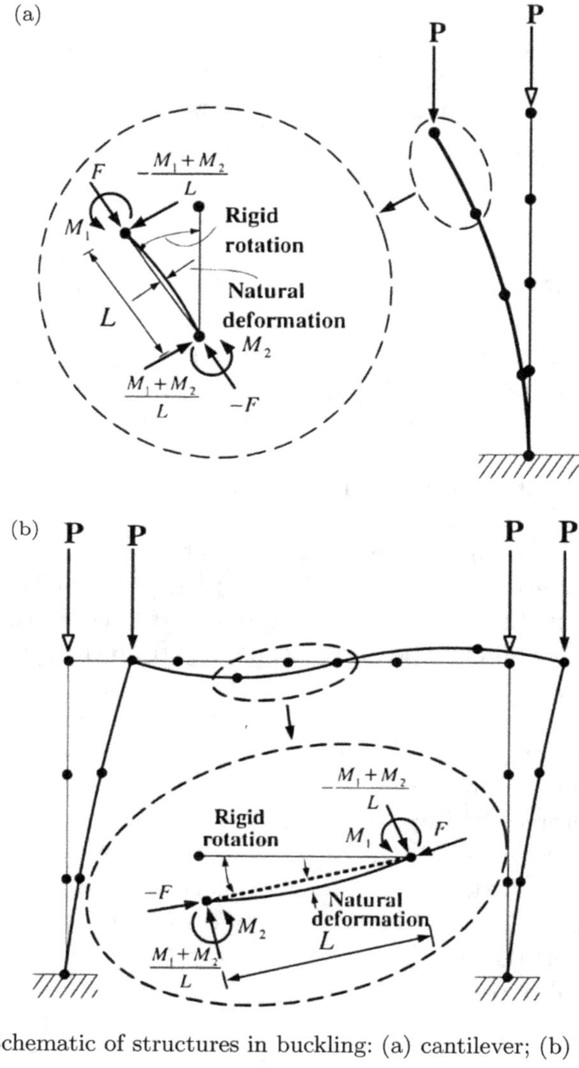

Fig. 3.7. Schematic of structures in buckling: (a) cantilever; (b) portal frame.

at C_1. Fortunately, the fact of large initial forces $\{^1_1f\}$ coupled with large rigid rotations $\{u\}_r$ can be easily dealt with by the rigid body rule in the force updating procedure, namely, by directly treating the initial forces $\{^1_1f\}$ at the current incremental step as the forces $\{^2_2f\}$ acting at the beginning of the next incremental step, according to the rigid body rule given in equation (3.51).

The fact that the natural deformations $\{u\}_n$ (which can be computed using the procedure in Yang and Chiou, 1987) are generally small allows us to compute the force increment $\{\Delta f\}$ generated during the incremental step from C_1 to C_2 by the following linearized equation:

$$\{\Delta f\} = [k_e]\{u\}_n = [k_e](\{u\} - \{u\}_r) = [k_e]\{u\} \tag{3.55}$$

where it is recognized that no actions will be generated by the elastic stiffness matrix $[k_e]$ when subjected to the rigid displacements $\{u\}_r$, i.e., $[k_e]\{u\}_r = \{0\}$. Consequently, the resulting forces $\{^2_2f\}$ acting at the beginning of the next incremental step can be given as follows:

$$\{^2_2f\} = \{^1_1f\} + [k_e]\{u\} \tag{3.56}$$

The preceding is the *force recovery* formula that will be used in updating the element forces for the new incremental step in nonlinear analysis. It is featured by the fact that the first term on the RHS accounts for the rigid rotation effect and the second term the elastic deformation effect.

Of interest to note is that the elastic stiffness matrix $[k_e]$ alone has been used in the force recovery and in the *predictor phase* (i.e., in constructing the structural stiffness matrix for solving the displacement increments) in incremental-iterative analysis (see Chapter 7 for further details). It was demonstrated that quite a number of nonlinear and postbuckling problems can be solved if used has been made of the *rigid body rule* and reliable incremental-iterative solution methods, such as Yang and Shieh's (1990) *generalized displacement control method* (Yang *et al.*, 2019).

3.10 Numerical Examples

In the solution of a nonlinear structure, it is required that the element stiffness equation (i.e., *stiffness matrix*) is qualified by the rigid body test, that a physically consistent *force recovery* procedure is adopted, and that a numerically stable *solution method* be used. This will be referred to as the three components of the *solution strategy* for nonlinear problems.

The plane frame element and stiffness matrices derived in this chapter will be employed to solve several elastic nonlinear frames, with the results compared with those available in the literature. The force recovery procedure adopted for each incremental (or iterative) step is the one given that is physically justified in equation (3.56). Such an approach is similar to the *natural deformation approach* adopted by Yang and Chiou (1987), in that the geometric stiffness additionally included by the latter in the calculation of the force increment is quite minor.

The *generalized displacement control method* devised by Yang and Shieh (1990), which is described in Chapter 7 with the algorithm given in Section 7.8, is adopted as the tool for tracing the nonlinear load-deflection curves of each problem. This method is featured by the fact that it is generally stable for tackling various instability phenomena such as snap-through and snap-back points that may create some computational difficulty in the incremental/iterative operation. With regard to the convergence criteria, iteration for equilibrium of the structure at each incremental step is performed whenever the norm of the unbalanced forces or the normal of the displacement increments for the iterative step is larger than a preset tolerance of 10^{-4}. Throughout the numerical examples to be presented in this chapter, it was found that generally three cycles of iteration are required at each incremental step, indicating that the three components of the solution strategy mentioned above, i.e., stiffness matrix, force recovery, and solution method, can perfectly work together to solve the problems of concern.

3.10.1 *Williams' toggle*

The following properties have been adopted for each member of the frame shown in Figure 3.8(a): $E = 10.3 \times 10^6$ ψ $(71.0 \times 10^6$ kPa), $A = 0.183\,\text{in}^2$ $(1.181\ \text{cm}^2)$, $I_z = 9.00 \times 10^{-4}$ $(0.0375\ \text{cm}^4)$. In analysis, each member of the frame is represented by ten frame elements. As can be seen from Figure 3.8(b), the solution obtained presently is in close agreement with the analytical solution given by Williams (1964). In addition, the computational time consumed by the present approach has been compared with that by Yang and Chiou (1987) using the natural deformation concept in Table 3.1. For this simple problem, the efficiency is roughly the same using both approaches.

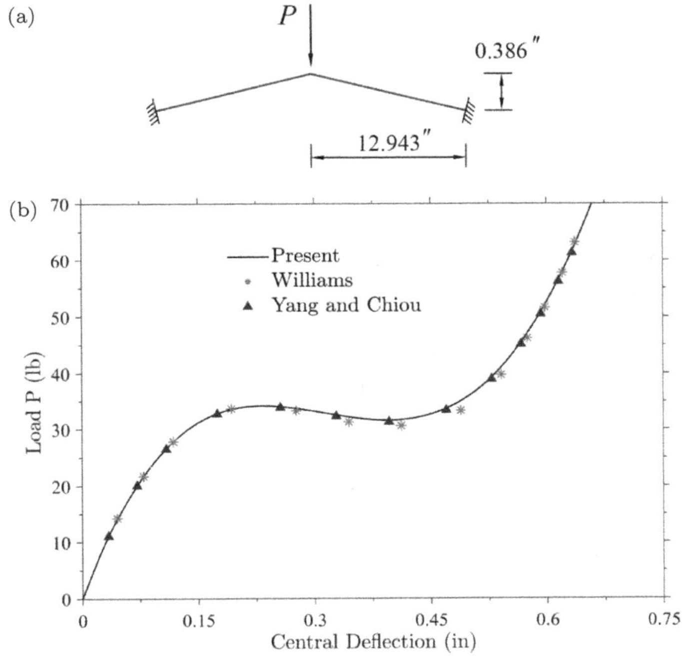

Fig. 3.8. Williams' toggle: (a) geometry; (b) load-deflection curve.

Table 3.1. Data of computation for Williams' toggle.

Issues of concern	Present	Yang and Chiou
Total time(s)	0.234	0.261
Total incremental steps	253	252
Average iterative steps per incremental step	2.72	2.00

3.10.2 *Axially compressed cantilever*

The buckling of an axially compressed cantilever, as shown in Figure 3.9(a), is of interest in that it involves very large rotations. To avoid numerical difficulties near the bifurcation point, i.e., when the load approaches the buckling load, an imperfection of the moment type has been introduced at the free end of the cantilever, to make the load-deflection curve deviate from the perfect buckling path that includes the bifurcation point. The cantilever was modeled by ten

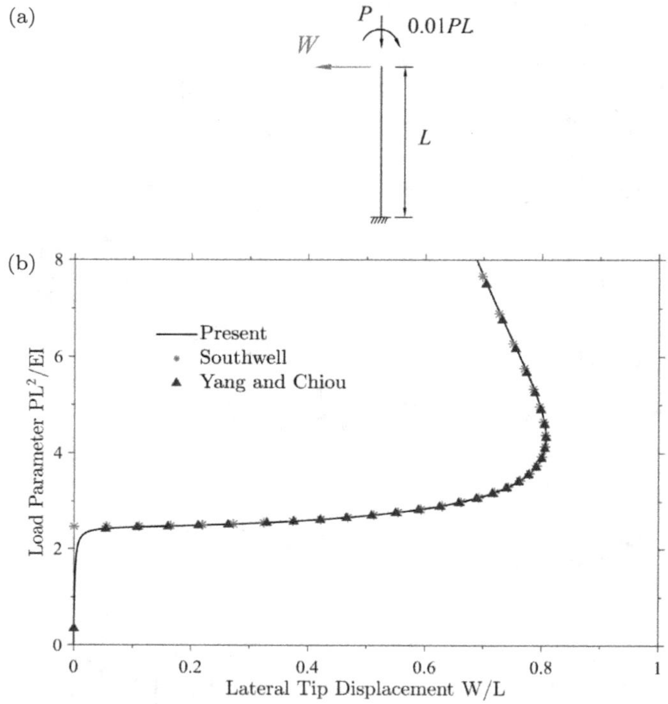

Fig. 3.9. Axially compressed cantilever: (a) geometry; (b) load-deflection curve.

Table 3.2. Data of computation for axially compressed cantilever.

Issues of concern	Present	Yang and Chiou
Total time(s)	15.616	14.491
Total incremental steps	82,961	83,244
Average iterative steps per incremental step	1.28	1.02

elements in the finite element analysis. The numerical result shown in Figure 3.9(b) compares very well with that based on Southwell (1941). From the computational times listed in Table 3.2 for the present approach and Yang and Chiou (1987) using the natural deformation concept, it can be seen that the efficiency is roughly the same.

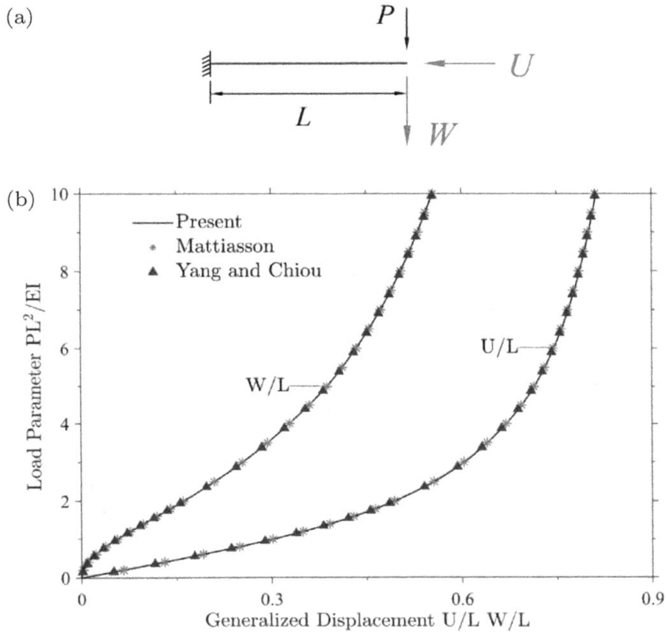

Fig. 3.10. Cantilever under shear load: (a) geometry; (b) load-deflection curve.

Table 3.3. Data of computation for cantilever under shear load.

Issues of concern	Present	Yang and Chiou
Total time(s)	0.394	0.330
Total incremental steps	806	808
Average iterative steps per incremental step	3.14	2.00

3.10.3 *Cantilever under shear load*

This example serves as a good benchmark for testing the capability of the derived element to solve problems that are primarily of the shear type. In the analysis, the cantilever shown in Figure 3.10(a) has been divided into twenty elements. Figure 3.10(b) indicates that good agreement has been achieved between the present solution and Mattiasson's (1981) solution. From the computational times listed in Table 3.3 for the present approach and Yang and Chiou (1987) using

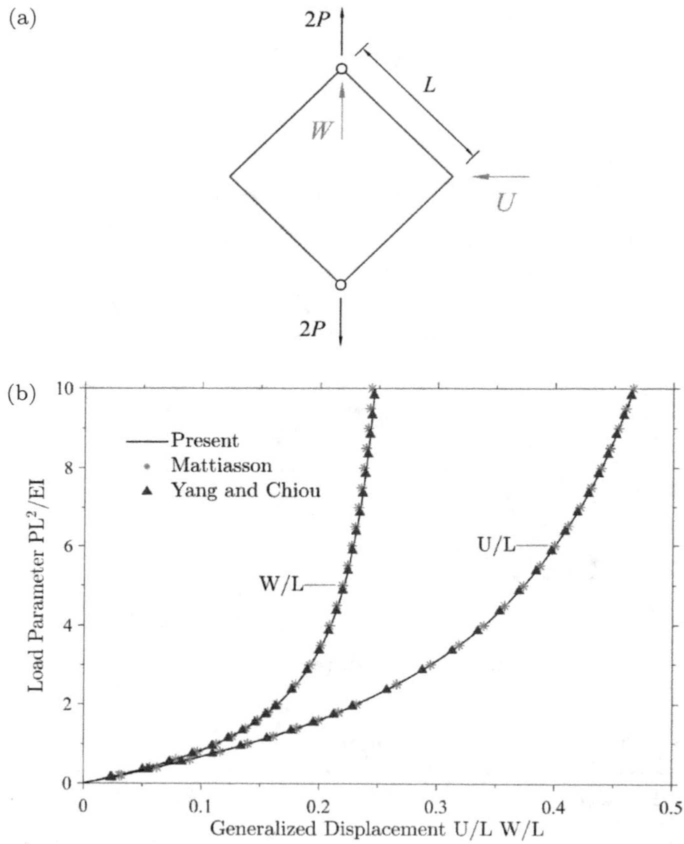

Fig. 3.11. Pinned-fixed square diamond frame in tension: (a) geometry; (b) load-deflection curve.

the natural deformation concept, it can be seen that the efficiency is roughly the same.

3.10.4 *Pinned-fixed square diamond frame*

A pinned-fixed square diamond frame is loaded either in tension or in compression, as shown in Figures 3.11(a) and 3.12(a). Due to the symmetry condition of the frame, only half of the diamond frame is modeled in the analysis. Each member of the frame is represented by ten elements. The load-deflection curves obtained for the frame

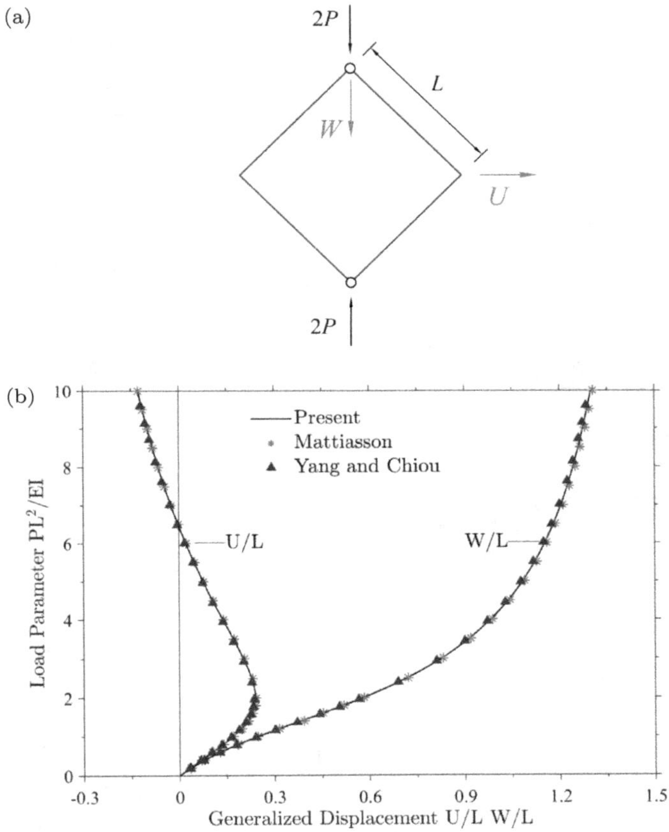

Fig. 3.12. Pinned-fixed square diamond frame in compression: (a) geometry; (b) load-deflection curve.

under both loading cases have been plotted in Figures 3.11(b) and 3.12(b), along with Mattiasson's (1981) analytical solutions. It can be seen that good agreement has been achieved for both cases. From the computational times listed in Tables 3.4 and 3.5, it can be seen that roughly the same efficiency was achieved by the present approach and Yang and Chiou (1987) using the natural deformation concept.

3.10.5 *Rigid-jointed square frame*

As a final example, a square frame loaded at the midpoint of the top and bottom members by a pair of tension and compression forces

Table 3.4. Data of computation for pinned-fixed square diamond frame in tension.

Issues of concern	Present	Yang and Chiou
Total time(s)	2.722	2.729
Total incremental steps	4,852	4,857
Average iterative steps per incremental step	1.97	1.72

Table 3.5. Data of computation for pinned-fixed square diamond frame in compression.

Issues of concern	Present	Yang and Chiou
Total time(s)	0.222	0.217
Total incremental steps	179	179
Average iterative steps per incremental step	3.55	2.71

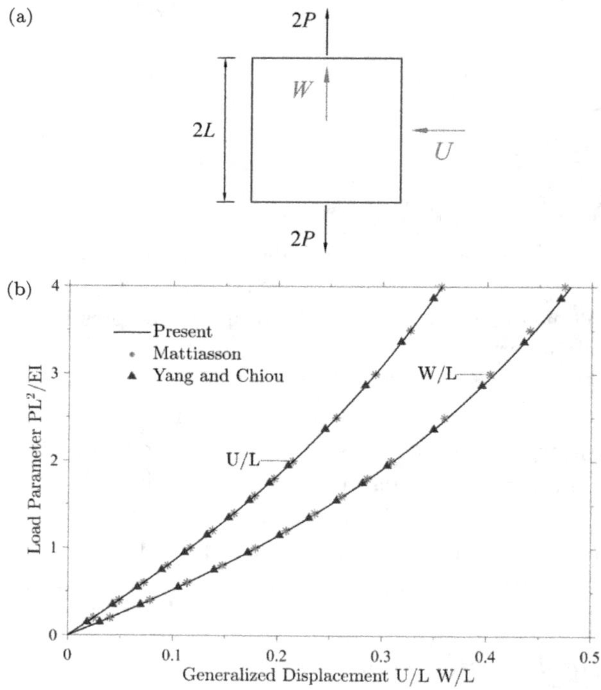

Fig. 3.13. Rigid-jointed square frame in tension: (a) geometry; (b) load-deflection curve.

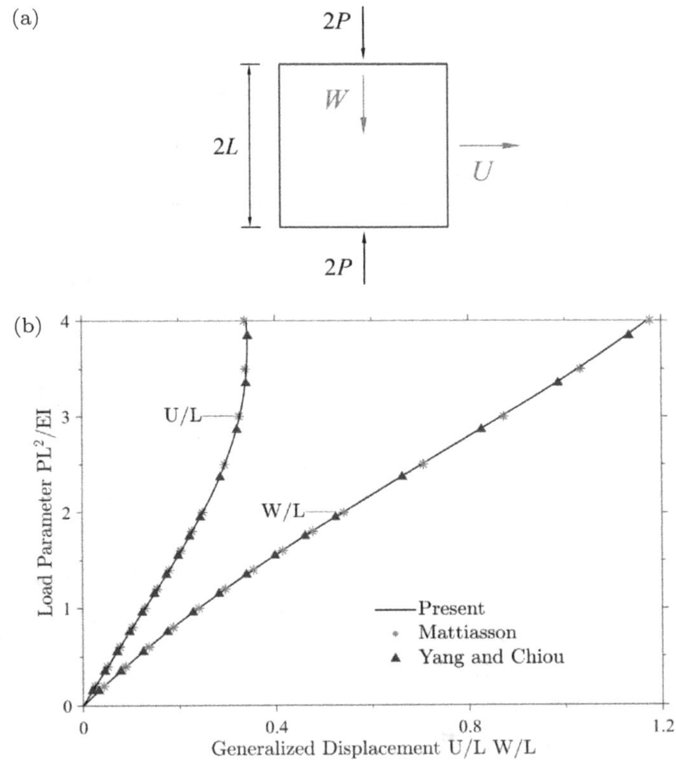

Fig. 3.14. Rigid-jointed square frame in compression: (a) geometry; (b) load-deflection curve.

is analyzed, as shown in Figures 3.13(a) and 3.14(a), respectively. Only half of the symmetric frame is adopted in the analysis, with each member divided into four elements. From Figures 3.13(b) and 3.14(b), it can be observed that for both loading cases, the solutions obtained here agree closely with those given by Mattiasson (1981).

From the numerical studies presented in this section, it is confirmed that the stiffness matrices derived in Section 3.4 based on the general theory, i.e., with the geometric stiffness matrix $[k_g]$ of equation (3.32) that considers all the nonlinear strain components, and the physically justified formula for recovering the element forces in equation (3.56), are reliable for solving the nonlinear problems, involving the snap-through and bifurcation points, as well as large rotations. One thing that remains to be further discussed is the

Table 3.6. Data of computation for rigid-jointed square frame in tension.

Issues of concern	Present	Yang and Chiou
Total time(s)	2.384	2.664
Total incremental steps	5,946	5,948
Average iterative steps per incremental step	1.78	1.75

Table 3.7. Data of computation for rigid-jointed square frame in compression.

Issues of concern	Present	Yang and Chiou
Total time(s)	0.917	0.812
Total incremental steps	1,427	1,439
Average iterative steps per incremental step	2.88	2.00

nonlinear-path solution scheme for dealing with the snap-through and snap-back points, e.g., the generalized displacement control method, which is exploited in Chapter 7. From the computational times listed in Tables 3.6 and 3.7, it can be seen that roughly the same efficiency was achieved by the present approach and Yang and Chiou (1987) using the natural deformation concept.

References

Bathe, K.J. and Bolourchi, S. (1979). Large displacement analysis of three-dimensional beam structures, *Int. J. Numer. Meth. Eng.*, 14, 961–986.

Bathe, K.J., Ramm, E., and Wilson, E.L. (1975). Finite element formulation for large deformation analysis, *Int. J. Numer. Meth. Eng.*, 9, 353–386.

Chajes, A. (1974). *Principles of Structural Stability Theory*, Prentice Hall, Englewood Cliffs.

Chen, W.F. and Lui, E.M. (1987). *Structural Stability – Theory and Implementation*, Elsevier, New York.

Cook, R.D., Malkus, D.S., and Plesha, M.E. (1989). *Concepts and Applications of Finite Element Analysis*, 3rd edn., John Wiley & Sons, Inc., New York.

Gattass, M. (1982). Large displacement, interactive-adaptive dynamic analysis of frames, thesis presented to Cornell University, in Ithaca, New York, in partial fulfillment of the requirements for the degree of Doctor of Philosophy.

Irons, B. and Ahmad, S. (1980). *Techniques of Finite Elements*, John Wiley & Sons, Inc., New York.

Jagannathan, D.S., Epstein, H.I., and Christiano, P. (1975). Fictitious strains due to rigid body rotation, *J. Struct. Div.*, (ASCE), 101(11), 2472–2476.

Mattiasson, K. (1981). Numerical results from large deflection beam and frame problems analyzed by means of elliptic integrals, *Int. J. Numer. Meth. Eng.*, 17(1), 145–153.

Porter, F.L. and Powell, G.H. (1971). Static and dynamic analysis of inelastic frame structures, Report No. EERC 71-3, Earthquake Engineering Research Center, University of California, Berkeley, California.

Przemieniecki, J.S. (1968). *Theory of Matrix Structural Analysis*, McGraw-Hill, New York.

Southwell, R.V. (1941). *An Introduction to the Theory of Elasticity for Engineers and Physicists*, 2nd edn., Oxford Press, Oxford.

Williams, F.W. (1964). An approach to the nonlinear behavior of the members of a rigid jointed framework with finite deflections, *Q. J. Mech. Appl. Math.*, 17(4), 451–469.

Yang, Y.B. and Chiou, H.T. (1987). Rigid body motion test for nonlinear analysis with beam elements, *J. Eng. Mech.*, (ASCE), 113(9), 1404–1419.

Yang, Y.B. and Kuo, S.R. (1994). *Theory of Nonlinear Framed Structures*, Prentice Hall, Englewood Cliffs.

Yang, Y.B. and McGuire, W. (1986a). Stiffness matrix for geometric nonlinear analysis, *J. Struct. Eng.* (ASCE), 112(4), 853–877.

Yang, Y.B. and McGuire, W. (1986b). Joint rotation and geometric nonlinear analysis, *J. Struct. Eng.* (ASCE), 112(4), 879–905.

Yang, Y.B. and Shieh, M.S. (1990). Solution method for nonlinear problems with multiple critical points, *AIAA J.*, 28(12), 2110–2116.

Yang, Y.B., Chen, A., Yan, Y., and Wang, Z.L. (2019). Using only elastic stiffness in nonlinear and postbuckling analysis of structures, *Int. J. Struct. Stab. Dyn.*, 19(10), 1950112.

Zienkiewicz, O.C. (1977). *The Finite Element Method*, 3rd edn., McGraw-Hill Book Co., Inc., New York.

Chapter 4

Truss: A Perfect Example for Rigid Body Rule

4.1 Introduction

The truss element is a simple and perfect structure for demonstrating the rigid body rule presented in Chapter 3 or Yang and Chiou (1987), concerning the rigid rotation of an element with initial nodal forces. The reason behind this is that *no kinematic hypothesis* needs to be imported for describing the cross-sectional behavior of the truss members, unlike the plane beam theory that is deeply rooted on the Bernoulli–Euler hypothesis of plane sections remaining plane after deformation. As such, the strain–displacement relations derived from elasticity remains perfectly valid for the truss elements. All the high-order terms encountered in the finite element formulation can be well interpreted physically, and they are as important as the low-order terms concerning the rigid body behavior.

The other matter in an incremental nonlinear analysis based on the updated Lagrangian formulation is the specification of *constitutive coefficients*. In this formulation, the constitutive coefficients are often set to be *constant* within each incremental step. A material specified in this sense can only be said to be an *incrementally linear* material and is in fact a nonlinear material. A really linear material is the one that is given in terms of the total second Piola–Kirchhoff stress and total Green–Lagrange strain. The difference between an incrementally linear and really linear material can be observed when the strains accumulated for the previous incremental steps are large.

Such a phenomenon has been exploited by Yang and Leu (1990, 1991) using some simple trusses as an example.

In an incremental nonlinear analysis that incorporates iterations at each incremental step for achieving the equilibrium of the structure, three different phases can be identified. The first or *predictor* phase involves the solution of displacement increments from the incremental equations of equilibrium for the structure. The second or *corrector* phase is concerned with the recovery of element force increments from the displacement increments obtained by the predictor. The resulting forces acting on each element at the end of each incremental step can therefore be obtained as the summation of the initial forces existing at the beginning of the incremental step (based on the rigid body rule) plus the force increments generated during the current step. In the third phase of analysis, the equilibrium of the structure is checked to ensure that convergence of iterations is achieved at the new deformed configuration. Namely, by summing the element forces at each node and comparing them with the external loads, the *unbalanced forces* for the structure can be calculated. Whenever the unbalanced forces appear to be nonnegligible, iteration for equilibrium of the structure involving the first and second phases should be repeated. In Chapter 7, a complete description of the step-by-step analysis procedure involving the three phases will be given.

Although numerous procedures have been proposed for performing the nonlinear analysis of structures, see for instance, Chajes and Churchill (1987), it seems that obscurities do exist in the literature on some essential parts of nonlinear analysis. This is especially true when the rigid body characteristics of the finite elements are concerned, for instance, in the treatment of higher order terms pertaining to the derivation of element stiffness matrices using the virtual work approach, which are central to the recovery and updating of element forces and, therefore, can affect the equilibrium checking of structures in an incremental-iterative nonlinear analysis.

To start, the finite element equations of equilibrium will be derived in Section 4.2 for a two-dimensional truss element based on the virtual work equation presented in Section 1.7.2, in particular, equation 1.131, using the updated Lagrangian formulation. With regard to the higher-order terms, it will be shown in Section 4.3 that the truncation of a higher-order term from the stiffness formulation depends not only on the order of the term itself but also on *how it affects the rigid body behavior*. In order not to violate the rigid body law, some matching terms must be treated as an entity and considered together throughout the derivation. Any attempt to retain part of them while neglecting the remaining will induce fictitious forces for elements undergoing rigid body rotations. Such defects can be carried over to the corrector phase to yield incorrect element forces, which will further result in an incorrect count of the unbalanced forces. Consequently, no guarantee can be made for the accuracy of the solution.

In the same section, the capability of the truss element with initial nodal forces to accommodate rigid rotations will be demonstrated. It can be seen that some high- and low-order terms should be treated as a matching pair in accommodating the rigid rotations, while a different combination of matching pars be used in interpreting the stretching behavior. From this point of view, we can observe the "beauty" or excellent mechanism of the elasticity-based strain components in describing the general physical behavior of a solid member.

In Section 4.4, it will be demonstrated that all the higher-order matrices derived in Section 4.2 can be presented in a symmetric form, which appears to be particularly convenient from the standpoint of computer programming. The procedure of formulation presented in Sections 4.2–4.4 for the plane truss element can be easily extended to the three-dimensional case, as will be shown in Section 4.5. Of particular interest is the consideration of rigid body characteristics of truss elements in the force recovery procedures in Section 4.6, along with the case studies of two examples in Section 4.7.

A large portion of the materials to be presented in the chapter has been obtained from Leu and Yang (1990). The materials acquired from the source paper have been completely digested, reorganized, and supplemented in such a manner that a smoother and more systematic presentation be maintained throughout the text based on the unified notation.

4.2 Formulation of Nonlinear Plane Truss Element

As was shown in Figure 4.1, the motion of a body can be described by three typical configurations in an incremental nonlinear analysis: the

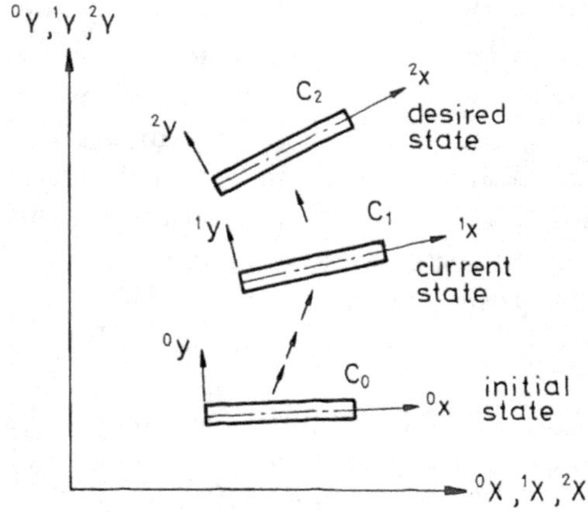

Fig. 4.1. Motion of truss member in space.

initial undeformed configuration C_0, the last calculated configuration C_1, and the current deformed configuration C_2. In the updated Lagrangian formulation, the last known configuration C_1 is used as the reference configuration for establishing the equation of equilibrium for the solid at the current configuration C_2. Based on the strain and stress measures presented in Sections 1.3 and 1.4, the virtual work equation of equilibrium can be written for the solid in a fully nonlinear incremental form as the one given in (1.131), i.e.,

$$\int_{{}^1V} {}_1C_{ijkl}\, {}_1\epsilon_{kl} \delta\, {}_1\epsilon_{ij}\, {}^1dV + \int_{{}^1V} {}^1\tau_{ij} \delta\, {}_1\eta_{ij}\, {}^1dV = {}^2_1R - {}^1_1R \tag{4.1}$$

where, as defined in Chapter 1, C_{ijkl} denotes the incremental constitutive coefficients, ${}^1\tau_{ij}$ the Cauchy (initial) stresses, ${}_1\epsilon_{kl}$ the updated Green strain increments, and ${}_1\eta_{ij}$ the nonlinear components of the strain increments ${}_1\epsilon_{kl}$. All these quantities have been referred to the C_1 configuration. In equation (4.1), ${}^2_1R$ and ${}^1_1R$ denote the virtual works done by the external loads acting on the solid at C_2 and C_1, respectively, i.e.,

$${}^2_1R = \int_{{}^1S} {}^2_1t_i\, \delta u_i\, {}^1dS + \int_{{}^1V} {}^2_1f_i\, \delta u_i\, {}^1dV \tag{4.2}$$

$${}^1_1R = \int_{{}^1V} {}^1\tau_{ij}\, \delta_1 e_{ij}\, {}^1dV \tag{4.3}$$

where ${}^2_1t_i$ and ${}^2_1f_i$ represent the surface tractions and body forces, respectively, acting on the solid at the C_2 configuration.

It should be noted that the domains of integration in the foregoing three equations, i.e., the volume ${}^1V$ and surface area ${}^1S$, have been defined for the solid at the C_1 configuration. For a solid in equilibrium at C_1, the external virtual work ${}^1_1R$ can be explicitly written as

$$ {}^1_1R = \int_{{}^1S} {}^1_1t_i \, \delta u_i \, {}^1dS + \int_{{}^1V} {}^1_1f_i \, \delta u_i \, {}^1dV \tag{4.4} $$

where ${}^1_1t_i$ and ${}^1_1f_i$ are the surface tractions and body forces, respectively, acting on the body at C_1. For the sake of simplicity in the present work, we shall neglect the effect of body forces ${}^2_1f_i$ and ${}^1_1f_i$. As a result, the external virtual works ${}^2_1R$ and ${}^1_1R$ reduce to

$$ {}^2_1R = \int_{{}^1S} {}^2_1t_i \, \delta u_i \, {}^1dS \tag{4.5} $$

$$ {}^1_1R = \int_{{}^1S} {}^1_1t_i \, \delta u_i \, {}^1dS \tag{4.6} $$

By decomposing the strain increments ${}_1\epsilon_{ij}$ into the linear and nonlinear components,

$$ {}_1\epsilon_{ij} = {}_1e_{ij} + {}_1\eta_{ij} \tag{4.7} $$

we can obtain from equation (4.1) the incremental equation of equilibrium for the solid at C_2, but referred to C_1 as follows:

$$ \begin{aligned} &\int_V C_{ijkl}\, e_{kl}\delta e_{ij} dV + \int_V C_{ijkl}\, \eta_{kl}\delta e_{ij} dV + \int_V C_{ijkl}\, e_{kl}\delta \eta_{ij} dV \\ &+ \int_V C_{ijkl}\, \eta_{kl}\delta \eta_{ij} dV + \int_V {}^1\tau_{ij}\, \delta \eta_{ij} dV + {}^1R = {}^2R \end{aligned} \tag{4.8} $$

where the left superscript "1" for V and the left subscripts "1" for C_{ijkl}, e_{ij}, η_{ij}, ${}^1R$, and ${}^2R$ have been dropped for clarity, because all these variables are known to be referred to the last calculated configuration C_1. It should be noted that the term ${}^1R$ has been relocated to the left-hand side of equation (4.8) to signify that it represents part of the effect of initial stresses or forces.

At this point, we like to emphasize that equation (4.8) remains an exact statement of the equilibrium of the solid at the current configuration C_2, with reference to the coordinates of the solid at configuration C_1, because no approximation has been made in the foregoing derivation. Naturally, this equation can be used as a valid basis for the formulation of a great variety of finite elements for solid mechanics problems. In the following, a plane truss element suitable for the incremental nonlinear analysis will be derived based on this equation. This element serves as a perfect example for us to physically illustrate the effects of stretching and rigid body rotations implied by the nonlinear strain components. The significance of such effects in nonlinear analysis will also be highlighted in the sections to follow.

For a truss element, we need to consider only the axial component of the stress and strain tensors. By denoting the initial axial stress as ${}^1\tau_{xx}$, axial strain increment as $\epsilon_{xx}(= e_{xx} + \eta_{xx})$, and axial stress increment as S_{xx}, the incremental constitutive law (1.62) reduces to the following for the plane truss element:

$$S_{xx} = E\epsilon_{xx} \tag{4.9}$$

where E denotes the modulus of elasticity of the material. Consequently, the incremental equation (4.8) reduces to

$$\int_V E\, e_{xx}\delta e_{xx} dV + \int_V E\, \eta_{xx}\delta e_{xx} dV + \int_V E\, e_{xx}\delta\eta_{xx} dV \\ + \int_V E\, \eta_{xx}\delta\eta_{xx} dV + \int_V {}^1\tau_{xx}\delta\eta_{xx} dV + {}^1R = {}^2R \tag{4.10}$$

Consider a truss element of length L in Figure 4.2, in which the C_1 coordinates x and y have been denoted with the left superscripts omitted for clarity. Let A and B denote the two ends of the element, and (u, v) the axial and transverse displacements of the element at

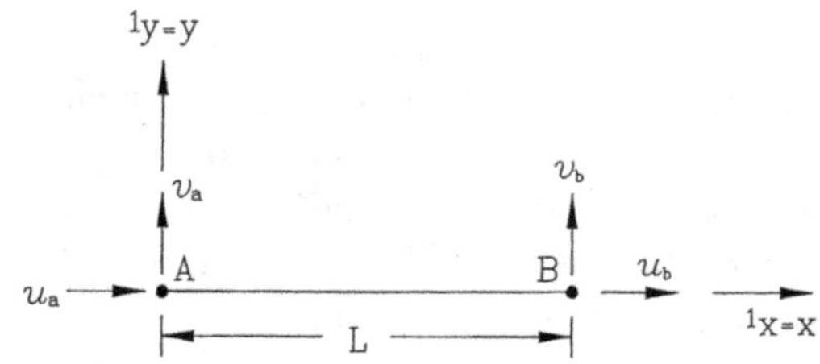

Fig. 4.2. Degrees of freedom for plane truss element.

section x, which have been generated during the incremental step from C_1 to C_2. The displacements (u, v) may be related to the nodal displacement by the linear interpolation functions as follows:

$$u = u_a \left(1 - \frac{x}{L}\right) + u_b \left(\frac{x}{L}\right) \tag{4.11}$$

$$v = v_a \left(1 - \frac{x}{L}\right) + v_b \left(\frac{x}{L}\right) \tag{4.12}$$

where (u_a, v_a) and (u_b, v_b) denote the displacements at the two ends of the element. According to equations (1.34) and (1.35), the linear and nonlinear parts of the axial strain ϵ_{xx} can be expressed as

$$e_{xx} = \frac{\partial u}{\partial x} = \frac{\Delta u}{L} \tag{4.13}$$

$$\eta_{xx} = \frac{1}{2}\left[\left(\frac{\partial u}{\partial x}\right)^2 + \left(\frac{\partial v}{\partial x}\right)^2\right] = \frac{1}{2}\left(\frac{\Delta u^2}{L^2} + \frac{\Delta v^2}{L^2}\right) \tag{4.14}$$

where the *relative nodal displacements* are

$$\Delta u = u_b - u_a \quad \text{and} \quad \Delta v = v_b - v_a \tag{4.15}$$

As shown in Figure 4.2, the nodal displacement vector for the truss element is

$$\{u\}^T = \langle u_a \quad v_a \quad u_b \quad v_b \rangle \tag{4.16}$$

Correspondingly, the element force vectors at C_1 and C_2 are

$$\{^1f\}^T = \langle {}^1F_{xa} \quad {}^1F_{ya} \quad {}^1F_{xb} \quad {}^1F_{yb} \rangle \tag{4.17}$$

$$\{^2f\}^T = \langle {}^2F_{xa} \quad {}^2F_{ya} \quad {}^2F_{xb} \quad {}^2F_{yb} \rangle \tag{4.18}$$

where it should be noted that for a truss element in equilibrium at the C_1 configuration, the transverse shear forces ${}^1F_{ya}$ and ${}^1F_{yb}$ are identically equal to zero, and the axial forces acting at the two ends of the element are equal in magnitude but opposite in direction, that is, ${}^1F_{xb} = -{}^1F_{xa}$. Moreover, the initial axial force can be related to the initial stress ${}^1\tau_{xx}$ by

$${}^1F_x = \int_A {}^1\tau_{xx}\, dA \tag{4.19}$$

where A denotes the cross-sectional area of the element at C_1.

For the sake of simplicity, we shall assume that only loadings of the concentrated type are allowed to act on the two ends of the truss element, while loadings of the distributed type will be considered only when they are replaced by statically equivalent nodal loads. In accordance, each of the terms in equation (4.10) can be derived as follows:

$$\int_V E e_{xx} \delta e_{xx} dV = \int_V E \frac{\Delta u}{L} \delta \left(\frac{\Delta u}{L} \right) dV = \{\delta u\}^T [k_e] \{u\} \tag{4.20}$$

$$\begin{aligned} \int_V E \eta_{xx} \delta e_{xx} dV &= \int_V \frac{1}{2} E \left(\frac{\Delta u^2}{L^2} + \frac{\Delta v^2}{L^2} \right) \delta \left(\frac{\Delta u}{L} \right) dV \\ &= \{\delta u\}^T [s_1] \{u\} \end{aligned} \tag{4.21}$$

$$\begin{aligned} \int_V E e_{xx} \delta \eta_{xx} dV &= \int_V E \frac{\Delta u}{L} \left[\frac{\Delta u}{L} \delta \left(\frac{\Delta u}{L} \right) + \frac{\Delta v}{L} \delta \left(\frac{\Delta v}{L} \right) \right] dV \\ &= \{\delta u\}^T [s_2] \{u\} \end{aligned} \tag{4.22}$$

$$\begin{aligned} \int_V E \eta_{xx} \delta \eta_{xx} dV &= \int_V \frac{1}{2} E \left(\frac{\Delta u^2}{L^2} + \frac{\Delta v^2}{L^2} \right) \left[\frac{\Delta u}{L} \delta \left(\frac{\Delta u}{L} \right) \right. \\ &\quad \left. + \frac{\Delta v}{L} \delta \left(\frac{\Delta v}{L} \right) \right] dV = \{\delta u\}^T [s_3] \{u\} \end{aligned} \tag{4.23}$$

$$\begin{aligned} \int_V {}^1\tau_{xx} \delta \eta_{xx} dV &= \int_V {}^1F_x \left[\frac{\Delta u}{L} \delta \left(\frac{\Delta u}{L} \right) + \frac{\Delta v}{L} \delta \left(\frac{\Delta v}{L} \right) \right] dV \\ &= \{\delta u\}^T [k_g] \{u\} \end{aligned} \tag{4.24}$$

$${}^1R = \int_S {}^1t_i \delta u_i dS = \{\delta u\}^T \{{}^1f\} \tag{4.25}$$

$${}^2R = \int_S {}^2t_i \delta u_i dS = \{\delta u\}^T \{{}^2f\} \tag{4.26}$$

where $dV = Adx$ and the integration for x is carried over the length L of the element, $[k_e]$ is the elastic stiffness matrix, $[k_g]$ the geometric stiffness matrix, and $[s_1]$, $[s_2]$, and $[s_3]$ the higher-order nonlinear stiffness matrices.

The elastic stiffness matrix $[k_e]$ has been given previously as the $[k]$ matrix in equation (2.100) for the plane truss element. The geometric stiffness matrix $[k_g]$ for the plane truss element can be given

as follows:

$$[k_g] = \frac{^1F_{xb}}{L} \cdot \begin{bmatrix} 1 & 0 & -1 & 0 \\ & 1 & 0 & -1 \\ & & 1 & 0 \\ Symm. & & & 1 \end{bmatrix} \tag{4.27}$$

and the higher-order stiffness matrices $[s_1]$, $[s_2]$, and $[s_3]$ can be expressed in terms of Δu and Δv as follows:

$$[s_1] = \frac{EA}{2L^2} \begin{bmatrix} \Delta u & \Delta v & -\Delta u & -\Delta v \\ 0 & 0 & 0 & 0 \\ -\Delta u & -\Delta v & \Delta u & \Delta v \\ 0 & 0 & 0 & 0 \end{bmatrix} \tag{4.28}$$

$$[s_2] = \frac{EA}{2L^2} \begin{bmatrix} 2\Delta u & 0 & -2\Delta u & 0 \\ \Delta v & \Delta u & -\Delta v & -\Delta u \\ -2\Delta u & 0 & 2\Delta u & 0 \\ -\Delta v & -\Delta u & \Delta v & \Delta u \end{bmatrix} \tag{4.29}$$

$$[s_3] = \frac{EA}{6L^3} \begin{bmatrix} 3\Delta u^2 + \Delta v^2 & 2\Delta u\Delta v & -3\Delta u^2 - \Delta v^2 & -2\Delta u\Delta v \\ & 3\Delta v^2 + \Delta u^2 & -2\Delta u\Delta v & -3\Delta v^2 - \Delta u^2 \\ & & 3\Delta u^2 + \Delta v^2 & 2\Delta u\Delta v \\ Symm. & & & 3\Delta v^2 + \Delta u^2 \end{bmatrix} \tag{4.30}$$

where the relation $^1F_x = {}^1F_{xb}$ has been used in deriving the $[k_g]$ matrix. In the literature, the sum of the two matrices $[s_1]$ and $[s_2]$ has been referred to as the $1/2[N_1]$ or $1/2[k_1]$ matrix, whereas the matrix $[s_3]$ has been called the $1/3[N_2]$ or $1/3[k_2]$ matrix (Mallett and Marcal, 1968; Chajes and Churchill, 1987). It will be shown in Section 4.3 that the use of three distinct matrices $[s_1]$, $[s_2]$, and $[s_3]$, plus the $[k_e]$ and $[k_g]$ matrices, enables us to analyze the stretching and rigid body characteristics of the truss element, while the above notation, though attractive for producing symmetric matrices, offers no such advantages.

Making use of the expressions in (4.20) through (4.26) and admitting the arbitrary nature of virtual displacements $\{\delta u\}$, we can derive from equation (4.10) the incremental stiffness equation for the two-dimensional truss element as

$$([k_e] + [k_g] + [s_1] + [s_2] + [s_3])\{u\} + \{^1f\} = \{^2f\} \tag{4.31}$$

in which the term $\{^1f\}$ on the left-hand side represents the initial forces acting on the element at C_1, the term $\{^2f\}$ on the right-hand side, the total forces acting on the element at C_2, and the remaining terms on the left-hand side denote the incremental forces generated by the resistance of the element against the external force increment, i.e., $\{^2f\} - \{^1f\}$. It should be noted that this equation continues to be exact for describing the behavior of the element from C_1 to C_2, on the condition that the material property can be described in an incremental manner following the constitutive law (4.9).

4.3 Physical Interpretation of Element Forces

The five stiffness matrices appearing in equation (4.31) represent five types of actions generated by the element undergoing the incremental displacement $\{u\}$ from C_1 to C_2. The following is a complete list of all the actions generated by each of the five stiffness matrices:

$$[k_e]\{u\} = \langle -F_{xl} \quad 0 \quad F_{xl} \quad 0 \rangle^T \tag{4.32}$$

$$[k_g]\{u\} = \langle -{}^1F_x\tfrac{\Delta u}{L} \quad -{}^1F_x\tfrac{\Delta v}{L} \quad {}^1F_x\tfrac{\Delta u}{L} \quad {}^1F_x\tfrac{\Delta v}{L} \rangle^T \tag{4.33}$$

$$[s_1]\{u\} = \langle -F_{xn} \quad 0 \quad F_{xn} \quad 0 \rangle^T \tag{4.34}$$

$$[s_2]\{u\} = \langle -F_{xl}\tfrac{\Delta u}{L} \quad -F_{xl}\tfrac{\Delta v}{L} \quad F_{xl}\tfrac{\Delta u}{L} \quad F_{xl}\tfrac{\Delta v}{L} \rangle^T \tag{4.35}$$

$$[s_3]\{u\} = \langle -F_{xn}\tfrac{\Delta u}{L} \quad -F_{xn}\tfrac{\Delta v}{L} \quad F_{xn}\tfrac{\Delta u}{L} \quad F_{xn}\tfrac{\Delta v}{L} \rangle^T \tag{4.36}$$

in which F_{xl} and F_{xn} denote the linear and nonlinear components of the force increments generated during the incremental step,

$$F_{xl} \equiv {}_1F_{xl} = EAe_{xx} = EA\frac{\Delta u}{L} \tag{4.37}$$

$$F_{xn} \equiv {}_1F_{xn} = EA\eta_{xx} = \frac{1}{2}EA\left(\frac{\Delta u^2}{L^2} + \frac{\Delta v^2}{L^2}\right) \tag{4.38}$$

where the subscript "1" is inserted to indicate that the forces are expressed with respect to C_1. In the following section, physical interpretation will be explored for each of the five types of actions given in equations (4.32)–(4.36).

4.3.1 *Element forces due to stretching*

In this section, we shall proceed to demonstrate how the terms on the left-hand side of equation (4.31) will interact with each other to represent the effects of stretching and rigid body rotation for a truss element during the movement from C_1 to C_2. First, as is shown in Figure 4.3(a), the forces generated by the $[s_2]$ matrix are in effect

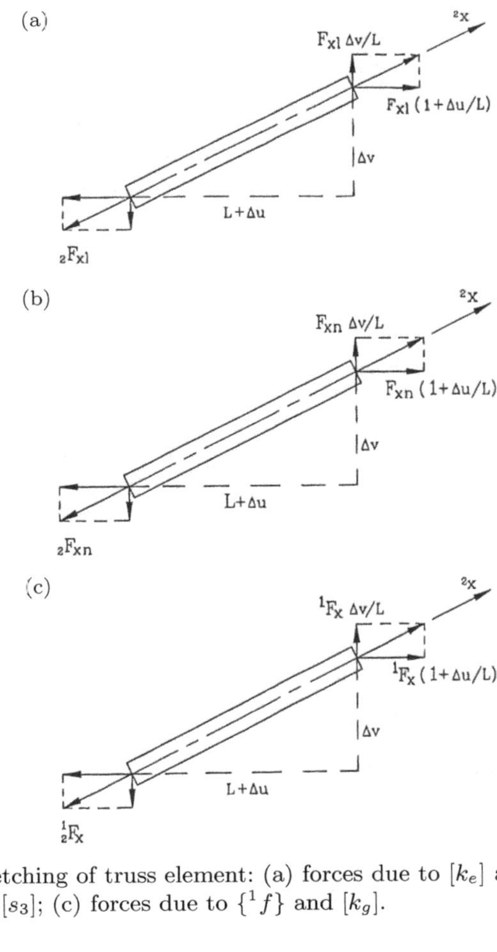

Fig. 4.3. Stretching of truss element: (a) forces due to $[k_e]$ and $[s_2]$; (b) forces due to $[s_1]$ and $[s_3]$; (c) forces due to $\{^1f\}$ and $[k_g]$.

to transform the force increment F_{xl} generated by the $[k_e]$ matrix from the element axis at C_1 to that at C_2. Collectively, the forces generated by the $[k_e]$ and $[s_2]$ matrices can be represented by a single force directed along the axis of the element at C_2 as

$$_2F_{xl} = F_{xl}(1 + {}_1e) = EA\frac{\Delta u}{L}(1 + {}_1e) \tag{4.39}$$

where the *stretch* ${}_1e$ is

$${}_1e = \frac{{}^2L - {}^1L}{{}^1L} \equiv \frac{{}^2L - L}{L} \tag{4.40}$$

and ${}^2L$ denotes the length of the member at C_2,

$${}^2L = \sqrt{(L + \Delta u)^2 + (\Delta v)^2} \tag{4.41}$$

Similarly, the forces generated by the $[s_3]$ matrix are in effect to transform the force increment F_{xn} generated by the $[s_1]$ matrix from the C_1 axis to the C_2 axis, as can be seen from Figure 4.3(b). The combined result can also be represented by a single force acting along the axis of the element at the C_2 configuration as follows:

$$_2F_{xn} = F_{xn}(1 + {}_1e) = \frac{1}{2}EA\left(\frac{\Delta u^2}{L^2} + \frac{\Delta v^2}{L^2}\right)(1 + {}_1e) \tag{4.42}$$

Finally, the $[k_g]$ matrix serves to transform the initial force ${}^1F_x$, which has been represented in the force vector $\{{}^1f\}$ as ${}^1F_{xb}$ [see equation (4.17) and the note following], from the initial axis to the deformed axis [see Figure 4.3(c)]. The resulting force acting along the axis of the element at C_2 is

$${}^1_2F_x = {}^1F_x(1 + {}_1e) \tag{4.43}$$

which is consistent with the rigid body rule.

Taking the sum of the above three effects, we can obtain the total force acting along the axis of the element at C_2 as

$${}^2_2F_x = {}_2F_{xl} + {}_2F_{xn} + {}^1_2F_x = (F_{xl} + F_{xn} + {}^1F_x)(1 + {}_1e) \tag{4.44}$$

or

$${}^2_2F_x = \left\{EA\left[\frac{\Delta u}{L} + \frac{1}{2}\left(\frac{\Delta u^2}{L^2} + \frac{\Delta v^2}{L^2}\right)\right] + {}^1F_x\right\}(1 + {}_1e) \tag{4.45}$$

which is as expected. The above derivation has demonstrated that the terms $[s_2]\{u\}$, $[s_3]\{u\}$, and $[k_g]\{u\}$ interact with the terms $[k_e]\{u\}$, $[s_1]\{u\}$, and $\{^1 f\}$, respectively, in accounting for the stretching effect of various orders for the truss element. Compared with the $[N_1]$ and $[N_2]$ matrices (or the $[k_1]$ and $[k_2]$ matrices) used by Mallett and Marcal (1968) and Chajes and Churchill (1987), the adoption of the three distinct matrices $[s_1]$, $[s_2]$, and $[s_3]$ has demonstrated its versatility in explaining the behavior of the truss element in stretching.

4.3.2 *Element forces due to rigid body rotation*

The effect of rigid body rotation can also be investigated using the present element matrices. For instance, consider the case where the element is rotated about its left end as a rigid body for a right angle of $\pi/2$ radians counterclockwise. For the element of length L, the element displacement vector $\{u\}_r$ is

$$\{u\}_r^T = \langle 0 \quad 0 \quad -L \quad L \rangle \tag{4.46}$$

By definition, we have $\Delta u = u_b - u_a = -L$ and $\Delta v = v_b - v_a = L$. According to equations (4.37) and (4.38), the linear and nonlinear force increments reduce to $F_{xl} = -EA$ and $F_{xn} = -EA$, respectively.

As can be seen from Figure 4.4(a), the forces generated by the matrices $[k_e]$ and $[s_1]$ counterbalance each other during the rigid body motion $\{u\}_r$. Similarly, the forces generated by the matrix $[s_2]$ also counterbalance those generated by the matrix $[s_3]$ when subjected to the rigid body motion [see Figure 4.4(b)]. Such relations remain valid regardless of the magnitude of the angle of rotation. On the other hand, by adding $[k_g]\{u\}_r$ to the initial forces $\{^1 f\}$, we observe that the axial force $^1F_x (= {^1F_{xb}})$, which was already acting on the element at C_1, will be rotated following the rigid rotation [see Figure 4.4(c)]. An overall result of the above three effects is that the initial forces have been directed along the rotated axis of the truss element, while their magnitudes remain unchanged, implying that the equilibrium of the element is preserved after the rigid body rotation. It should be added that the above observation regarding the rigid body characteristics of the geometric stiffness matrix $[k_g]$ is consistent with the rigid body rule presented previously in Section 3.5.

In fact, by letting $\{u\}_r$ represent any rigid body motion, it is possible to show that the strain increment ϵ_{xx} will remain equal to zero.

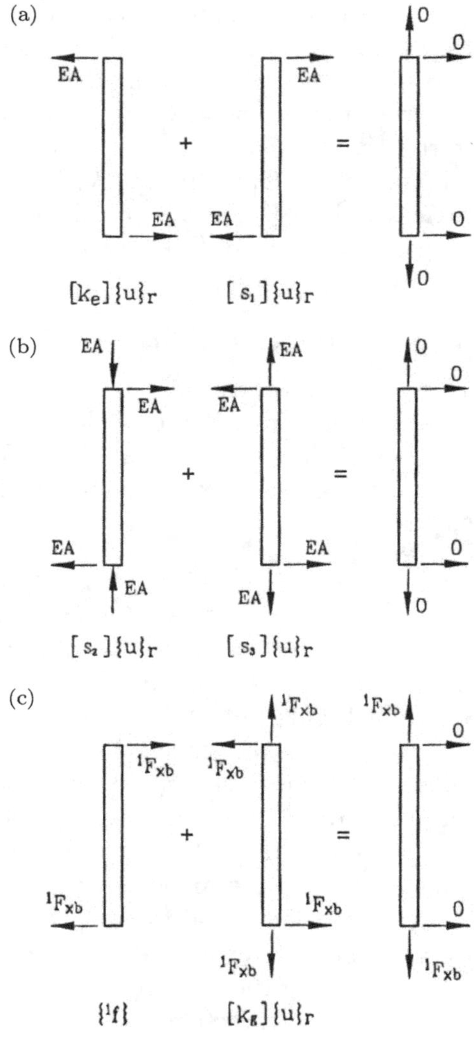

Fig. 4.4. Rigid body rotation of truss element: (a) forces due to $[k_e]$ and $[s_1]$; (b) forces due to $[s_2]$ and $[s_3]$; (c) forces due to $\{^1f\}$ and $[k_g]$.

Thus, we can write $\epsilon_{xx} = 0$ or

$$e_{xx} = -\eta_{xx} \tag{4.47}$$

From equations (4.37) and (4.38), it is known that

$$F_{xl} = -F_{xn} \quad \text{or} \quad F_{xl} + F_{xn} = 0 \tag{4.48}$$

Substituting the preceding equation into equation (4.44) yields

$$ {}_2^2F_x = {}^1F_x(1 + {}_1e) \equiv {}_1^1F_x \tag{4.49}$$

since it is realized that ${}_1e = ({}^2L - L)/L = 0$ for a bar undergoing rigid body rotations. Again, we have shown that for a truss element subjected to any rigid body rotation, the initial axial force ${}^1F_x$ acting on the element will always be directed along the (rotated) member axis with no change on the magnitude of the acting force.

Before closing this section, we like to point out that the $[k_g]$ matrix works with the initial forces vector $\{{}^1f\}$ in dealing with the rigid body behaviors of the initial forces acting on the element. It should not be regarded merely as a matrix similar to the $[k_e]$ matrix in computing the force increment. Rather, the $[k_g]$ matrix or the effects it represents should always be included in procedures for updating the initial nodal forces (based on the rigid body rule) in an incremental-iterative nonlinear analysis. Moreover, in order not to introduce any fictitious forces due to rigid body motions, both the $[k_e]$ and $[s_1]$ should be considered together in a force recovery procedure. The same is also true for the two matrices $[s_2]$ and $[s_3]$.

4.4 Higher-Order Stiffness Matrices in Symmetric Form

The higher-order stiffness matrices $[s_1]$ and $[s_2]$ presented in Section 4.2 appear to be asymmetric, while the matrix $[s_3]$ is advantageously shown to be symmetric. Whether a matrix is symmetric or not can affect the efficiency in implementation and execution of a computer analysis program. In the literature, the sum of the two matrices $[s_1]$ and $[s_2]$ has been denoted as the $1/2[N_1]$ or $1/2[k_1]$ matrix, which is known to be symmetric (Mallett and Marcal, 1968; Chajes and Churchill, 1987). In this section, it will be shown that *equivalent stiffness matrices* $[s_2]_{\text{eq}}$ and $[s_3]_{\text{eq}}$ that possess the particular feature of symmetry can also be derived for the $[s_2]$ and $[s_3]$ matrices respectively, based on the common characteristics of the three matrices: $[s_2]$, $[s_3]$, and $[k_g]$. By subtracting the stiffness matrix $[s_2]_{\text{eq}}$ from the sum of the $[s_1]$ and $[s_2]$ matrices, an equivalent stiffness matrix $[s_1]_{\text{eq}}$ which appears to be symmetric, can also be derived for the $[s_1]$ matrix.

A comparison of equation (4.33) with equations (4.35) and (4.36) shows that the force components generated by the $[s_2]$ and $[s_3]$ matrices are identical in form to those by the $[k_g]$ matrix, if the force parameter ${}^1F_x$ associated with the $[k_g]$ matrix is replaced by the force increments F_{xl} and F_{xn} associated with the $[s_2]$ and $[s_3]$ matrices, respectively. Such an analogy can also be observed from the discussion in the preceding section on the stretching of truss members, where it is known that the effect of the $[k_g]$, $[s_2]$ and $[s_3]$ matrices are in effect to transform the force components ${}^1F_x$, $F_{xl}(=[k_e]\{u\})$, and $F_{xn}(=[s_1]\{u\})$, respectively, from the C_1 axis to the C_2 axis, and to multiply them by the stretch factor $(1+{}_1e)$. Of interest to note is the fact that the force parameters ${}^1F_x$, $F_{xl}(=[k_e]\{u\})$, and $F_{xn}(=[s_1]\{u\})$ are all related to the virtual linear strain term δe_{xx} in the virtual work expressions [see equations (4.20), (4.21), and (4.25), along with (4.3)], while the $[k_g]$, $[s_2]$ and $[s_3]$ matrices are all related to the virtual nonlinear strain term $\delta\eta_{xx}$ [see equations (4.22)–(4.24)].

Based on the analogy between the $[k_g]$, $[s_2]$ and $[s_3]$ matrices, we can replace the force parameter ${}^1F_x = ({}^1F_{xb})$ in the geometric stiffness matrix $[k_g]$ of Eq. (4.27) by the force increments F_{xl} and F_{xn}, and obtain, respectively, equivalent expressions for the $[s_2]$ and $[s_3]$ matrices as follows:

$$[s_2]_{\text{eq}} = \frac{F_{xl}}{L}\begin{bmatrix} 1 & 0 & -1 & 0 \\ & 1 & 0 & -1 \\ & & 1 & 0 \\ \text{Symm.} & & & 1 \end{bmatrix} \tag{4.50}$$

$$[s_3]_{\text{eq}} = \frac{F_{xn}}{L}\begin{bmatrix} 1 & 0 & -1 & 0 \\ & 1 & 0 & -1 \\ & & 1 & 0 \\ \text{Symm.} & & & 1 \end{bmatrix} \tag{4.51}$$

in which the symmetry of the stiffness matrices has been restored. Since the matrices $[s_2]_{\text{eq}}$ and $[s_3]_{\text{eq}}$ play a role very similar to the $[k_g]$ matrix in the calculation of element forces, and since they represent an effect of an order higher than the $[k_g]$ matrix, the sum of the two matrices $[s_2]_{\text{eq}}$ and $[s_3]_{\text{eq}}$ has sometimes been denoted as the $\Delta[k_g]$

matrix,

$$\Delta[k_g] = [s_2]_{\text{eq}} + [s_3]_{\text{eq}} \tag{4.52}$$

On the other hand, by subtracting the $[s_2]_{\text{eq}}$ matrix from the sum of the $[s_1]$ and $[s_2]$ matrices in (4.28) and (4.29), an alternative expression can be obtained for the $[s_1]$ matrix:

$$[s_1]_{\text{eq}} = \frac{EA}{2L^2}\begin{bmatrix} \Delta u & \Delta v & -\Delta u & -\Delta v \\ & -\Delta u & -\Delta v & \Delta u \\ & & \Delta u & \Delta v \\ \text{Symm.} & & 0 & -\Delta u \end{bmatrix} \tag{4.53}$$

which again appears to be symmetric.

Using the concept of equivalent stiffness matrices, we have demonstrated in this section that the stiffness matrices $[s_2]_{\text{eq}}$ and $[s_3]_{\text{eq}}$ are identical in form to the geometric stiffness matrix $[k_g]$, and that equivalent symmetric form also exists for the stiffness matrix $[s_1]$. Such characteristics enable us to include all the higher-order effects of the truss element in a nonlinear analysis in a rather straightforward manner. There is no doubt that the use of the symmetric stiffness matrices $[s_1]_{\text{eq}}$, $[s_2]_{\text{eq}}$, and $[s_3]_{\text{eq}}$ can result in greater efficiency of computation, in comparison with the use of the original asymmetric stiffness matrices $[s_1]$, $[s_2]$ and $[s_3]$.

4.5 Space Truss Element: A Three-Dimensional Generalization

The procedure presented in Sections 4.2–4.4 has been restricted to the plane truss element. It can be easily extended to the three-dimensional case to arrive at the *space truss element*, if consideration of the contribution from the third dimension is necessary, as will be demonstrated in this section.

First of all, for the space truss element shown in Figure 4.5, the displacements (u, v, w) of a generic cross section x can be related to the displacements (u_a, v_a, w_a) and (u_b, v_b, w_b) at the two ends A and

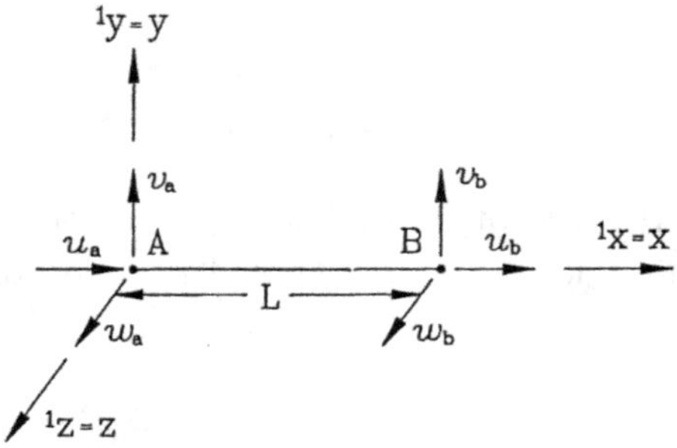

Fig. 4.5. Degrees of freedom for space truss element.

B of the element by the linear interpolation functions as

$$u = u_a\left(1 - \frac{x}{L}\right) + u_b\left(\frac{x}{L}\right) \tag{4.54}$$

$$v = v_a\left(1 - \frac{x}{L}\right) + v_b\left(\frac{x}{L}\right) \tag{4.55}$$

$$w = w_a\left(1 - \frac{x}{L}\right) + w_b\left(\frac{x}{L}\right) \tag{4.56}$$

According to equations (1.34) and (1.35), the linear and nonlinear components of the axial strain ϵ_{xx}, i.e., e_{xx} and η_{xx} can be augmented from equations (4.13) and (4.14) as follows:

$$e_{xx} = \frac{\partial u}{\partial x} - \frac{\Delta u}{L} \tag{4.57}$$

$$\eta_{xx} = \frac{1}{2}\left[\left(\frac{\partial u}{\partial x}\right)^2 + \left(\frac{\partial v}{\partial x}\right)^2 + \left(\frac{\partial w}{\partial x}\right)^2\right] = \frac{1}{2}\left(\frac{\Delta u^2}{L^2} + \frac{\Delta v^2}{L^2} + \frac{\Delta w^2}{L^2}\right) \tag{4.58}$$

where the *relative nodal displacements* are

$$\Delta u = u_b - u_a \quad \Delta v = v_b - v_a \quad \text{and} \quad \Delta w = w_b - w_a \tag{4.59}$$

For the space truss element shown in Figure 4.5, the element displacement vector consists of three translational degrees of freedom at each end, i.e.,

$$\{u\}^T = \langle u_a \quad v_a \quad w_a \quad u_b \quad v_b \quad w_b \rangle \tag{4.60}$$

Correspondingly, the force vectors at C_1 and C_2 should consist of three direct forces at each end of the element, i.e.,

$$\{{}^1f\}^T = \langle {}^1F_{xa} \quad {}^1F_{ya} \quad {}^1F_{za} \quad {}^1F_{xb} \quad {}^1F_{yb} \quad {}^1F_{zb} \rangle \tag{4.61}$$

$$\{{}^2f\}^T = \langle {}^2F_{xa} \quad {}^2F_{ya} \quad {}^2F_{za} \quad {}^2F_{xb} \quad {}^2F_{yb} \quad {}^2F_{zb} \rangle \tag{4.62}$$

As was noted previously, for a truss element in equilibrium at C_1, all the transverse shear forces vanish, i.e., ${}^1F_{ya} = {}^1F_{za} = {}^1F_{yb} = {}^1F_{zb} = 0$, and the axial forces on the two ends of the bar are equal in magnitude but opposite in sign, i.e., ${}^1F_{xb} = -{}^1F_{xa}$.

For the present three-dimensional case, each of the terms in equation (4.10) can be derived as follows:

$$\int_V Ee_{xx}\delta e_{xx}dV = \int_V E\frac{\Delta u}{L}\delta\left(\frac{\Delta u}{L}\right)dV = \{\delta u\}^T[k_e]\{u\} \tag{4.63}$$

$$\begin{aligned}\int_V E\eta_{xx}\delta e_{xx}dV &= \int_V \frac{1}{2}E\left(\frac{\Delta u^2}{L^2}+\frac{\Delta v^2}{L^2}+\frac{\Delta w^2}{L^2}\right)\delta\left(\frac{\Delta u}{L}\right)dV \\ &= \{\delta u\}^T[s_1]\{u\}\end{aligned} \tag{4.64}$$

$$\begin{aligned}\int_V Ee_{xx}\delta\eta_{xx}dV &= \int_V E\frac{\Delta u}{L}\left[\frac{\Delta u}{L}\delta\left(\frac{\Delta u}{L}\right)+\frac{\Delta v}{L}\delta\left(\frac{\Delta v}{L}\right)\right. \\ &\quad \left.+\frac{\Delta w}{L}\delta\left(\frac{\Delta w}{L}\right)\right]dV = \{\delta u\}^T[s_2]\{u\}\end{aligned} \tag{4.65}$$

$$\begin{aligned}\int_V E\eta_{xx}\delta\eta_{xx}dV &= \int_V \frac{1}{2}E\left(\frac{\Delta u^2}{L^2}+\frac{\Delta v^2}{L^2}+\frac{\Delta w^2}{L^2}\right) \\ &\quad \times\left[\frac{\Delta u}{L}\delta\left(\frac{\Delta u}{L}\right)+\frac{\Delta v}{L}\delta\left(\frac{\Delta v}{L}\right)+\frac{\Delta w}{L}\delta\left(\frac{\Delta w}{L}\right)\right]dV \\ &= \{\delta u\}^T[s_3]\{u\}\end{aligned} \tag{4.66}$$

$$\begin{aligned}\int_V {}^1\tau_{xx}\delta\eta_{xx}dV &= \int_V {}^1F_x\left[\frac{\Delta u}{L}\delta\left(\frac{\Delta u}{L}\right)+\frac{\Delta v}{L}\delta\left(\frac{\Delta v}{L}\right)\right. \\ &\quad \left.+\frac{\Delta w}{L}\delta\left(\frac{\Delta w}{L}\right)\right]dV = \{\delta u\}^T[k_g]\{u\}\end{aligned} \tag{4.67}$$

$$ {}^1R = \int_S {}^1t_i \, \delta u_i dS = \{\delta u\}^T \{{}^1f\} \tag{4.68} $$

$$ {}^2R = \int_S {}^2t_i \, \delta u_i dS = \{\delta u\}^T \{{}^2f\} \tag{4.69} $$

where the matrices $[k_e]$, $[k_g]$, $[s_1]$, $[s_2]$, and $[s_3]$ are known to have a dimension of 6×6 for the space truss element. For the present case, the elastic stiffness matrix $[k_e]$ has been given in equation (2.103) as the $[k]$ matrix, and the geometric stiffness matrix $[k_g]$ is

$$ [k_g] = \frac{{}^1F_{xb}}{L} \begin{bmatrix} 1 & 0 & 0 & -1 & 0 & 0 \\ & 1 & 0 & 0 & -1 & 0 \\ & & 1 & 0 & 0 & -1 \\ & & & 1 & 0 & 0 \\ & & & & 1 & 0 \\ \text{Symm.} & & & & & 1 \end{bmatrix} \tag{4.70} $$

and the higher-order stiffness matrices $[s_1]$, $[s_2]$, and $[s_3]$ can be expressed as follows:

$$ [s_1] = \frac{EA}{2L^2} \begin{bmatrix} \Delta u & \Delta v & \Delta w & -\Delta u & -\Delta v & -\Delta w \\ 0 & 0 & 0 & 0 & 0 & 0 \\ 0 & 0 & 0 & 0 & 0 & 0 \\ -\Delta u & -\Delta v & -\Delta w & \Delta u & \Delta v & \Delta w \\ 0 & 0 & 0 & 0 & 0 & 0 \\ 0 & 0 & 0 & 0 & 0 & 0 \end{bmatrix} \tag{4.71} $$

$$ [s_2] = \frac{EA}{2L^2} \begin{bmatrix} 2\Delta u & 0 & 0 & -2\Delta u & 0 & 0 \\ \Delta v & \Delta u & 0 & -\Delta v & -\Delta u & 0 \\ \Delta w & 0 & \Delta u & -\Delta w & 0 & -\Delta u \\ -2\Delta u & 0 & 0 & 2\Delta u & 0 & 0 \\ -\Delta v & -\Delta u & 0 & \Delta v & \Delta u & 0 \\ -\Delta w & 0 & -\Delta u & \Delta w & 0 & \Delta u \end{bmatrix} \tag{4.72} $$

$$ [s_3] = \frac{EA}{6L^3} \begin{bmatrix} [h] & -[h] \\ -[h] & [h] \end{bmatrix} \tag{4.73} $$

where

$$[h] = \begin{bmatrix} 3\Delta u^2 + \Delta v^2 + \Delta w^2 & 2\Delta u\Delta v & 2\Delta u\Delta w \\ 2\Delta u\Delta v & 3\Delta v^2 + \Delta u^2 + \Delta w^2 & 2\Delta v\Delta w \\ 2\Delta u\Delta w & 2\Delta v\Delta w & 3\Delta w^2 + \Delta u^2 + \Delta v^2 \end{bmatrix} \tag{4.74}$$

Again, making use of the expressions in (4.63)–(4.69) and admitting the arbitrary nature of the virtual displacements $\{\delta u\}$ we can derive from equation (4.10) the incremental equation of equilibrium for the space truss element, which is identical in form to the one presented in (4.31) for the two-dimensional case.

At this moment, it is sufficient to note that the procedures presented in Section 4.3 for exploiting the rigid body and stretching characteristics of the truss element, in terms of the stiffness matrices $[k_e]$, $[k_g]$, $[s_1]$, $[s_2]$, and $[s_3]$ and those in Section 4.4 for deriving the *equivalent stiffness* expressions can be easily extended to the present three-dimensional case. For instance, the stiffness matrices $[s_2]$, and $[s_3]$ can be recast in a form identical to the geometric stiffness matrix $[k_g]$ as follows:

$$[s_2]_{\text{eq}} = \frac{F_{xl}}{L} \begin{bmatrix} 1 & 0 & 0 & -1 & 0 & 0 \\ & 1 & 0 & 0 & -1 & 0 \\ & & 1 & 0 & 0 & -1 \\ & & & 1 & 0 & 0 \\ & & & & 1 & 0 \\ \text{Symm.} & & & & & 1 \end{bmatrix} \tag{4.75}$$

$$[s_3]_{\text{eq}} = \frac{F_{xn}}{L} \begin{bmatrix} 1 & 0 & 0 & -1 & 0 & 0 \\ & 1 & 0 & 0 & -1 & 0 \\ & & 1 & 0 & 0 & -1 \\ & & & 1 & 0 & 0 \\ & & & & 1 & 0 \\ \text{Symm.} & & & & & 1 \end{bmatrix} \tag{4.76}$$

Similarly, the equivalent matrix $[s_1]_{\text{eq}}$ can be obtained by subtracting the $[s_2]_{\text{eq}}$ matrix from the sum of the two matrices $[s_1]$ and $[s_2]$ as follows:

$$[s_1]_{\text{eq}} = \frac{EA}{2L^2}\begin{bmatrix} \Delta u & \Delta v & \Delta w & -\Delta u & -\Delta v & -\Delta w \\ & -\Delta u & 0 & -\Delta v & \Delta u & 0 \\ & & -\Delta u & -\Delta w & 0 & \Delta u \\ & & & \Delta u & \Delta v & \Delta w \\ & & & & -\Delta u & 0 \\ \text{Symm.} & & & & & -\Delta u \end{bmatrix} \tag{4.77}$$

which possesses the desired property of symmetry.

4.6 Force Recovery Procedures Based on Matrix Equations

As was stated in Section 4.1, an incremental-iterative nonlinear analysis may be functionally divided into three major phases, i.e., the *predictor*, *corrector*, and *equilibrium-checking* phases. A complete description of the procedure for nonlinear analysis is given in Chapter 7. Here, we like to emphasize that it is the corrector phase that governs the accuracy of the solution of a nonlinear analysis. For a solution to be acceptable, the corrector or the procedure for recovering the element forces has to be exact or exact in the limit sense, while the predictor can be more or less approximate, because it affects only the speed of convergence or the number of iterations.

For the purpose of recovering the element forces, the incremental element equation as given in (4.31) can be rewritten,

$$\{^2f\} = ([k_e] + [k_g] + [s_1] + [s_2] + [s_3])\{u\} + \{^1f\} \tag{4.78}$$

This equation, or its origin given in equation (4.31), has been derived for the truss element with due account taken of the stretching and rigid body effects. Such an equation can be said to be exact within the context of continuum mechanics principles, because no approximation has been made in the derivation, except that the material should obey the law given in (4.9). Thus, this equation is qualified for recovery of the element forces in the analysis of nonlinear trusses by incremental-iterative procedures, if the material law of

(4.9) is followed. It should be noted that all the quantities in equation (4.78) have been expressed with respect to the axes of the last calculated configuration C_1 of the element.

Of interest is the case when the strain increment ϵ_{xx} may be considered small for each incremental step of the nonlinear analysis. For this special case, the *incremental constitutive law* of (4.9) reduces to

$$S_{xx} = Ee_{xx} \tag{4.79}$$

In accordance, the terms that account for the effects of nonlinear strain η_{xx}, i.e., the matrices $[s_1]$ and $[s_3]$ given in equations (4.21) and (4.23) for the plane element or in equations (4.64) and (4.66) for the space element can be omitted from equation (4.78),

$$\{^2f\} = ([k_e] + [k_g] + [s_2])\{u\} + \{^1f\} \tag{4.80}$$

One particular drawback with this equation is that the rule of rigid body motions has been violated, because the fictitious forces generated by the $[k_e]$ and $[s_2]$ matrices upon rigid body rotations can no longer be counterbalanced (see Section 4.3.2 for interpretation). Consequently, this equation is strictly valid for incremental steps that are small not only in deformations, but also in rotations. It should not be used in the postbuckling analysis of trusses of which the rotations can be very large at certain nodal points.

It is still possible to obtain a simplified but fairly accurate equation for recovering the element forces. For instance, since the effect of the $[k_g]$ matrix is to let the initial forces $\{^1f\}$ move with the rigid body motion, we may therefore include such an effect simply by regarding the forces $\{^1f\}$, which are acting at C_1 and referred to the C_1 axes, as the ones acting at C_2 and directed along the axis of the bar at C_2, according to the rigid body rule (Section 3.5 or Yang and Chiou, 1987). Also, it has been demonstrated in Section 4.3.2 that the forces generated by the $[k_e]$ and $[s_1]$ matrices during a rigid body motion are always in equilibrium (i.e., they counterbalance each other upon rigid body rotations). To account for the rigid body rotation of the bar from C_1 to C_2, the forces generated by the $[k_e]$ and $[s_1]$ matrices can be directly regarded as the ones acting at C_2, and directed along the axis of the bar at C_2, also according to the rigid body rule. It follows that the resulting forces acting on the bar at C_2 can be expressed with respect to the same configuration as

$$\{{}_2^2f\} = ([k_e] + [s_1])\{u\} + \{{}^1f\} \tag{4.81}$$

where the higher-order stiffness matrices are associated with the effect of stretching, as represented by the term ${}_1e$ in equation (4.44), have been neglected. This equation can be applied to problems involving incremental steps that are large in rotations, but small in deformations.

4.7 Case Studies by Finite Element Analysis

The objective of this section is to numerically evaluate the performance of the element derived and the procedures presented in the preceding sections of this chapter for large-displacement analysis of trusses. The solution scheme adopted herein for solving the nonlinear stiffness equations is the *generalized displacement control method* to be presented in Chapter 7 or Yang and Shieh (1990). This method is chosen because of its self-adaptive capability in solving problems involving various critical points, such as the limit points and snap-back points. Besides, it has the advantage that the step size can be adjusted according to the variation in the degree of nonlinearity of the stiffness of the structure under consideration.

4.7.1 *Two-member truss*

The two-member truss shown in Figure 4.6 is taken as the first example. Each of the two members of the truss is assumed to have the same cross-sectional area A, length, and elastic modulus E. From the results given in Figure 4.7, we observe that very large errors occur in the solution when the small-strain equation (4.80) is used for calculating the element forces, which do not obey the rigid body law, and that the errors decrease as smaller load increments are used. In contrast, good agreement has been made between Papadrakakis' (1981) solution and the present solution obtained using either the exact version of (4.78) or the approximate version of (4.81). One feature of equations (4.78) and (4.81) is that they both forbid the occurrence of fictitious forces upon rigid body motions. From this example, it

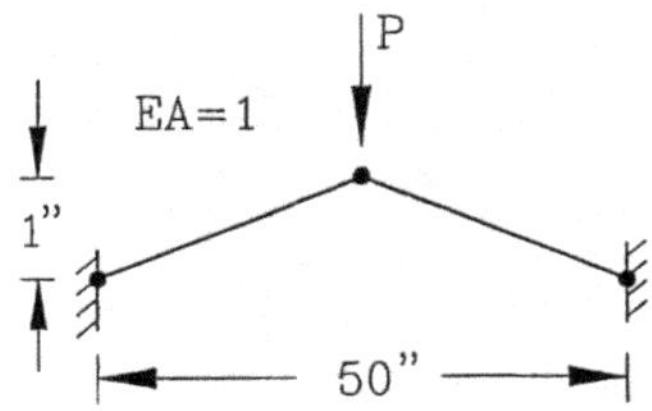

Fig. 4.6. Two-member truss.

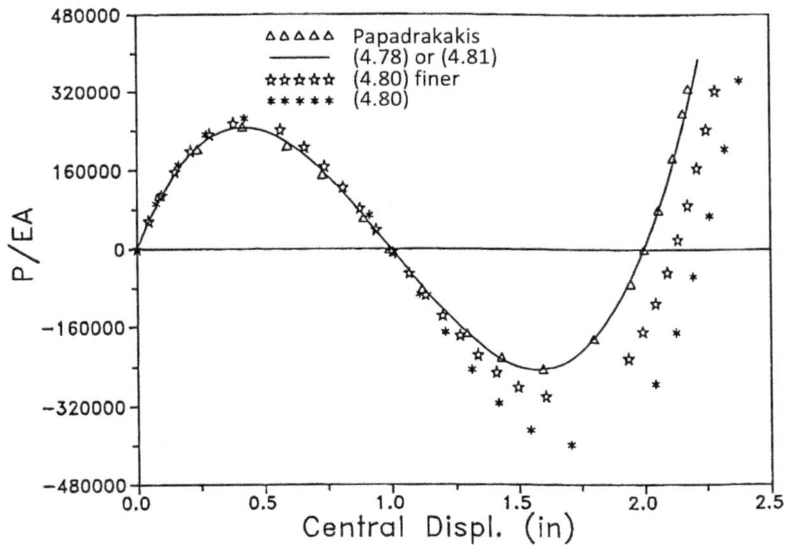

Fig. 4.7. Load-deflection curves of two-member truss.

is concluded that the ability of a force recovery procedure to incorporate rigid body modes can affect significantly the accuracy of the nonlinear analysis.

4.7.2 *24-Member shallow dome*

The second example studied here is the 24-member shallow dome shown in Figure 4.8, which has been commonly used as a benchmark for testing the accuracy of various element formulations and solution algorithms (Hangai and Kawamata, 1971; Jagannathan *et al.*, 1975; Holzer *et al.*, 1980; Papadrakakis, 1981). The dome is assumed to be

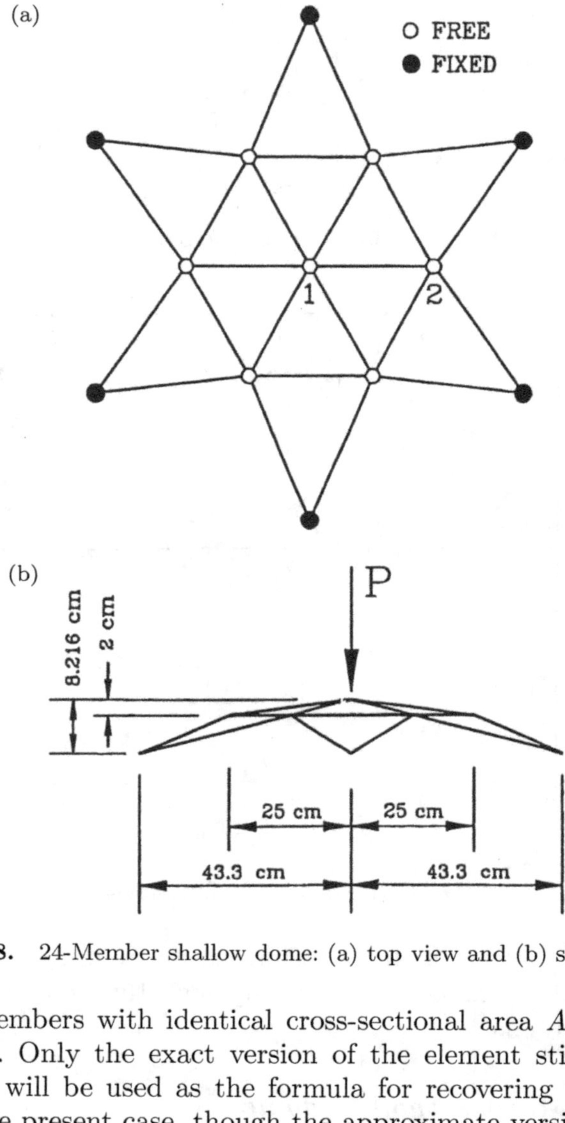

Fig. 4.8. 24-Member shallow dome: (a) top view and (b) side view.

made of members with identical cross-sectional area A and elastic modulus E. Only the exact version of the element stiffness equation (4.78) will be used as the formula for recovering the element forces in the present case, though the approximate version in (4.81) can be used with equal success.

For this particular example, the accuracy of solutions in Figure 4.9 can be evaluated by two zero load states due to the geometric constraints. In the first zero load state, node 1 lies in the plane of the internal ring nodes. As a result, the vertical deflection of joint 1

(a)

(b)

(c)

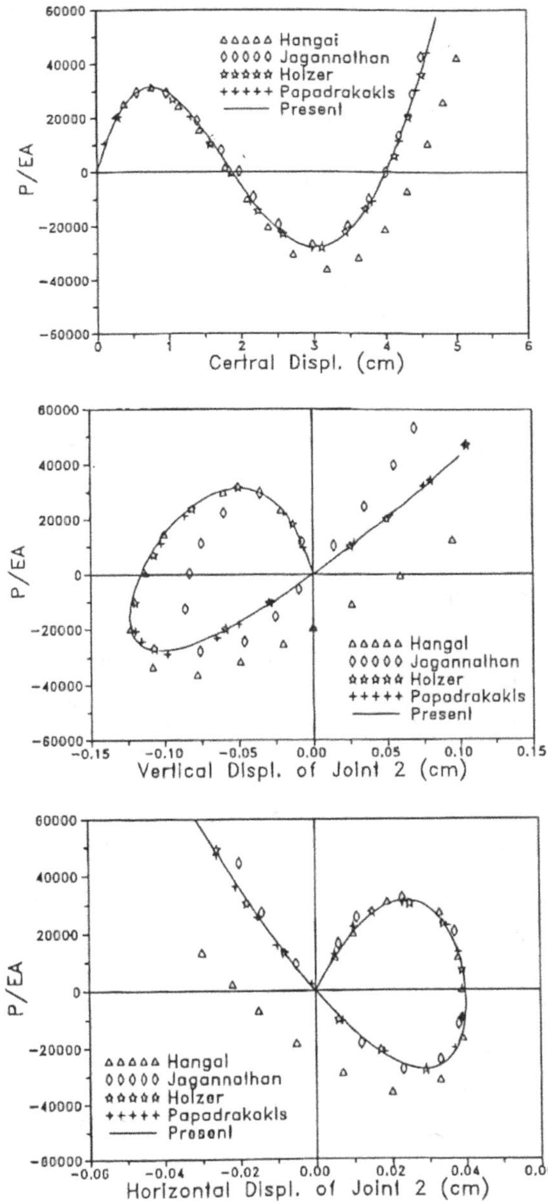

Fig. 4.9. Load–deflection curves for 24-member shallow dome: (a) vertical deflection of joint 1; (b) vertical deflection of joint 2; and (c) horizontal deflection of joint 2.

should be equal to 2 cm plus the vertical deflection of joint 2. The corresponding vertical deflections obtained by the present study for joints 1 and 2 are 1.91178 cm (downward) and 0.08823 cm (upward), respectively. This has the indication that the geometric condition is satisfied with extremely high precision. It should be noted that solutions obtained by Jagannathan *et al.* (1975), Holzer *et al.* (1980), and Papadrakakis (1981) seem to reflect the same characteristics as well. The second zero load state is reached when the vertical position of node 1 is a mirror image of the initial shape relative to the internal ring nodes. From this condition, the vertical deflection of node 1 should be equal to 4 cm and all deflection components of node 2 should sum up to zero. It is clear from Figures 4.9(a)–4.9(c) that this condition is again satisfied by the present solution and all previous ones, except that of Hangai and Kawamata (1971).

References

Chajes, A. and Churchill, J.E. (1987). Nonlinear frame analysis by finite element method, *J. Struct. Eng.*, ASCE, 113(6), 1221–1235.

Hangai, Y. and Kawamata, S. (1971). Nonlinear analysis of space frames and snap-through buckling of reticulated shell structures, *Proc. Int. Association for Shel and Spatial Structures (IASS) Pacific Symp., Part I on Tension Structures and Space Frames*, Tokyo and Kyoto, Japan, September, pp. 9-4-1–9-4-12.

Holzer, S.M., Plaut, R.H., and Somers, A.E., Jr. (1980). Stability of lattice structures under combined loads, *J. Eng. Mech. Div.*, ASCE, 106(2), 289–305.

Jagannathan, D.S., Epstein, H.I., and Christiano, P. (1975). Snap-through buckling of reticulated shells, *Proc. Institution of Civil Engineers, Part 2*, 59, 727–742.

Leu, L.J. and Yang, Y.B. (1990). Effects of rigid body and stretching on nonlinear analysis of trusses, *J. Struct. Eng.*, ASCE, 116(10), 2582–2598.

Mallett, R.H. and Marcal, P.V. (1968). Finite element analysis of nonlinear structures, *J. Struct. Div.*, ASCE, 94(9), 2081–2105.

Papadrakakis, M. (1981). Post-buckling analysis of spatial structures by vector iteration methods, *Comput. Struct.*, 14(5–6), 393–402.

Yang, Y. B. and Chiou, H. T. (1987). Rigid body motion test for nonlinear analysis with beam elements, *J. Eng. Mech.*, ASCE, 113(9), 1404–1419.

Yang, Y.B. and Leu, L.J. (1990). Postbuckling analysis of trusses with various Lagrangian formulations, *AIAA J.*, 28(5), 946–948.

Yang, Y.B. and Leu, L.J. (1991). Constitutive laws and force recovery procedures in nonlinear analysis of trusses, *Comp. Meth. Appl. Mech. Eng.*, 92, 121–131.

Yang, Y.B. and Shieh, M.S. (1990). Solution method for nonlinear problems with multiple critical points, *AIAA J.*, 28(12), 2110–2116.

Chapter 5

Fundamentals of Nonlinear Theory of Space Frames

5.1 Introduction

Previous research on the buckling of structural frames using analytical approaches has been directed in general toward the torsional-flexural buckling of single members or the in-plane buckling of plane frames (Timoshenko and Gere, 1961; Vlasov, 1961; Chajes, 1974; Simitses, 1976; Chen and Lui, 1987). As can be seen from the literature, very few analytical studies have been conducted on the out-of-plane buckling of plane frames or on the buckling of space frames that involve three-dimensional rotations. This is in part due to the mathematical difficulties involved in the solution of the associated boundary-value problem, imagining that we are faced with the problem of solving a large set of simultaneous differential equations along with a large number of continuity conditions and boundary conditions. One side effect of this mathematical hindrance has been the oversight of some fundamental aspects that cannot be directly extended from the buckling theories of single members or plane frames restrained against out-of-plane deformations.

It is true that the mathematical difficulties encountered in the analytical study of space frames can be greatly eased through the use of numerical approaches, such as the finite element method, along with the aid of modern high-performance computers. However, in adopting any numerical method, we should be very careful since it offers nothing but an approximation for representing the underlying governing differential equations, continuity conditions, and boundary conditions of the problem considered. Our question herein is whether the numerical model or the finite element model used, can consistently reproduce the behavior of the structure under consideration. From the theoretical point of view, if we are able to describe a particular problem by a finite element model, we should be able to describe the same problem by an equivalent set of governing differential equations, continuity conditions, and boundary conditions. Before we apply some criteria, say, those presented in Chapter 3, to test the legitimacy of the finite element model, we should be able to apply the same criteria to test the legitimacy of the governing differential equations, continuity conditions, and (natural) boundary conditions that underlies the finite element. In other words, we are saying that the arguments existing at the element formulation level, if there are any, should be settled first at the fundamental mechanics equations level, and that only when we can be sure of the consistency of the underlying mechanics equations adopted, can it be meaningful to talk about the legitimacy of the numerical model to be used as an approximate substitute.

Based on the consideration that the finite element method is nothing but a numerical substitute of the fundamental mechanics equations, there exists a need to establish some benchmark problems using the most fundamental mechanics equations. It is anticipated that the procedures and equations adopted in the analytical study can be directly translated into the finite element procedures, thereby providing some clues for clarifying the ambiguities existing in the finite element simulation. Moreover, the solutions obtained analytically can be used as the basis for calibrating numerical solutions obtained by other approaches. As part of this effort, a series of analytical studies have been conducted by Yang and Kuo (1991a,b) on the buckling of some simple plane frames that are allowed to buckle laterally or out of plane of the frame. Some single-member buckling problems will be presented in Section 5.7, while the two-member buckling problems are reserved for Chapter 9.

The materials to be presented in this chapter follow basically the same lines as those presented by the senior author and co-workers in a number of technical papers. However, they are not merely reproduction of those previous works. Efforts have been undertaken in this chapter to rigorously present the static and kinematic relations, and the equations for cross-sectional forces, for adoption in the formulation of the buckling theory for three-dimensional solid beams. With these relations and equations, a number of nonlinear virtual work terms that were previously regarded merely as of higher orders and discarded from the formulation can now be more properly interpreted.

The approach adopted will be referred to as the *engineering approach* in that only *three* components of strains (and stresses) are included in the formulation based on the engineers judgment. With proper assumptions adopted for the bending moments and torques in the deformed configuration, the final virtual work equation (and associated finite element) can be derived. This approach will be detailed in Sections 5.3–5.5. We believe that engineers will prefer to adopt the engineering approach for its straightforwardness and elegance in derivation.

There is another approach called the *elasticity approach* or *general theory* in that all the *six* components of strains (and stresses) are included in the formulation. With this approach, basically, all the higher-order terms generated either in the integrals or by the boundary conditions will combine to yield some physically meaningful terms or just cancel out with each other. Such an approach has been presented in detail in the original version of our book (Yang and Kuo, 1994). However, in preparing this revised version, we choose to abandon the elasticity approach because the lengthy mathematical operations involved in the formulation are likely to discourage most engineers while making itself not very welcome. Fortunately, with the aid of some proper assumptions, the engineering approach can be relied on to derive virtual work equations and finite element equations that are in full agreement with those by the elasticity approach.

One feature that makes the buckling of space frames or the lateral buckling of plane frames much more complicated than that of single members or plane frames considering only in-plane deformations has been the involvement of three-dimensional rotations. The static and kinematic relations presented in Section 5.2 and the equations of equilibrium for cross-sectional forces (especially moments and torques) derived in Section 5.3 are the prerequisite to the updated Lagrangian formulation to be presented later for the three-dimensional solid beam. In Section 5.4, we shall identify the various torsional and bending moments according to the way they are generated in the virtual work formulation. Such a classification of moment properties is necessary for the three-dimensional analysis, in which distinction has to be made between the initial and buckling configurations of structural members.

After the presentation of the engineering approach in Section 5.4, the buckling differential equations and boundary conditions will be derived for the solid beam by the method of variation in Section 5.5. For the purpose of illustration, we shall demonstrate in Section 5.6 that the natural boundary conditions derived for the three-dimensional beam are suitable candidates for testing the quality of rigid body rotations.

All the mechanics equations derived have been conveniently presented in the incremental form. We shall show in Section 5.7 that with this type of equation, a two-stage analysis can be conducted for the buckling of structures, with the first stage referring to the *prebuckling* behavior and the second stage the *buckling* behavior of structures. It is worth noting that the force-displacement relations for structural members have been established duly for the buckling position.

To focus on the flow of presentation leading to the finite element matrices for the space frame element (Chapter 6) and nonlinear solution algorithm (Chapter 7), further analytical applications of the rigid body rule to the derivation of some new elements (such as the *triangular plate element*) will be deferred until Chapter 8. The closed-form solutions to be given in Chapter 9 serve as the benchmarks for calibrating the finite element solutions to be presented in Chapter 6.

5.2 Statics and Kinematics

In describing the behavior of a solid beam under the action of applied loads, two typical stages can be identified. One is called the *prebuckling* stage, in which the deformations of the beam are assumed to be generally small, for which the effects of change in geometry may be neglected. The other is called the *buckling* stage, which is characterized by the occurrence of large deformations in directions not parallel to those of the prebuckling stage in response to a small increase in external loadings. In this section, the static and kinematic relations in connection with each stage that are necessary to the derivation of the buckling theory for three-dimensional solid beams will first be presented.

5.2.1 *Prebuckling stage*

In this section, we shall focus on the static behavior of a solid beam in the prebuckling stage, as characterized by the application of the external loadings from zero at the C_0 configuration to a reference load level smaller than the critical value at the C_1 configuration. Since the deformations of the beam at this stage are generally small, the effects of change in the geometry of the beam can be neglected. It becomes immaterial whether configuration C_0 or C_1 is selected as the reference, as is typical in a *linear analysis*. For the purpose of conducting the buckling analysis in later sections, it will be more convenient to relate all physical parameters to the C_1 configuration. By doing so, al the relations derived here can be readily applied in the formulation of the buckling theory for the three-dimensional beam in Section 5.4.

As shown in Figure 5.1, we shall use the x-axis to denote the centroidal axis of the beam, and y and z the principal directions of the cross-section. Let u_x, u_y, and u_z denote the displacements of a generic point N at section x of the beam along the three axes x, y, and z. The strains for the generic point can be written as

$$
{}_1e_{xx} = u_{x,x} \tag{5.1}
$$

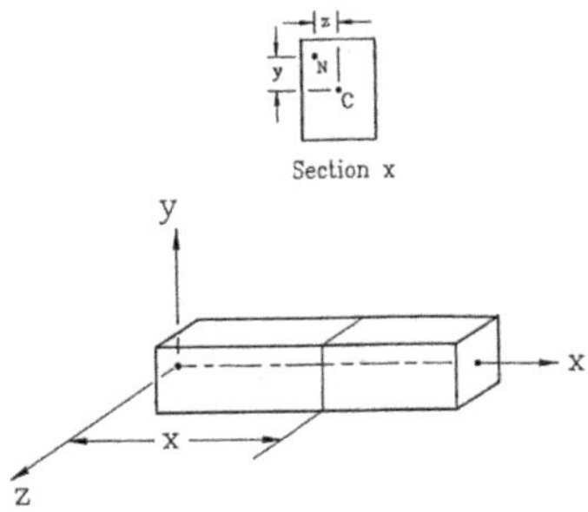

Fig. 5.1. Three-dimensional solid beam.

$$ {}_1e_{xy} = \frac{1}{2}(u_{x,y} + u_{y,x}) \tag{5.2} $$

$$ {}_1e_{xz} = \frac{1}{2}(u_{x,z} + u_{z,x}) \tag{5.3} $$

where a comma denotes differentiation with respect to the coordinate following. The other components that are of little significance in the linear analysis are

$$ {}_1e_{yy} = u_{y,y} \tag{5.4} $$

$$ {}_1e_{yz} = \frac{1}{2}(u_{y,z} + u_{z,y}) \tag{5.5} $$

$$ {}_1e_{zz} = u_{z,z} \tag{5.6} $$

Based on the *Bernoulli–Euler hypothesis* of plane sections remaining plane after deformation, the displacements (u_x, u_y, u_z) of a generic point N at section x with coordinates (y, z) can be related to the displacements (u, v, w) of the centroid C of the same section as follows [see Figure 3.2 and equations (3.9a) and (3.9b) for the two-dimensional case]:

$$ u_x = u - yv' - zw' \tag{5.7} $$

$$ u_y = v - z\theta_x \tag{5.8} $$

$$ u_z = w + y\theta_x \tag{5.9} $$

Substituting the preceding expressions for the displacements (u_x, u_y, u_z) into equations (5.1)–(5.6) for the strain components yields the following:

$$ {}_1e_{xx} = u' - yv'' - zw'' \tag{5.10} $$

$$ {}_1e_{xy} = -\frac{1}{2}z\theta'_x \tag{5.11} $$

$$ {}_1e_{xz} = \frac{1}{2}y\theta'_x \tag{5.12} $$

$$ {}_1e_{yy} = {}_1e_{yz} = {}_1e_{zz} = 0 \tag{5.13} $$

As can be seen, the strains ${}_1e_{yy}$, ${}_1e_{yz}$, and ${}_1e_{zz}$ vanish due to the assumption of undistorted cross sections, as implied by equations (5.8) and (5.9).

According to *Hooke's law*, the stresses can be related to the strains as follows:

$$^1\tau_{xx} = E_1 e_{xx} \tag{5.14}$$

$$^1\tau_{xy} = G(2_1 e_{xy}) \tag{5.15}$$

$$^1\tau_{xz} = G(2_1 e_{xz}) \tag{5.16}$$

where E and G denote the moduli of elasticity and rigidity, respectively. Consequently,

$$^1\tau_{xx} = E(u' - yv'' - zw'') \tag{5.17}$$

$$^1\tau_{xy} = -Gz\theta'_x \tag{5.18}$$

$$^1\tau_{xz} = Gy\theta'_x \tag{5.19}$$

Based on the conditions of equilibrium, we can integrate the stresses over the cross-section of the beam to obtain the following *stress resultants*: the axial force ^1F_x, transverse shears ^1F_y, and ^1F_z, bending moments ^1M_y and ^1M_z, and torsion ^1M_x:

$$^1F_x = \int_A {}^1\tau_{xx} dA \tag{5.20}$$

$$^1F_y = \int_A {}^1\tau_{xy} dA \tag{5.21}$$

$$^1F_z = \int_A {}^1\tau_{xz} dA \tag{5.22}$$

$$^1M_x = \int_A ({}^1\tau_{xz} y - {}^1\tau_{xy} z) dA \tag{5.23}$$

$$^1M_y = \int_A {}^1\tau_{xx} z dA \tag{5.24}$$

$$^1M_z = -\int_A {}^1\tau_{xx} y dA \tag{5.25}$$

where A denotes the cross-sectional area of the beam.

Substituting the expression for $^1\tau_{xx}$ in (5.17) into (5.20), (5.24), and (5.25), we can express the *generalized stresses* ^1F_x, ^1M_y, and

${}^1M_z$ for a cross-section of the solid beam in terms of the *generalized strains* u', v'', and w'' as

$$ {}^1F_x = EAu' \tag{5.26}$$

$$ {}^1M_y = -EI_y w'' \tag{5.27}$$

$$ {}^1M_z = EI_z v'' \tag{5.28}$$

where I_y and I_z represent the moments of inertia about the y- and z-axes, respectively, of the cross-section,

$$ I_y = \int_A z^2 dA \tag{5.29}$$

$$ I_z = \int_A y^2 dA \tag{5.30}$$

In arriving at (5.26)–(5.28), the following orthogonality conditions for the *principal centroidal axes* have been adopted:

$$ \int_A y dA = \int_A z dA = \int_A yz dA = 0 \tag{5.31}$$

Solving (5.26)–(5.28) for u', v'', and w'' and substituting them into (5.17) yields

$$ {}^1\tau_{xx} = \frac{{}^1F_x}{A} + \frac{{}^1M_y z}{I_y} - \frac{{}^1M_z y}{I_z} \tag{5.32}$$

This is exactly the formula for calculating the axial stress ${}^1\tau_{xx}$ based on the linear theory of solid beams.

5.2.2 *Buckling stage*

In this sub-section, we shall talk about the buckling behavior of a three-dimensional solid beam, as typified by the process from the C_1 to the C_2, configuration. Let us cut a beam at section x and consider its left-hand portion as shown in Figure 5.2. The stresses acting on section x are exactly those exerted by the right-hand portion of the beam. For the present purposes, we shall attach two sets of coordinates to the centroid C of the cross-section. The first set is referred to as the η-ζ coordinates, of which η and ζ denote the two principal

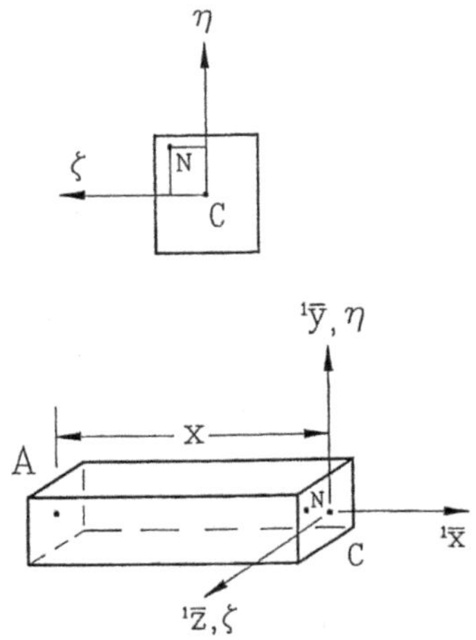

Fig. 5.2. Coordinates of beam element at C_1.

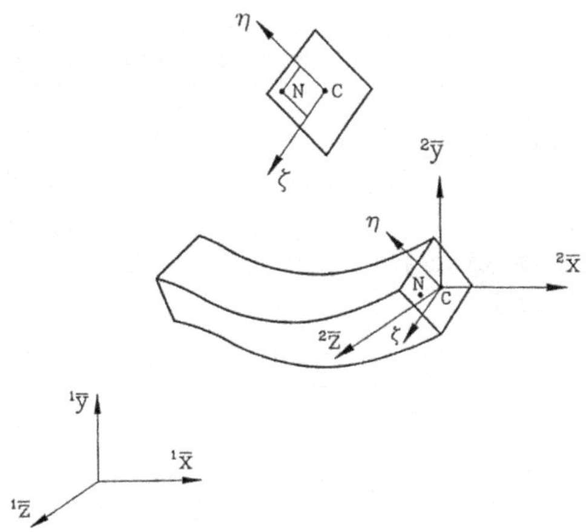

Fig. 5.3. Coordinates of beam element at C_2.

axes of the cross-section. Note that this set of coordinates is assumed to be embedded at the cross-section, which may rotate as the beam deforms. The second set of coordinates is assumed to have an origin fixed at the centroid C of the cross-section. It is referred to as the $^1\bar{x}$- $^1\bar{y}$- $^1\bar{z}$ axes at C_1, with the $^1\bar{y}$- and $^1\bar{z}$ axes coincident with the η- and ζ-axes (see Figure 5.2), and as the $^2\bar{x}$- $^2\bar{y}$- $^2\bar{z}$ axes at C_2, with the $^2\bar{y}$- and $^2\bar{z}$-axes parallel to the $^1\bar{y}$- and $^1\bar{z}$-axes (see Figure 5.3).

Since the embedded coordinates (η, ζ) are coincident with the $({}^1\bar{y}, {}^1\bar{z})$ axes at C_1, we can write

$$ {}^1\bar{x} = 0, \quad {}^1\bar{y} = \eta = y, \quad {}^1\bar{z} = \zeta = z \tag{5.33} $$

where y and z denote the principal axes of the cross-section of the beam at C_1. In view of the preceding relations, the displacement increments u_x, u_y, and u_z of a generic point N at section x occurring during the incremental step from C_1 to C_2 are identical in form to those given in (5.7)–(5.9), based on the *Bernoulli–Euler hypothesis* of plane sections remaining plane after deformation. From equations (5.7)–(5.9), the coordinates of point N can be referred to as the displaced axes at the C_2 configuration (see Figure 5.3) as

$$ {}^2\bar{x} = -y\theta_z + z\theta_y \tag{5.34} $$

$$ {}^2\bar{y} = y - z\theta_x \tag{5.35} $$

$$ {}^2\bar{z} = z + y\theta_x \tag{5.36} $$

where it should be noted that the angles of rotation are defined as: $\theta_y = -w'$ and $\theta_z = v'$.

By the conditions of equilibrium, the axial force $({}^2F_x)$, shear forces $({}^2F_y, {}^2F_z)$, torque $({}^2M_x)$, and bending moments $({}^2M_y, {}^2M_z)$ acting on section x at the C_2 configuration can be related to the stresses acting on the same section at the same configuration but with reference to the initial C_1 configuration as follows:

$$ {}^2F_x = \int_A {}^2S_{xx} dA \tag{5.37} $$

$$ {}^2F_y = \int_A {}^2S_{xy} dA \tag{5.38} $$

$$ {}^2F_z = \int_A {}^2S_{xz} dA \tag{5.39} $$

$$ {}^2M_x = \int_A ({}^2S_{xz} \cdot {}^2\bar{y} - {}^2S_{xy} \cdot {}^2\bar{z}) dA \tag{5.40} $$

$$ {}^2M_y = \int_A ({}^2S_{xx} \cdot {}^2\bar{z} - {}^2S_{xz} \cdot {}^2\bar{y}) dA \tag{5.41} $$

$$ {}^2M_z = \int_A ({}^2S_{xy} \cdot {}^2\bar{x} - {}^2S_{xx} \cdot {}^2\bar{y}) dA \tag{5.42} $$

where ${}^2S_{xx}$, ${}^2S_{xy}$, and ${}^2S_{xz}$ denote the *second Piola–Kirchhoff stresses* acting on section x at C_2, with reference to the C_1 configuration. In contrast, the *stress resultants* acting on the section at the C_1 configuration have been given in (5.20)–(5.25).

One particular property of the stress resultants at the C_2 configuration is that the moments ${}^2M_x$, ${}^2M_y$, and ${}^2M_z$ will be affected by the rotations θ_x, θ_y, and θ_z as implied by the coordinates ${}^2\bar{x}$, ${}^2\bar{y}$, and ${}^2\bar{z}$. For instance, by substituting the expressions (5.34)–(5.36) for ${}^2\bar{x}$, ${}^2\bar{y}$, and ${}^2\bar{z}$ into equations (5.40)–(5.42), we obtain

$$ {}^2M_x = \int_A ({}^2S_{xz} \cdot y - {}^2S_{xy} \cdot z)dA \tag{5.43} $$

$$ {}^2M_y = \int_A {}^2S_{xx} \cdot zdA + \int_A ({}^2S_{xx} y\theta_x + {}^2S_{xz} y\theta_z)dA \tag{5.44} $$

$$ {}^2M_z = -\int_A {}^2S_{xx} \cdot ydA + \int_A ({}^2S_{xx} z\theta_x + {}^2S_{xy} z\theta_y)dA \tag{5.45} $$

where some higher-order terms that cannot be physically justified have been omitted.

For convenience, the stresses ${}^2S_i$ acting on the cross-section at C_2, but referred to C_1, can be decomposed as

$$ {}^2S_i = {}^1\tau_i + S_i \tag{5.46} $$

where the subscript i can be xx, xy, or xz, ${}^1\tau_i$, denotes the *Cauchy stresses* existing on the section at C_1 and S_i the *updated Kirchhoff stress increments.* Substituting the expression for ${}^2S_i$ into equations (5.43)–(5.45), and recalling the definitions for the moments at C_1 in equations (5.23)–(5.25), we can express the moments that are acting on section x at C_2 as follows:

$$ {}^2M_x = \int_A ({}^2S_{xz} \cdot y - {}^2S_{xy} \cdot z)dA \tag{5.47} $$

$$ {}^2M_y = \int_A {}^2S_{xx} \cdot zdA - {}^1M_z\theta_x + \left(\int_A {}^1\tau_{xz} ydA\right)\theta_z \tag{5.48} $$

$$ {}^2M_z = -\int_A {}^2S_{xx} \cdot ydA + {}^1M_y\theta_x + \left(\int_A {}^1\tau_{xy} zdA\right)\theta_y \tag{5.49} $$

which are in consistence with those given by Yang and Kuo (1992) previously.

At this point, we like to point out that the preceding expressions for the moments acting on section x of the beam at C_2 have been derived based merely on the *Bernoulli–Euler hypothesis* for cross-sectional displacements, as implied by (5.7)–(5.9), and the elasticity definitions (5.40)–(5.42) for *stress resultants* over the cross-section. The proceeding expressions are crucial to the derivation of the moment increments induced by nodal moments upon three-dimensional rotations in the incremental formulation.

5.3 Rotational Properties of Nodal Moments

The moments to be considered in this text can be either *external moments* applied to the structure by mechanical devices or *internal moments* generated as stress resultants over the cross-section of each element of the structure. Conventionally, a *conservative direct force* is defined as the one whose direction and magnitude remain unchanged when the structure on which the force acts is displaced. Hence, the vector expression for the conservative force in the displaced configuration C_2 is identical to the one in the initial configuration C_1, when both are referred to the same coordinate system. However, the same is not true with a *conservative moment*. As we shall show in this section, the direction and magnitude of a conservative moment in general will change as the structure on which the moment acts is subjected to three-dimensional rotations. It follows that the vector expression for the conservative moment in the displaced configuration C_2 will be generally different from the one in the C_1 configuration.

Previously, various types of moments generated by different mechanisms have been identified by Ziegler (1977). Each type of moments is characterized by the manner in which the vector that defines the moment rotates following the deformation of the member to which the moment is applied. For instance, Figure 5.4 shows four types of moments in the displaced position, all of which are initially of magnitude ${}^1M_x$ and acting along the x-axis at C_1. In this text, the moments generated by the rotation of the initial moment about the other axes will be referred to as the *induced moments*.

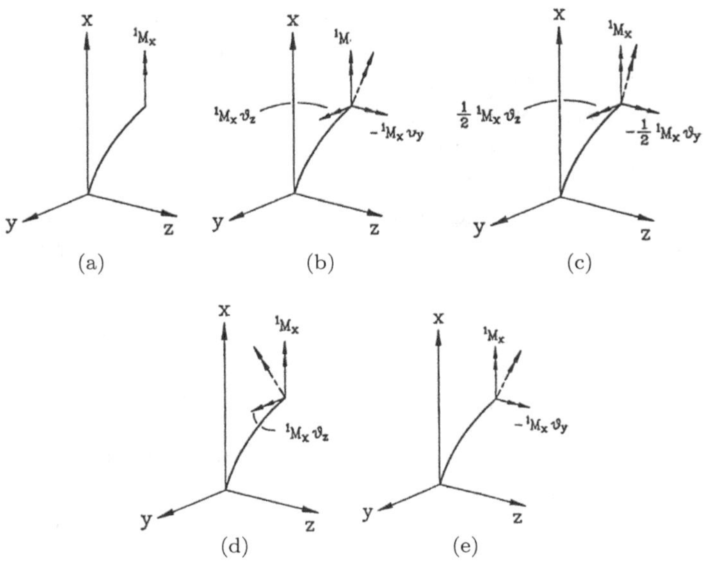

Fig. 5.4. Various types of moments: (a) axial; (b) tangential; (c) semitangential; (d) first-kind quasitangential; and (e) second-kind quasitangential.

A moment is defined as the *axial moment* if the direction and magnitude of the moment remain unchanged in the presence of finite rotations about the other axes (Figure 5.4(a)), which implies that zero moments will be induced. In contrast, a moment is defined as a *tangential* or *follower moment* if it follows completely the rotations in the three-dimensional space (Figure 5.4(b)). In this case, the moments induced about the axes perpendicular to the initial axis of the moment are equal to the tangential components.

It is possible to define a third type of moment called the *semitangential moment*, which is featured by the fact that the induced moments generated upon a particular rotation are equal to half of those generated by a tangential moment of equal magnitude undergoing the same rotation (Figure 5.4(c)). The moments in Figures 5.4(d) and 5.4(e) have been termed the *quasitangential moments* of the first and second kind, respectively, since only a single moment will be induced about one of the axes perpendicular to the initial moment vector. The reason for identifying the two types of quasitangential moments will be given later when dealing with the rotational properties of bending moments on a member.

Conventionally, a conservative force is defined in the sense that the work done by that force is path-independent. It can be shown that the work done by a semi- or quasi-tangential moment is path-independent if an appropriate definition is adopted for the rotation corresponding to each type of moment (Argyris *et al.*, 1978). Thus they are *conservative moments*. It can also be shown that both the axial and tangential moments are *nonconservative moments*. A thorough discussion of the conservative or nonconservative nature of various types of moments can be found in Ziegler (1977) and Argyris *et al.* (1978).

5.3.1 *Moments generated as stress resultants*

Ziegler (1977) and Argyris *et al.* (1978) have identified a bending moment as the quasitangential moment and a *St. Venant torque* as the *semitangential moment* when either is generated as a stress resultant at each cross section of the structural members. Representations of such moments are shown in Figures 5.5 and 5.6. In each case the action is represented by a distributed stress, and alternatively, by a statically equivalent couple or couples of forces.

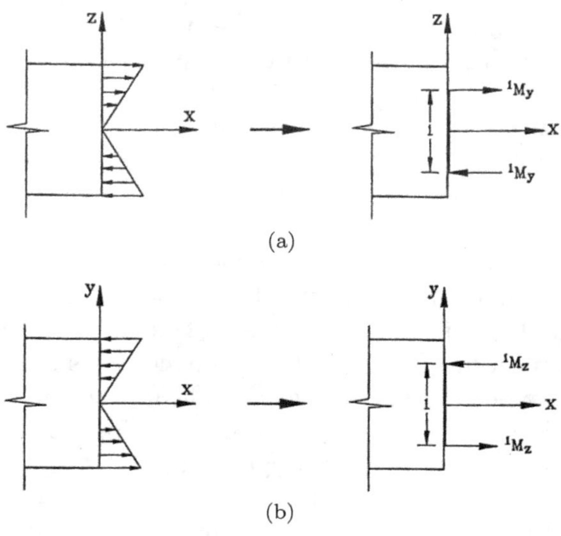

Fig. 5.5. Quasitangential moments: (a) ${}^1M_y$ and (b) ${}^1M_z$.

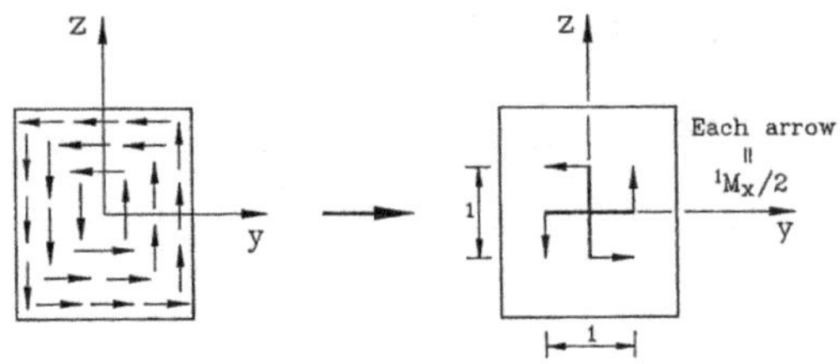

Fig. 5.6. Semitangential torque.

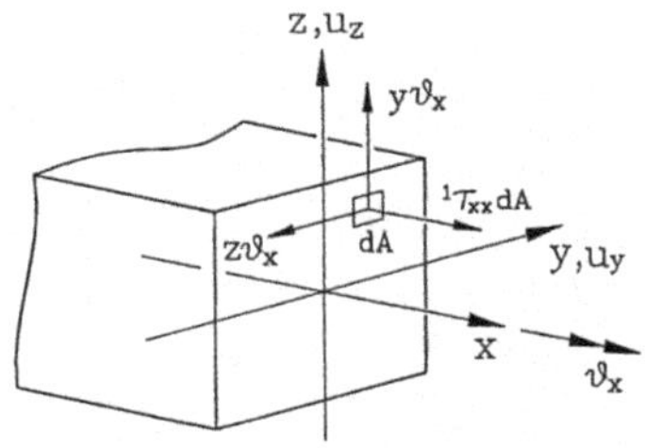

Fig. 5.7. Flexural stresses undergoing rotation.

Based on the *stress resultant* definition, the bending moments ${}^1M_y$ and ${}^1M_z$ acting on each cross-section of a beam at the C_1 configuration have been defined in (5.24) and (5.25), i.e.,

$$ {}^1M_y = \int_A {}^1\tau_{xx} z dA \tag{5.50} $$

$$ {}^1M_z = -\int_A {}^1\tau_{xx} y dA \tag{5.51} $$

where, as was shown in Figure 5.7, ${}^1\tau_{xx}$ denotes the axial stress, A the cross-sectional area, and y and z the principal axes of the cross-section of the beam at C_1. To calculate the *induced moments* for the bending moments ${}^1M_y$ and ${}^1M_z$ undergoing three-dimensional rotations, let us consider a small element of area dA of a bisymmetrical section on which the axial stress ${}^1\tau_{xx}$ is acting in Figure 5.7. When the element is subjected to a torsional rotation θ_x, the following differential moments will be induced:

$$ d(\Delta M_z) = {}^1\tau_{xx}\theta_x z dA \tag{5.52} $$

$$ d(\Delta M_y) = {}^1\tau_{xx}\theta_x y dA \tag{5.53} $$

which can be integrated to yield

$$\Delta M_z = {}^1M_y\theta_x \quad \text{(due to } {}^1M_y\text{)} \tag{5.54}$$

$$\Delta M_y = -{}^1M_z\theta_x \quad \text{(due to } {}^1M_z\text{)} \tag{5.55}$$

It is easy to show that the same moments as those given in equations (5.54) and (5.55) can be induced by the alternative mechanisms shown on the right-hand side of Figures 5.5(a) and 5.5(b) undergoing the same rotation. Bending moments that are generated as *stress resultants*, which will induce moments of the type shown in equations (5.54) and (5.55), are referred to as *quasitangential moments of the first kind* (QT-1 moments). Later, it will be shown that there exists a second kind of quasitangential moments which cannot be interpreted as conventional stress resultants.

The *St. Venant torque* ${}^1M_x$ resulting from the cross-sectional shear stresses at C_1 has been defined in (5.23), i.e.,

$${}^1M_x = \int_A ({}^1\tau_{xz}y - {}^1\tau_{xy}z)dA \tag{5.56}$$

By the condition of equilibrium,

$$\frac{\partial^1\tau_{xy}}{\partial y} + \frac{\partial^1\tau_{xz}}{\partial z} = 0 \tag{5.57}$$

it is easy to show that the following relation holds:

$$\int_A {}^1\tau_{xz}ydA = -\int_A {}^1\tau_{xy}zdA = \frac{1}{2}{}^1M_x \tag{5.58}$$

To obtain the induced moments for a St. Venant torque undergoing rotations, consider Figure 5.8. If the cross-section rotates about the z-axis, a moment will be induced about the y-axis:

$$d(\Delta M_y) = {}^1\tau_{xz}\theta_z ydA \tag{5.59}$$

Similarly, a rotation about the y-axis leads to

$$d(\Delta M_z) = {}^1\tau_{xy}\theta_y zdA \tag{5.60}$$

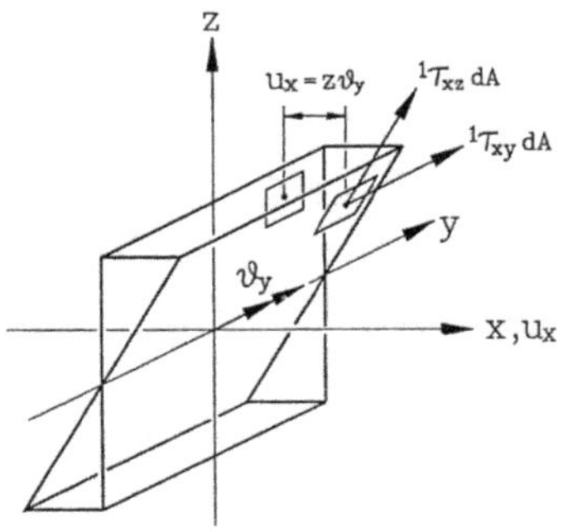

Fig. 5.8. Torsional shear stresses undergoing rotations.

Integrating the preceding two equations over the cross-section yields the induced moments as:

$$\Delta M_y = \frac{1}{2}{}^1M_x\theta_z \tag{5.61}$$

$$\Delta M_z = -\frac{1}{2}{}^1M_x\theta_y \tag{5.62}$$

The moments induced by the St. Venant torque are equal to half of those generated by the tangential torque shown in Figure 5.4(b). For this reason, a St. Venant torque has been referred to as the *semi-tangential torque* (ST torque). Evidently, identical moments ΔM_y and ΔM_z can be induced by the semitangential torque mechanism shown in Figure 5.6 undergoing the same rotations, if the direct forces constituting the couples are assumed to remain unchanged both in magnitude and direction upon rotations.

5.3.2 *Moments generated by external devices*

Conceptually, moments and torques need not be viewed as the conventional stress resultants. They can also be generated by the conservative mechanisms shown in Figures 5.9 and 5.10. For example, Figure 5.9 shows that a torque can be generated by a quasitangential mechanism that consists of only a couple of direct forces with a lever arm directed along either the y- or z-axis. In the presence of finite rotations, the *first-kind quasitangential torque* (QT-1 torque)

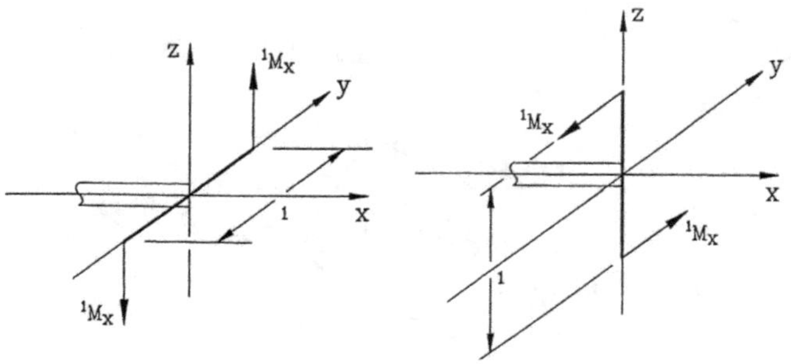

Fig. 5.9. Quasitangential torque: (a) first kind and (b) second kind.

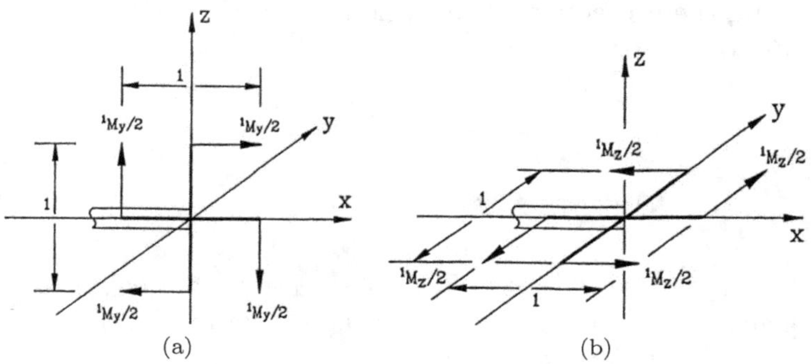

Fig. 5.10. Semitangential bending moments: (a) ^1M_y and (b) ^1M_z.

of Figure 5.9(a) will induce the following moments:

$$\Delta M_y = {}^1M_x\theta_z \tag{5.63}$$

$$\Delta M_z = 0 \tag{5.64}$$

Similarly, the moments induced by the *second-kind quasitangential torque* (QT-2 torque) of Figure 5.9(b) undergoing rotations are

$$\Delta M_y = 0 \tag{5.65}$$

$$\Delta M_z = -{}^1M_x\theta_y \tag{5.66}$$

Bending moments are quasitangential moments of the first kind when treated as conventional stress resultants. If the stress resultant restriction is removed, they can be generated as quasitangential moments of the second kind (QT-2) or as semitangential moments. When based on the semitangential definition, as shown in Figure 5.10, the following moments are to be induced by the bending moments undergoing three-dimensional rotations:

$$\Delta M_z = \frac{1}{2}{}^1M_y\theta_x; \quad \Delta M_x = -\frac{1}{2}{}^1M_y\theta_z \quad (\text{due to } {}^1M_y) \tag{5.67}$$

$$\Delta M_y = -\frac{1}{2}{}^1M_z\theta_x; \quad \Delta M_x = \frac{1}{2}{}^1M_z\theta_y \quad (\text{due to } {}^1M_z) \tag{5.68}$$

A list of the moments induced by various types of moment mechanisms has been listed in Table 5.1, in which the ST property for the torque ${}^1M_x$ and the QT-1 properties for the bending moments ${}^1M_y$ and ${}^1M_z$ have been based on the stress resultant definitions. The rotational properties presented in this section for various moment mechanisms should not be regarded as researches of merely of theoretical curiosity. They have to be included in the establishment of equilibrium conditions for structural joints connecting non-collinear members and in the specification of natural boundary conditions for structural nodes directly under the action of applied loads since it is realized that in a buckling analysis, all the physical quantities and relations should be set up for the structure in the buckling (C_2) configuration.

Table 5.1. Moments induced by various moment mechanisms undergoing rotations.

Origin		Rotations		
		θ_x	$\theta_y(=-w')$	$\theta_z(=v')$
${}^1M_x$	ST		$\Delta M_z = -1/2{}^1M_x\theta_y$	$\Delta M_y = 1/2{}^1M_x\theta_z$
	QT-1		$\Delta M_z = 0$	$\Delta M_y = {}^1M_x\theta_z$
	QT-2		$\Delta M_z = -{}^1M_x\theta_y$	$\Delta M_y = 0$
${}^1M_y$	QT-1	$\Delta M_z = {}^1M_y\theta_x$		$\Delta M_x = 0$
	QT-2	$\Delta M_z = 0$		$\Delta M_x = -{}^1M_y\theta_z$
	ST	$\Delta M_z = 1/2{}^1M_y\theta_x$		$\Delta M_x = -1/2{}^1M_y\theta_z$
${}^1M_z$	QT-1	$\Delta M_y = -{}^1M_z\theta_x$	$\Delta M_x = 0$	
	QT-2	$\Delta M_y = 0$	$\Delta M_x = {}^1M_z\theta_y$	
	ST	$\Delta M_y = -1/2{}^1M_z\theta_x$	$\Delta M_x = 1/2{}^1M_z\theta_y$	

5.4 Engineering Approach for Formulation of the Three-Dimensional Beam Element

As was stated in Chapter 1, in an updated Lagrangian formulation all quantities occurring within each incremental step of a nonlinear analysis are referred to the last known configuration C_1. Based on the principle of virtual displacements, the equation of equilibrium for a solid at the current configuration C_2 is the one given in equation (1.131). For cases where the strain increments ${}_1\epsilon_{ij}$ occurring within each incremental step can be considered reasonably small, this equation can be linearized as follows:

$$\int_{^1V} {}_1C_{ijkl}e_{kl}\delta_1 e_{ij}\,{}^1dV + \int_{^1V} {}^1\tau_{ij}\delta_1\eta_{ij}\,{}^1dV = {}^2_1R - {}^1_1R \tag{5.69}$$

where, as defined in previous chapters, ${}^1V$ is the volume of the solid at C_1, ${}_1e_{ij}$ and ${}_1\eta_{ij}$ are the linear and nonlinear components of the *Green–Lagrange strain increments*, ${}^1\tau_{ij}$ denotes the *Cauchy stresses* at C_1, and ${}_1C_{ijkl}$ the *incremental constitutive coefficients*. The terms on the right-hand side of equation (5.69), i.e., ${}^2_1R$ and ${}^1_1R$, denote the virtual works done by the external loads acting on the solid at C_2 and C_1, respectively,

$$ {}^2_1R = \int_{^1S} {}^2_1t_i\delta u_i\,{}^1dS + \int_{^1v} {}^2_1f_i\delta u_i\,{}^1dV \tag{5.70}$$

$$ {}^1_1R = \int_{^1V} {}^1\tau_{ij}\delta_1 e_{ij}\,{}^1dV \tag{5.71}$$

where ${}^1S$ is the surface area of the solid, and ${}^2_1t_i$ and ${}^2_1f_i$ denote the surface tractions and body forces, respectively, acting on the solid at the current configuration C_2, but referred to the last calculated configuration C_1.

For a solid in equilibrium at C_1, the external virtual work ${}^1_1R$ can also be written as

$$ {}^1_1R = \int_{^1S} {}^1_1t_i\delta u_i\,{}^1dS + \int_{^1v} {}^1_1f_i\delta u_i\,{}^1dV \tag{5.72}$$

where ${}^1_1t_i$ and ${}^1_1f_i$ denote the surface tractions and body forces, respectively, acting on the solid at C_1. In this text, the effect of body forces ${}^2_1f_i$ and ${}^1_1f_i$ will be excluded for simplicity. Consequently, the external virtual work terms ${}^2_1R$ and ${}^1_1R$ reduce to

$$ {}^2_1R = \int_{{}^1S} {}^2_1t_i \delta u_i \, {}^1dS \tag{5.73} $$

$$ {}^1_1R = \int_{{}^1S} {}^1_1t_i \delta u_i \, {}^1dS \tag{5.74} $$

Conventionally, it is assumed that only three components of stresses and three corresponding components of strains are significant for a three-dimensional beam. With the updated Lagrangian formulation, the three components of stresses considered are those of the *Cauchy stresses* existing at the C_1 configuration, ${}^1\tau_{xx}$, ${}^1\tau_{xy}$ and ${}^1\tau_{xz}$, and the three corresponding strain increments are those of the *updated Green strain increments*, ${}_1\epsilon_{xx}$, ${}_1\epsilon_{xy}$ and ${}_1\epsilon_{xz}$.

Let us decompose the three strains of concern into the linear and nonlinear parts:

$$ {}_1\epsilon_{xx} = {}_1e_{xx} + {}_1\eta_{xx} \tag{5.75} $$

$$ {}_1\epsilon_{xy} = {}_1e_{xy} + {}_1\eta_{xy} \tag{5.76} $$

$$ {}_1\epsilon_{xz} = {}_1e_{xz} + {}_1\eta_{xz} \tag{5.77} $$

Consequently, the virtual work equation (5.69) reduces to

$$ \begin{aligned} &\int_V (Ee_{xx}\delta e_{xx} + 4Ge_{xy}\delta e_{xy} + 4Ge_{xz}\delta e_{xz})dV \\ &\quad + \int_V ({}^1\tau_{xx}\delta\eta_{xx} + 2{}^1\tau_{xy}\delta\eta_{xy} + 2{}^1\tau_{xz}\delta\eta_{xz})dV \\ &= {}^2R - {}^1R \end{aligned} \tag{5.78} $$

in which E and G denote the *moduli of elasticity and rigidity* respectively, and the factors of 4 and 2 have been added to account for the symmetry of shear strains, i.e., $e_{xy} = e_{yx}$, $e_{xz} = e_{zx}$, $\eta_{xy} = \eta_{yx}$, $\eta_{xz} = \eta_{zx}$, and so on. In equation 5.78, the left superscript "1" for V and the left subscripts "1" for the strain components and virtual work terms have been dropped, since it is known that all these variables have been expressed with respect to the last calculated configuration C_1. Physical meanings for equation (5.78) can be given as follows: The external virtual work increment, ${}^2R - {}^1R$, caused by the increase in surface tractions from ${}^1_1t_i$ to ${}^2_1t_i$, will be transformed into the *strain energy* and *potential energy* of the beam, as represented by the first and second integrals on the left-hand side of the equation, thereby resulting in the deformation of the beam from the C_1 to the C_2 configuration.

5.4.1 *Strain energy due to linear strains*

The linear strain components e_{xx}, e_{xy}, and e_{xz} have been expressed in terms of the cross-sectional displacements u_x, u_y, and u_z in (5.10)–(5.12). Substituting these expressions for the strain components into the first integral of (5.78) yields

$$\begin{aligned}&\int_V (Ee_{xx}\delta e_{xx} + 4Ge_{xy}\delta e_{xy} + 4Ge_{xz}\delta e_{xz})dV \\ &\quad = \frac{1}{2}\int_0^L \int_A E\delta \left(u'^2 + y^2 v''^2 + z^2 w''^2 - 2yu'v'' - 2zu'w''\right. \\ &\qquad \left. + 2yzv''w''\right) dAdx + \frac{1}{2}\int_0^L \int_A G\delta(y^2 + z^2)\theta_x'^2 dAdx \qquad (5.79)\end{aligned}$$

where L denotes the length of the beam. In view of the conditions in (5.13) for the *principal centroidal coordinates* (y, z), the definitions in (5.29) and (5.30) for the moments of inertia I_y and I_z, and the following for the torsional constant J:

$$J = \int_A (y^2 + z^2)dA = I_z + I_y \qquad (5.80)$$

the preceding equation reduces to

$$\begin{aligned}&\int_V (Ee_{xx}\delta e_{xx} + 4Ge_{xy}\delta e_{xy} + 4Ge_{xz}\delta e_{xz})dV \\ &\quad = \int_0^L (EAu'\delta u' + EI_y w''\delta w'' + EI_z v''\delta v'' + GJ\theta_x'\delta\theta_x')dx \qquad (5.81)\end{aligned}$$

in which EAu' denotes the axial shortening effect, $EI_y w''$ and $EI_z v''$ the flexural deformation effect, and $GJ\theta_x'$ the torsional deformation effect.

5.4.2 *Potential energy due to initial stresses*

Based on the theory of elasticity, the nonlinear components of strains at a generic point of the cross-section can be related to the displacements u_x, u_y, and u_z at the same point as

$$\eta_{xx} = \frac{1}{2}(u_{x,x}^2 + u_{y,x}^2 + u_{z,x}^2) \tag{5.82}$$

$$\eta_{xy} = \frac{1}{2}(u_{x,y}u_{x,x} + u_{y,y}u_{y,x} + u_{z,y}u_{z,x}) \tag{5.83}$$

$$\eta_{xz} = \frac{1}{2}(u_{x,z}u_{x,x} + u_{y,z}u_{y,x} + u_{z,z}u_{z,x}) \tag{5.84}$$

where a comma denotes differentiation with respect to the coordinate following. Substituting the expressions for the displacements u_x, u_y, and u_z in (5.7)–(5.9) into the preceding equations yields

$$\begin{aligned}\eta_{xx} = \frac{1}{2}\Big(&u'^2 + y^2 v''^2 + z^2 w''^2 - 2yu'v'' - 2zu'w'' + 2yzv''w'' \\ &+ v'^2 - 2zv'\theta_x' + z^2\theta_x'^2 + w'^2 + 2yw'\theta_x' + y^2\theta_x'^2\Big)\end{aligned} \tag{5.85}$$

$$\eta_{xy} = \frac{1}{2}(-u'v' + yv'v'' + zv'w'' + w'\theta_x + y\theta_x\theta_x') \tag{5.86}$$

$$\eta_{xz} = \frac{1}{2}(-u'w' + yv''w' + zw'w'' + v'\theta_x + z\theta_x\theta_x') \tag{5.87}$$

Previously, the axial shortening term $u_{x,x}^2$ for the nonlinear strain η_{xx} in equation (5.82) has been regarded as of higher order and excluded in the derivation of the beam theory (Argyris *et al.*, 1979; Washizu, 1982). Such an approach was referred to as the *simplified theory* in the previous version of our book (Yang and Kuo, 1994). From the point of mathematical derivation, there is no difficulty in including the term $u_{x,x}^2$ in the derivation of the beam theory, as will be demonstrated in this book.

As far as computer programming is concerned, it is suggested that all the nonlinear strain terms be duly included in the formulation. This will result in only a marginal increase in the cost of computation, as caused by the variations in stiffness matrices, but will definitely add to the overall rationality of the numerical model considered.

In the following, we shall derive the potential energy (in variational form) associated with each component of the initial stresses in the second integral of (5.78). Let us consider first the contribution by the axial stress ${}^1\tau_{xx}$. With the aid of the expression for ${}^1\tau_{xx}$ in (5.32), the one for η_{xx} in (5.85), and those for the stress resultants in (5.20)–(5.25), the following can be derived:

$$\begin{aligned}
&\int_V {}^1\tau_{xx}\delta\eta_{xx}dV \\
&\quad = \frac{1}{2}\int_0^L \left[{}^1F_x\delta(u'^2 + v'^2 + w'^2)\right. \\
&\quad + {}^1F_x\left(\frac{I_y}{A}\delta w''^2 + \frac{I_z}{A}\delta v''^2\right) + \left.\bar{K}\delta\theta_x'^2\right] dx \\
&\quad + \int_0^L [-{}^1M_z\delta(w'\theta_x') - {}^1M_y\delta(v'\theta_x') \\
&\quad - {}^1M_y\delta(u'w'') + {}^1M_z\delta(u'v'')]\, dx
\end{aligned} \tag{5.88}$$

in which $\bar{K}$ denotes the *Wagner coefficient*,

$$\bar{K} = {}^1F_x\frac{I_y + I_z}{A} = {}^1F_x\frac{J}{A} \tag{5.89}$$

In arriving at equation (5.88), the cross-section of the solid beam has been assumed to be bisymmetrical, in the sense that the conditions given below plus those in (5.31) are satisfied by the coordinate axes:

$$\int_A y^2 z dA = \int_A z^2 y dA = \int_A y^3 dA = \int_A z^3 dA = 0 \tag{5.90}$$

Next, let us consider the potential energy associated with the shear stresses ${}^1\tau_{xy}$ and ${}^1\tau_{xz}$. With the substitution of (5.86) and (5.87) for the nonlinear shear strains and the definitions for stress resultants in (5.21)–(5.23), we can write

$$\begin{aligned}
&\int_V (2{}^1\tau_{xy}\delta\eta_{xy} + 2{}^1\tau_{xz}\delta\eta_{xz})dV \\
&\quad = \int_0^L [{}^1F_y\delta(w'\theta_x - u'v') - {}^1F_z\delta(v'\theta_x + u'w')]dx \\
&\quad + \int_0^L [\alpha{}^1M_x\delta(v''w') - (1-\alpha){}^1M_x\delta(v'w'')]dx
\end{aligned} \tag{5.91}$$

where terms that are not physically meaningful, i.e., higher-order terms, have been neglected and the *torsional parameter* α is defined as (Yang and McGuire, 1986a, 1986b):

$$\alpha = \frac{\int_A {}^1\tau_{xz} y dA}{{}^1M_x} = 1 - \frac{\int_A (-{}^1\tau_{xy} z) dA}{{}^1M_x} \tag{5.92}$$

By letting $\alpha = 1/2$ for bisymmetrical cross-sections, equation (5.91) reduces to

$$\begin{aligned} &\int_V (2{}^1\tau_{xy}\delta\eta_{xy} + 2{}^1\tau_{xz}\delta\eta_{xz})dV \\ &= \int_0^L [{}^1F_y\delta(w'\theta_x - u'v') - {}^1F_z\delta(v'\theta_x + u'w')]dx \\ &+ \frac{1}{2}\int_0^L [{}^1M_x\delta(v''w') - {}^1M_x\delta(v'w'')]dx \end{aligned} \tag{5.93}$$

So far, we have derived the potential energy associated with each component of stresses for the three-dimensional beam. Collectively, we can combine the *potential energy* due to the three components of stresses in equations (5.88) and (5.93) as follows:

$$\begin{aligned} &\int_V ({}^1\tau_{xx}\delta\eta_{xx} + 2{}^1\tau_{xy}\delta\eta_{xy} + 2{}^1\tau_{xz}\delta\eta_{xz})dV \\ &= \frac{1}{2}\int_0^L \Bigg[{}^1F_x\delta(u'^2 + v'^2 + w'^2) \\ &+ {}^1F_x\left(\frac{I_y}{A}\delta w''^2 + \frac{I_z}{A}\delta v''^2\right) + \bar{K}\delta\theta_x'^2\Bigg] dx \\ &+ \int_0^L [-{}^1M_z\delta(w'\theta_x') - {}^1M_y\delta(v'\theta_x')) \\ &- {}^1M_y\delta(u'w'' + {}^1M_z\delta(u'v'')]\, dx \\ &+ \int_0^L [{}^1F_y\delta(w'\theta_x - u'v') - {}^1F_z\delta(v'\theta_x + u'w')]dx \\ &+ \frac{1}{2}\int_0^L [{}^1M_x\delta(v''w') - {}^1M_x\delta(v'w'')]dx \end{aligned} \tag{5.94}$$

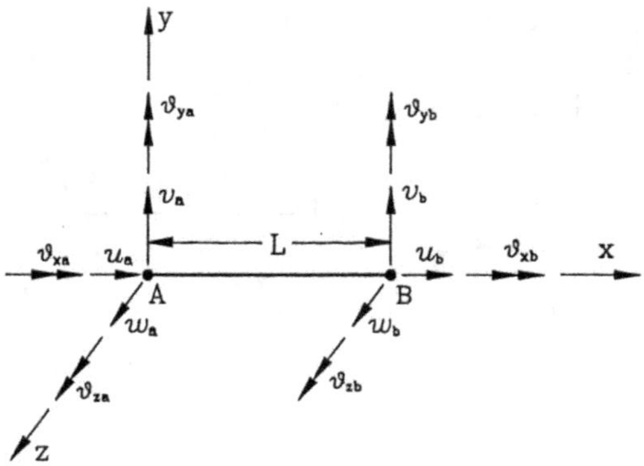

Fig. 5.11. Three-dimensional beam element.

The preceding expression for the three-dimensional beam is featured by the fact that the instability effect due to all the member actions, i.e., ${}^1F_x$, ${}^1F_y$, ${}^1F_z$, ${}^1M_x$, ${}^1M_y$, ${}^1M_z$, are duly taken into account. This serves as the rational basis for deriving the finite element to come.

5.4.3 *External virtual work increment*

For a three-dimensional solid beam, the element displacement vector $\{u\}$ consists of six degrees of freedom at each end of the element (Figure 5.11). In total, there will be twelve degrees of freedom associated with the two ends A and B of the element,

$$\{u\}^T = \langle u_a \quad v_a \quad w_a \quad \theta_{xa} \quad \theta_{ya} \quad \theta_{za} \quad u_b \quad v_b \quad w_b \quad \theta_{xb} \quad \theta_{yb} \quad \theta_{zb} \rangle \tag{5.95}$$

As in previous chapters, we shall assume that the surface tractions ${}^1t_i$ are allowed to act on the two end cross sections of the solid beam. Consider first the virtual work done by the surface tractions only at end B of the beam at C_2. From equation (5.70),

$${}^2_1R = \int_{A_b} {}^2t_i \delta u_i dA = \int_{A_b} ({}^2t_x \delta u_x + {}^2t_y \delta u_y + {}^2t_z \delta u_z) dA \tag{5.96}$$

where the subscript "1" has been omitted, since the reference configuration is known to be C_1, and $A_b = A$ denotes the area of the cross-section of end B. Treating the surface tractions (${}^2t_x$, ${}^2t_y$, ${}^2t_z$) as the second Piola–Kirchhoff stresses $({}^2S_{xx}, {}^2S_{xy}, {}^2S_{xz})$, then substituting the expressions (5.7)–(5.9) for the displacements (u_x, u_y, u_z) into the preceding equation, and recalling the definitions for moments $({}^2M_x, {}^2M_y, {}^2M_z)$ at C_2, in (5.47)–(5.49) and those for the forces $({}^2F_x, {}^2F_y, {}^2F_z)$ in (5.37)–(5.39), we can show the external work for end B at C_2, but with reference to C_1, as follows:

$$\begin{aligned}
{}^2R_b &= {}^2F_{xb} \cdot \delta u_b + {}^2F_{yb} \cdot \delta v_b + {}^2F_{zb} \cdot \delta w_b \\
&\quad + \left\{ {}^2M_{xb} + \left[\int_{A_b} ({}^1\tau_{xz} z + {}^1\tau_{xy} y) dA \right] \theta_{xb} \right\} \cdot \delta\theta_{xb} \\
&\quad + \left[{}^2M_{yb} + {}^1M_{zb}\theta_{xb} - \left(\int_{A_b} {}^1\tau_{xz} y dA \right) \theta_{zb} \right] \cdot \delta\theta_{yb} \\
&\quad + \left[{}^2M_{zb} - {}^1M_{yb}\theta_{xb} - \left(\int_{A_b} {}^1\tau_{xy} z dA \right) \theta_{yb} \right] \cdot \delta\theta_{zb}
\end{aligned} \tag{5.97}$$

where terms that are not physically meaningful, i.e., higher-order terms, have been neglected, the relations $\theta_z = v'$, $\theta_y = -w'$ have been used, and the subscript b has been added for node B. Rearranging,

$$\begin{aligned}
{}^2R_b &= {}^2F_{xb} \cdot \delta u_b + {}^2F_{yb} \cdot \delta v_b + {}^2F_{zb} \cdot \delta w_b + {}^2M_{xb} \cdot \delta\theta_{xb} + {}^2M_{yb} \cdot \delta\theta_{yb} \\
&\quad + {}^2M_{zb} \cdot \delta\theta_{zb} + \left({}^1M_{zb}\theta_{xb} - \frac{1}{2} {}^1M_{xb}\theta_{zb} \right) \cdot \delta\theta_{yb} \\
&\quad + \left(-{}^1M_{yb}\theta_{xb} + \frac{1}{2} {}^1M_{xb}\theta_{yb} \right) \cdot \delta\theta_{zb}
\end{aligned} \tag{5.98}$$

where $\alpha = 1/2$ has been used for bisymmetrical cross-sections.

By switching the subscripts in the preceding equation from b to a, we can obtain the virtual work ${}^2R_a$ for node A of the beam at C_2, but with reference to C_1. Summing up the two effects, we can obtain the total virtual work ${}^2R$ done by the surface tractions on the beam at C_2, but with reference to C_1 as follows:

$$\begin{aligned}
{}^2R &= {}^2R_a + {}^2R_b \\
&= \langle \delta u \rangle \{ {}^2f \} + \left[\begin{array}{c} \left({}^1M_z\theta_x - \frac{1}{2} {}^1M_x\theta_z \right) \cdot \delta\theta_y \\ + \left(-{}^1M_y\theta_x + \frac{1}{2} {}^1M_x\theta_y \right) \cdot \delta\theta_z \end{array} \right] \Bigg|_0^L
\end{aligned} \tag{5.99}$$

where the element displacement vector $\{u\}$ has been given in (5.95) and the nodal force vector $\{^2f\}$ for the three-dimensional beam is defined as

$$\{^2f\}^T = \langle {}^2F_{xa} \quad {}^2F_{ya} \quad {}^2F_{za} \quad {}^2M_{xa} \quad {}^2M_{ya} \quad {}^2M_{za} \\ {}^2F_{xb} \quad {}^2F_{yb} \quad {}^2F_{zb} \quad {}^2M_{xb} \quad {}^2M_{yb} \quad {}^2M_{zb}\rangle \tag{5.100}$$

On the other hand, the external virtual work ^1R done by the surface tractions on the beam has been given in (5.74). By the same procedure, the external virtual work ^1R at C_1 can be expressed with reference to C_1 as follows:

$$^1R = \langle \delta u\rangle\{^1f\} \tag{5.101}$$

where the initial force vector $\{^1f\}$ can be obtained by switching the superscripts in equation (5.100) from "2" to "1", i.e.,

$$\{^1f\}^T = \langle {}^1F_{xa} \quad {}^1F_{ya} \quad {}^1F_{za} \quad {}^1M_{xa} \quad {}^1M_{ya} \quad {}^1M_{za} \\ {}^1F_{xb} \quad {}^1F_{yb} \quad {}^1F_{zb} \quad {}^1M_{xb} \quad {}^1M_{yb} \quad {}^1M_{zb}\rangle \tag{5.102}$$

5.4.4 *Incremental virtual work equation*

In this book, the *engineering approach* is exclusively adopted in that only three components of strains and stresses are included in the formulation. Particularly, the expression of the nonlinear axial strain η_{xx} will be completely included in the formulation. Previously, the nonlinear term $u_{x,x}^2$ in the axial strain η_{xx} was excluded by Argyris *et al.* (1979) and Washizu (1982). From the point of derivation of general finite elements, it is suggested that all the nonlinear strain terms be duly included in the formulation. Although this will result in a marginal increase in the cost of computation, due to the inclusion of additional nonlinear strain terms in the stiffness matrices, the overall rationality of the numerical model derived is enhanced, as it allows us to deal with a wider range of problems encountered in practice.

Substituting all the expressions derived above for the strain energy, potential energy, and external virtual works, i.e., (5.81),

(5.94), (5.99), and (5.101), into the incremental virtual work equation (5.78), one can obtain the following equation for the three-dimensional beam:

$$
\begin{aligned}
&\int_0^L (EAu'\delta u' + EI_y w''\delta w'' + EI_z v''\delta v'' + GJ\theta'_x\delta\theta'_x)dx \\
&\quad + \frac{1}{2}\int_0^L \Big[{}^1F_x\delta(u'^2 + v'^2 + w'^2) \\
&\quad + {}^1F_x\left(\frac{I_y}{A}\delta w''^2 + \frac{I_z}{A}\delta v''^2\right) + \bar{K}\delta\theta'^2_x\Big]\, dx \\
&\quad + \int_0^L \big[-{}^1M_z\delta(w'\theta'_x) - {}^1M_y\delta(v'\theta'_x) \\
&\quad - {}^1M_y\delta(u'w'') + {}^1M_z\delta(u'v'')\big]\, dx \\
&\quad + \int_0^L [{}^1F_y\delta(w'\theta_x - u'v') - {}^1F_z\delta(v'\theta_x + u'w')]dx \\
&\quad + \frac{1}{2}\int_0^L [{}^1M_x\delta(v''w') - {}^1M_x\delta(v'w'')]dx \\
&\quad = \langle\delta u\rangle(\{{}^2f\} - \{{}^1f\}) + \left[\begin{array}{c} ({}^1M_z\theta_x - \frac{1}{2}{}^1M_x\theta_z)\cdot\delta\theta_y \\ + (-{}^1M_y\theta_x + \frac{1}{2}{}^1M_x\theta_y)\cdot\delta\theta_z \end{array}\right]\Bigg|_0^L
\end{aligned}
\tag{5.103}
$$

where it should be noted that $\theta_z = v'$ and $\theta_y = -w'$ and, in treating the terms associated with the two ends of the beam, a distinction should be made between the cross-sectional forces based on the deformation sign convention (see Figure 2.3) and those based on the finite element sign convention (e.g., see Figure 2.2). For instance, at node A of the beam (i.e., at $x = 0$), we have ${}^1M_x = -{}^1M_{xa}$, ${}^1M_y = -{}^1M_{ya}$, and so on, but for node B (i.e., at $x = L$), we have instead ${}^1M_x = {}^1M_{xb}$, ${}^1M_y = {}^1M_{yb}$, and so on.

In equation (5.103), as a natural result of the derivation, the *semitangential* property of the torque ${}^1M_x$ and the *quasitangential* property of the bending moments (${}^1M_y, {}^1M_z$) can be readily observed from the boundary terms on the right-hand side.

Compared with the existing works in the literature for three-dimensional beam elements, the present approach is featured by the fact that all the nonlinear strain components have been duly included in the formulation. This has resulted in the complete inclusion of all the member actions, i.e., ${}^1F_x$, ${}^1F_y$, ${}^1F_z$, ${}^1M_x$, ${}^1M_y$, and ${}^1M_z$, in the potential energy of the beam, which has been the source of instability for the structural element. Such a formulation is believed to be the most *complete* and *rational* among existing comparable theories. It should be noted that the virtual work equation of equilibrium as given in equation (5.103) is a *linearized* one, implying that the equilibrium of the beam is valid up to, but not including, terms of the order of the products or squares of the displacement increments u, v, and w.

The functional presented in equation (5.103) for the 3D beam element is important, since it offers the basis for deriving the governing differential equations and natural boundary condition if the variational approach is adopted (see Section 5.5), or as the basis for deriving the finite element stiffness matrices, if use is made of the interpolation functions (see Chapter 6). In Section 5.6, it will be shown that the present theory is rational and physically meaningful, in the sense that the natural boundary conditions derived can pass the rigid body test.

5.5 Differential Equations and Boundary Conditions

In this section, the *method of variation* will be employed to derive the governing differential equations and natural boundary conditions (or member-force equations) from the virtual work equations presented as equation (5.103) in Section 5.4 for the three-dimensional beam based on the engineering approach.

Following the procedure of Section 3.3 for the two-dimensional beam, we may first integrate each of the terms on the left-hand side of equation (5.103) by parts to obtain the integral and boundary terms containing the virtual displacements δu, δv, and δw. By taking the arbitrary nature of virtual displacements δu, δv, and δw, we can obtain from the integral terms the *buckling differential equations* as the *Euler–Lagrange equations* of the functional:

$$EAu'' + {}^1F_x u'' - {}^1F_y v'' - {}^1F_z w'' - ({}^1M_y w'')' + ({}^1M_z v'')' = 0 \tag{5.104}$$

$$EI_y w'''' - {}^1F_x w'' + {}^1F_x \frac{I_y}{A} w'''' - {}^1F_y \theta_x' + ({}^1M_z \theta_x')' + {}^1F_z u''$$
$$- {}^1M_x v''' - ({}^1M_y u')'' = 0 \tag{5.105}$$

$$EI_z v'''' - {}^1F_x v'' + {}^1F_x \frac{I_z}{A} v'''' + {}^1F_z \theta_x' + ({}^1M_y \theta_x')' + {}^1F_y u''$$
$$+ {}^1M_x w''' + ({}^1M_z u')'' = 0 \tag{5.106}$$

$$GJ\theta_x'' + \bar{K}\theta_x'' - {}^1M_z w'' - {}^1M_y v'' = 0 \tag{5.107}$$

where the relations ${}^1M_y' = {}^1F_z$ and ${}^1M_z' = -{}^1F_y$ have ben used (see Figure 5.12 for the sign convention of the forces and bending moments). One assumption of the foregoing buckling differential equations is that only concentrated loads are allowed to be applied at the two ends of the beam, which excludes the consideration of distributed loads of any type along the axis of the beam.

As a byproduct of the variational procedure, two sets of boundary conditions can be derived for the beam at its two ends A and B (with $x = 0, L$). The *natural boundary conditions* for the beam are

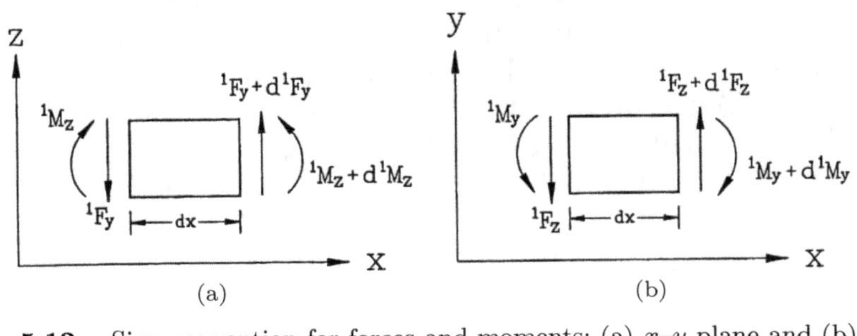

Fig. 5.12. Sign convention for forces and moments: (a) x–y plane and (b) x–z plane.

as follows:

$$EAu' + {}^1F_x u' - {}^1F_y v' - {}^1F_z w' - {}^1M_y w'' + {}^1M_z v'' = {}^2F_x - {}^1F_x \tag{5.108}$$

$$-EI_z v''' + {}^1F_x v' - {}^1F_x \frac{I_z}{A} v''' - {}^1F_z \theta_x - {}^1M_y \theta_x' - {}^1M_x w'' - {}^1M_z u'' = {}^2F_y - {}^1F_y \tag{5.109}$$

$$-EI_y w''' + {}^1F_x w' - {}^1F_x \frac{I_y}{A} w''' + {}^1F_y \theta_x - {}^1M_z \theta_x' + {}^1M_x v'' - {}^1M_y u'' = {}^2F_z - {}^1F_z \tag{5.110}$$

$$GJ\theta_x' + \bar{K}\theta_x' - {}^1M_z w' - {}^1M_y v' = {}^2M_x - {}^1M_x \tag{5.111}$$

$$-EI_y w'' - {}^1F_x \frac{I_y}{A} w'' + \frac{1}{2}{}^1M_x v' + {}^1M_y u' = {}^2M_y - {}^1M_y + \left({}^1M_z \theta_x - \frac{1}{2}{}^1M_x \theta_z\right) \tag{5.112}$$

$$EI_z v'' + {}^1F_x \frac{I_z}{A} v'' + \frac{1}{2}{}^1M_x w' + {}^1M_z u' = {}^2M_z - {}^1M_z + \left(-{}^1M_y \theta_x + \frac{1}{2}{}^1M_x \theta_y\right) \tag{5.113}$$

where the relations $\theta_z = v'$ and $\theta_y = -w'$ have been utilized. Of interest are the *semitangential* moments induced by the nodal torque ${}^1M_x$ and the *quasitangential* moments induced by the nodal bending moments ${}^1M_y$ and ${}^1M_y$ upon three-dimensional rotations, which have been included in the parenthesized terms on the right-hand side of equations (5.112) and (5.113).

Correspondingly, the *geometric boundary conditions* at the two ends A and B (with $x = 0,\ L$) of the beam are as follows:

$$\delta u = 0 \tag{5.114}$$

$$\delta v = 0 \tag{5.115}$$

$$\delta w = 0 \tag{5.116}$$

$$\delta \theta_x = 0 \tag{5.117}$$

$$\delta\theta_y = -\delta w' = 0 \tag{5.118}$$

$$\delta\theta_z = \delta v' = 0 \tag{5.119}$$

As was noted previously, the geometric boundary conditions correspond to prescribed displacements and rotations of the beam, while the natural boundary conditions correspond to prescribed boundary forces and moments. Either the geometric or natural boundary condition, but not both, must be prescribed for each kinematic degree of freedom at the two ends A and B of the beam for the boundary-value problem considered.

The buckling differential equations and natural boundary conditions as derived above will be referred to those of the *engineering approach* for three-dimensional beams, in that the axial strain and two shear strains have been included in the formulation of the beam theory based on physically justified assumptions. This approach is attractive in that less mathematical operations are required in the formulation, compared with the *elasticity approach* that includes all the six components of strains and requires basically no assumptions, as presented in the original version of the book by Yang and Kuo (1994). However, the final equations and finite elements derived by both approaches are the same.

The natural boundary conditions as derived above may appear to be of limited use, as they are manipulable only for the buckling analysis of certain simple problems, such as simply supported beams and cantilevers subjected to in-plane loads. This is due to the fact that for problems of this sort, the natural boundary conditions reduce to those for the linear problems and that rigid body motions are prevented because of the imposition of geometric boundary conditions. It should be noted, however, that such an impression is absolutely not true for problems other than trivial ones, such as for frames that may undergo out-of-plane rotations, which certainly include rigid rotations.

For structures of which the effect of torsion cannot be neglected, a correct specification of the natural boundary conditions for structural members in the buckling position is very important. One simple example of this is the buckling of structural members subjected to torsional moments, of which the critical loads are dependent upon the property of applied moments undergoing three-dimensional rotations. Other examples falling in this category include the buckling of planar frames that are not restricted against lateral or out-of-plane deformations, the buckling of three-dimensional frames, and the buckling of curved beams modeled by straight beam elements (Yang *et al.*, 1991). For problems of this sort, the natural boundary conditions have to be included in the establishment of the conditions of equilibrium for structural joints in the buckling position (Yang and Kuo, 1991a, 1991b), as will be demonstrated in Chapter 9.

For solid beams of which the effect of cross-sectional warping can be neglected, the differential equations presented here are more general than the classical theories available in the literature in that the effects of instability caused by various member actions, such as the axial forces, bending moments, shear forces, and torsions have all been taken into account. In contrast, the classical equations derived by Bleich (1952), Vlasov (1961), and Timoshenko and Gere (1961) consider only the effects of axial forces and bending moments, while the instability of a bar subjected to torsional loadings was treated separately by Ziegler (1977). Either of these classical equations can be regarded as a special cases of the present theories with certain terms omitted.

5.6 Test of the Theory by Rigid Body Rule

Following the procedure of Section 3.6, we shall demonstrate in this section that the natural boundary conditions derived previously for the three-dimensional beam can pass the rigid body test. First, we assume that the beam is initially in equilibrium under the action of a set of nodal forces at C_1 and is subjected to a rigid body rotation θ_{zr} on the x–y plane of the following form:

$$u = u_r \tag{5.120}$$

$$v = v_r + x\theta_{zr} \tag{5.121}$$

$$w = \theta_y = \theta_x = 0 \tag{5.122}$$

$$\theta_z = v' = \theta_{zr} \tag{5.123}$$

where u_r and v_r denote the rigid translations along the x- and y-axis respectively, and θ_{zr} the rigid rotation about the z-axis (assumed to be small). With these expressions, we can obtain from the natural boundary conditions in equations (5.108)–(5.113) the following for the forces acting on the two ends of the beam after the rigid body motion:

$$^2F_x = {}^1F_x - {}^1F_y\theta_{zr} \tag{5.124}$$

$$^2F_y = {}^1F_y + {}^1F_x\theta_{zr} \tag{5.125}$$

$$^2F_z = {}^1F_z \tag{5.126}$$

$$^2M_x = {}^1M_x - {}^1M_y\theta_{zr} \tag{5.127}$$

$$^2M_y = {}^1M_y + {}^1M_x\theta_{zr} \tag{5.128}$$

$$^2M_z = {}^1M_z \tag{5.129}$$

which have also been plotted in Figure 5.13. This figure indicates that the initial forces acting on the beam have rotated following the rigid body rotation, while their magnitudes remain unchanged. Thus, the test has been passed for this particular rigid body mode.

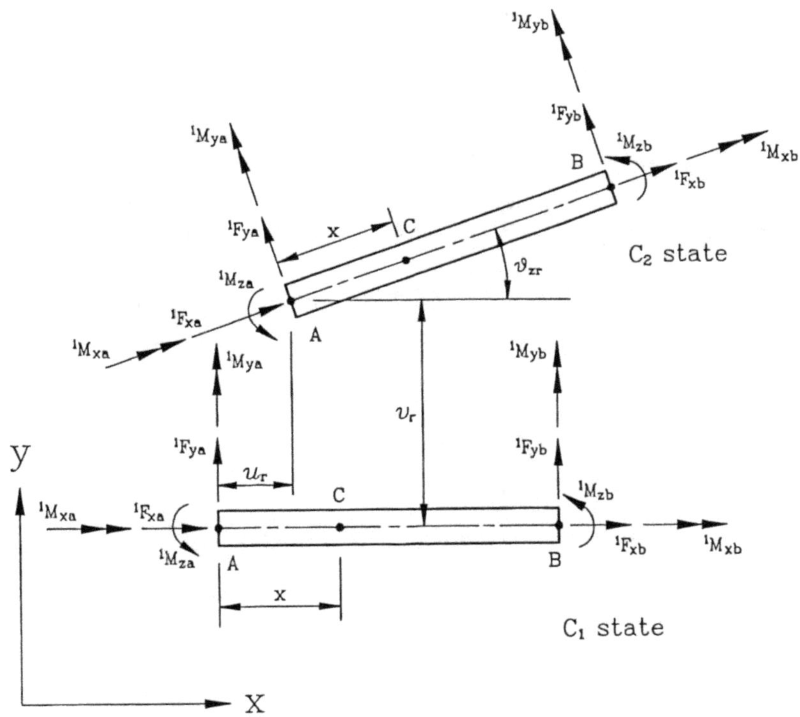

Fig. 5.13. Rigid body motion in x–y plane.

Next, let us assume that the initially stressed beam is subjected to a rigid body rotation θ_{yr} about the y-axis and rigid body translations u_r and w_r along the x- and z-axes, respectively. By letting

$$u = u_r \tag{5.130}$$

$$v = \theta_z = \theta_x = 0 \tag{5.131}$$

$$w = w_r - x\theta_{yr} \tag{5.132}$$

$$\theta_y = -w' = \theta_{yr} \tag{5.133}$$

the forces acting on the beam after the rigid body motion can be obtained from equations (5.108)–(5.113) as follows:

$$^2F_x = {}^1F_x + {}^1F_z\theta_{yr} \tag{5.134}$$

$$^2F_y = {}^1F_y \tag{5.135}$$

$$^2F_z = {}^1F_z - {}^1F_x\theta_{yr} \tag{5.136}$$

$$^2M_x = {}^1M_x + {}^1M_z\theta_{yr} \tag{5.137}$$

$$^2M_y = {}^1M_y \tag{5.138}$$

$$^2M_z = {}^1M_z - {}^1M_x\theta_{yr} \tag{5.139}$$

From the results depicted in Figure 5.14, it is evident that the present equations for beams based on the present theory can adequately account for the rigid body motions on the x–z plane.

Finally, let us assume that the beam is subjected instead to a rigid rotation θ_{xr} about the axis of the beam and rigid translations v_r and w_r along the y- and z-axes, respectively, that is,

$$u = \theta_y = \theta_z = 0 \tag{5.140}$$

$$v = v_r \tag{5.141}$$

$$w = w_r \tag{5.142}$$

$$\theta_x = \theta_{xr} \tag{5.143}$$

For this particular rigid body mode, the following forces can be obtained from equations (5.108) (5.113):

$$^2F_x = {}^1F_x \tag{5.144}$$

$$^2F_y = {}^1F_y - {}^1F_z\theta_{xr} \tag{5.145}$$

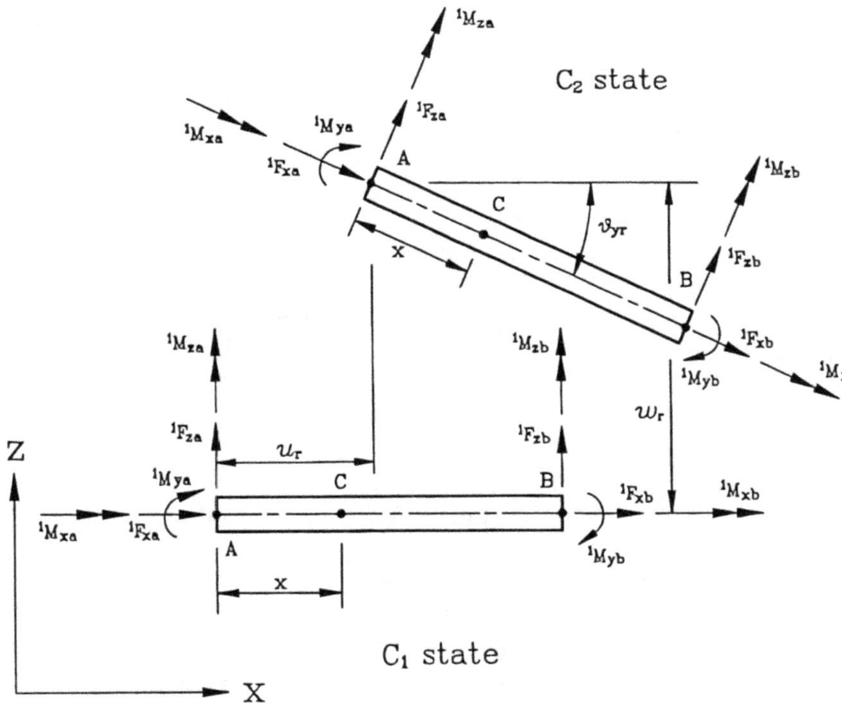

Fig. 5.14. Rigid body motion in x–z plane.

$$^2F_z = {}^1F_z + {}^1F_y\theta_{xr} \tag{5.146}$$

$$^2M_x = {}^1M_x \tag{5.147}$$

$$^2M_y = {}^1M_y - {}^1M_z\theta_{xr} \tag{5.148}$$

$$^2M_z = {}^1M_z + {}^1M_y\theta_{xr} \tag{5.149}$$

which have also ben plotted in Figure 5.15. From this figure, it is certain that the natural boundary conditions derived previously based on the present theory can adequately account for the rigid rotation about the axis of the beam.

As can be seen from the above test, whether a set of natural boundary conditions can pass the rigid body test or not depends entirely on the terms containing the rotation angles or the first derivatives of displacements, that is, θ_x, θ_y $(= -w')$, and θ_z $(= v')$. We like to mention that not all the rigid body-related terms, i.e.,

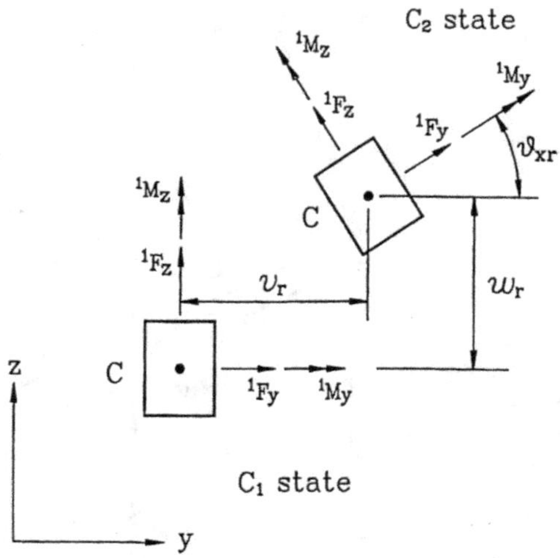

Fig. 5.15. Rigid body motion in y–z plane.

those containing the rotation angles θ_x, θ_y $(= -w')$, and θ_z $(= v')$, have been consistently included in the formulation of classical beam theories, such as those by Bleich (1952), Vlasov (1961), Timoshenko and Gere (1961), and Gjelsvik (1981). It is true that for certain problems for which the rigid body motions are completely restrained, such as single members, equations not fully qualified for rigid body modes can still be used to yield good results. However, care must be exerted if we intend to derive finite elements from equations of this sort and use them in a general analysis program for solving a wide range of three-dimensional frames involving the nonlinear and postbuckling responses. In this case, proper measures should be taken to eliminate the fictitious forces that are likely to be induced by elements incapable of dealing with rigid body motions.

One important lesson from the above rigid body test is that the moments induced by the initial end moments undergoing three-dimensional rotations, as represented by the last boundary term (in brackets) on the right-hand side of the source equation (5.103), should always be taken into account in writing the natural boundary conditions or in deriving the finite elements for a three-dimensional beam. Omission of any such terms from the natural boundary conditions, or the effects they represent from the source equation, will result in violation of the rigid body rule, which obviously is not acceptable. Such a fact seems to have not received enough attention from previous researchers.

5.7 Application of the Incremental Theory

The buckling differential equations and natural boundary conditions for the three-dimensional beam derived in Section 5.5 have been expressed in incremental form. One advantage with equations of this form is that they can be directly employed in the formulation of non-linear finite elements, in which case it is preferable to have all the related equations expressed in incremental form. Another advantage is the possibility of conducting a two-stage analysis for analyzing the critical loads of structures that do not exhibit large deformations prior to buckling, i.e., with negligible prebuckling deformations.

The first stage relates to the deformation of the structure from its initial C_0 configuration to the C_1 configuration, in response to the application of external loadings from zero to a certain level. This stage has also been referred to as the *prebuckling stage*, in which the deformations of the structure are assumed to be so small that they can be neglected from the second-stage analysis to follow. Based on the smallness of deformations, the internal forces of each member of the structure corresponding to a reference load level can be determined by the conventional linear analysis procedure. On the other hand, the second stage refers to *bifurcation* of the structure from the initially stressed configuration (C_1) to the final buckling configuration (C_2). It is characterized by the fact that the external loadings acting on the structure remain generally constant, while displacements of large magnitude can occur in directions not parallel to those of the prebuckling displacements. The second stage is also known as the *buckling stage*. Of major concern in a buckling analysis is to determine the *critical load* that may result in the (bifurcation) buckling of the structure, or in the transition of the structure from the prebuckling stage to the buckling stage.

5.7.1 *Axially compressed columns*

The following is an illustration of the procedure for two-stage analysis. Consider a structural member subjected to a compressive force P

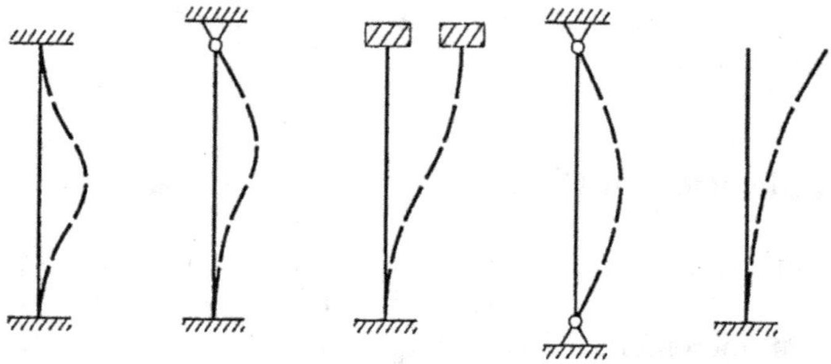

Fig. 5.16. Axially compressed columns.

in Figure 5.16. At the first stage (or step) from C_0 to C_1, the external compressive load is increased from 0 to P. For this particular step, all the initial forces are equal to zero, i.e., ${}^1F_x = {}^1F_y = {}^1F_z = {}^1M_x = {}^1M_y = {}^1M_z = 0$, and the only external load is ${}^2F_x = -P$. Accordingly, the natural boundary conditions in equations (5.108)–(5.113) reduce to: $EAu' = -P$, $-EI_zv''' = -EI_yw''' = GJ\theta_x' = -EI_yw'' = -EI_zv'' = 0$, which represents exactly the forces acting on the column at the end of the stage. Here u, v, w, and θ_x should be interpreted as displacements of the column occurring at the first stage.

From the results obtained at the end of the first stage, we have the following information available at the beginning of the second stage (or step), which again is referred to as the C_1 configuration: ${}^1F_x = -P$ and ${}^1F_y = {}^1F_z = {}^1M_x = {}^1M_y = {}^1M_z = 0$. Since the external loads remain constant during the buckling process, we can therefore express the external load at the end of the step (i.e., at C_2) as: ${}^2F_x = {}^1F_x = -P$. The buckling differential equations (5.104)–(5.107) reduce to the following ones:

$$EI_yw'''' + Pw'' - P\frac{I_y}{A}w'''' = 0 \tag{5.150}$$

$$EI_zv'''' + Pv'' - P\frac{I_z}{A}v'''' = 0 \tag{5.151}$$

$$GJ\theta_x'' - P\frac{I_y + I_z}{A}\theta_x'' = 0 \tag{5.152}$$

where v, w, and θ_x should now be interpreted as the buckling deformations of the structure occurring at the second stage. The present theory differs from the existing classical ones in the existence of the terms $P\frac{I_y}{A}w''''$ and $P\frac{I_z}{A}v''''$ in equations (5.150) and (5.151), respectively.

Since the three buckling differential equations (5.150)–(5.152) are uncoupled, they can be solved independently. First of all, the critical load for the column to buckle in the torsional mode can be obtained from equation (5.152):

$$P_{cr} = GJ\frac{A}{I_y + I_z} \tag{5.153}$$

which is independent of the torsional boundary conditions of the column.

The general solutions to (5.50) and (5.151) are

$$w = c_1 \sin k_y x + c_2 \cos k_y x + c_3 x + c_4 \tag{5.154}$$

$$v = d_1 \sin k_z x + d_2 \cos k_z x + d_3 x + d_4 \tag{5.155}$$

where

$$k_y = \sqrt{\frac{P}{EI_y\left(1 - \frac{P}{EA}\right)}}, \quad k_z = \sqrt{\frac{P}{EI_z\left(1 - \frac{P}{EA}\right)}}, \tag{5.156a,b}$$

and $c_1, c_2, \ldots, d_1, d_1, \ldots$ are integration constants to be determined. Once the boundary conditions are given for the two ends of the column, the critical modes can be solved from the preceding two equations for each of the cases shown in Figure 5.16.

For example, let us assume that the column is simply supported at both ends. For this case, the essential boundary conditions are

$$v = 0 \quad \text{at } x = 0, L \tag{5.157}$$

$$w = 0 \quad \text{at } x = 0, L \tag{5.158}$$

and the (nontrivial) natural boundary conditions can be obtained from equations (5.109)–(5.110) and (5.112)–(5.113) as follows:

$$^2M_y = -EI_y w'' + P\frac{I_y}{A}w'' = 0 \quad \text{at } x = 0, L \tag{5.159}$$

$$^2M_z = EI_z v'' - P\frac{I_z}{A}v'' = 0 \quad \text{at } x = 0, L \tag{5.160}$$

By enforcing the displacement v in equation (5.155) to satisfy the conditions given in equations (5.157) and (5.160), the following characteristic equation can be derived:

$$d_1 \sin k_z L = 0 \tag{5.161}$$

from which the buckling coefficient $k_z L$ is obtained as

$$k_z L = \pi \tag{5.162}$$

and the *critical load* P_{cz} for the column to buckle on the x–y plane is

$$P_{cz} = \frac{1}{1 + \frac{\pi^2 I_z}{AL^2}} \frac{\pi^2 EI_z}{L^2} \tag{5.163}$$

The preceding expression for P_{cz} differs from the well-known conventional one given below:

$$P_{cz} = \frac{\pi^2 EI_z}{L^2} \tag{5.164}$$

This is due to the inclusion of $(\pi^2 I_z)/(AL^2)$ in the denominator of the coefficient in equation (5.163), which originates from the inclusion of the nonlinear effect $u_{x,x}^2$ in the nonlinear strain component $\eta_{x,x}$. Such an effect has been carried over through the whole virtual work formulation. In general, the effect of $(\pi^2 I_z)/(AL^2)$ is very small for most practical problems and can be neglected in practice.

Similarly, by enforcing the displacement v in equation (5.154) to satisfy the boundary conditions given in equations (5.158) and (5.159), we can derive another critical load P_{cy} for the x–z plane as follows:

$$P_{cy} = \frac{1}{1 + \frac{\pi^2 I_y}{AL^2}} \frac{\pi^2 EI_y}{L^2} \tag{5.165}$$

which is basically no different from the conventional one given as follows:

$$P_{cy} = \frac{\pi^2 EI_y}{L^2} \tag{5.166}$$

Under this load, the column will buckle in the flexural mode on the x–z plane.

The above derivation serves merely as an illustration of the procedure of application of the incremental equations. Solutions for problems with other boundary conditions in Figure 5.16 can be found from almost any textbook on stability (e.g., Timoshenko and Gere, 1961), neglecting the axial shortening effect $(\pi^2 I_z)/(AL^2)$ or $(\pi^2 I_y)/(AL^2)$ as presented herein based on the present theory of three-dimensional beams, which will not be recapitulated here.

5.7.2 *Torsionally loaded columns*

Next, let us consider the column subjected to a torque T at the free end in Figure 5.17(a). Also, let us assume that the torque T is generated be a semitangential mechanism that consists of a couple of direct forces as the one shown in Figure 5.17(b). For the first stage, all the initial forces are equal to zero, i.e., ${}^1F_x = {}^1F_y = {}^1F_z = {}^1M_x = {}^1M_y = {}^1M_z = 0$, and the only external load is ${}^2M_x = T$. The natural boundary conditions in equations (5.108)–(5.113) reduce to: $GJ\theta'_x = T$ and $EAu' = -EI_z v''' = -EI_y w''' = -EI_y w'' = -EI_z v'' = 0$. These results can be used as the loading condition for the second-stage analysis. Namely, the member forces existing at the beginning

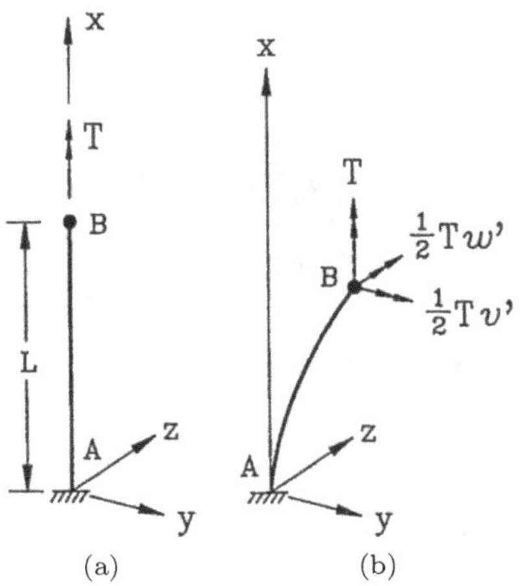

Fig. 5.17. Torsionally loaded column: (a) before buckling and (b) after buckling.

of the second stage are: ${}^1M_x = T$ and ${}^1F_x = {}^1F_y = {}^1F_z = {}^1M_x = {}^1M_y = {}^1M_z = 0$. The buckling differential equations in (5.104)–(5.107) reduce to

$$EI_y w'''' - Tv''' = 0 \tag{5.167}$$

$$EI_Z v'''' + Tw''' = 0 \tag{5.168}$$

Since the applied torque T is of the semitangential type, according to Table 5.1, it will induce a moment $\Delta M_z = -(1/2)T\theta_y = (1/2)Tw'$ about the z-axis and another moment $\Delta M_y = (1/2)T\theta_z = (1/2)Tv'$ about the y-axis, when subjected to three-dimensional rotations in the buckling stage. Such induced moments have to be considered in setting up the natural boundary conditions for the free end B of the column. Consider the free body of joint B of the column in the buckling position in the xy-plane and xz-plane in Figures 5.18(a) and 5.18(b), respectively. The moments acting on the upper side of the joint are generated *externally* by the moment mechanism on the two planes as

$${}^2M_y = \Delta M_y = \frac{1}{2}Tv' \tag{5.169}$$

$${}^2M_z = \Delta M_z = \frac{1}{2}Tw' \tag{5.170}$$

while those acting on the lower side are generated *internally* as the stress resultants of the cross-section,

$${}^2M_y = -EI_y w'' + Tv' \tag{5.171}$$

$${}^2M_z = EI_z v'' + Tw' \tag{5.172}$$

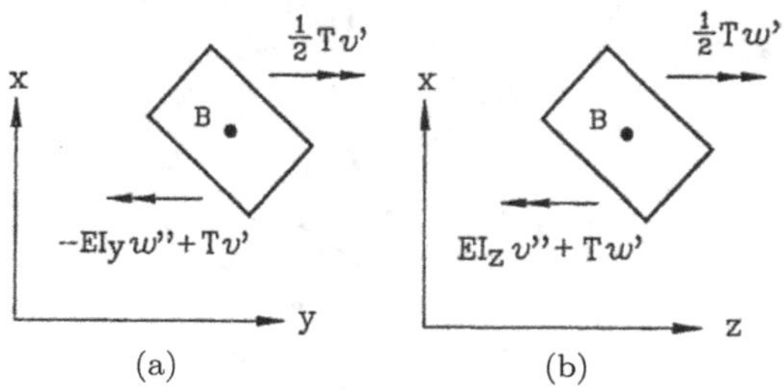

Fig. 5.18. Moment equilibrium conditions for joint B: (a) xy-plane and (b) xz-plane.

readily available through equations (5.112) and (5.113) in recognition of the following relations: ${}^1M_x = T$, ${}^1M_y = {}^1M_z = 0$, $\theta_z = v'$, and $\theta_y = -w'$. From the conditions of equilibrium for the moments acting at joint B, the following can be written:

$$-EI_y w'' + Tv' = \frac{1}{2}Tv' \quad \text{at } x = L \tag{5.173}$$

$$EI_z v'' + Tw' = \frac{1}{2}Tw' \quad \text{at } x = L \tag{5.174}$$

The natural boundary conditions for the shear forces at joint B can be obtained from equations (5.109) and (5.110) as

$$-EI_z v''' - Tw'' = 0 \quad \text{at } x = L \tag{5.175}$$

$$-EI_y w''' + Tv'' = 0 \quad \text{at } x = L \tag{5.176}$$

by noting that ${}^2F_y = {}^1F_y = 0$ and ${}^2F_z = {}^1F_z = 0$. The essential boundary conditions for the fixed end of the column are simply as follows:

$$v = v' = 0 \quad \text{at } x = 0 \tag{5.177}$$

$$w = w' = 0 \quad \text{at } x = 0 \tag{5.178}$$

By differentiating equation (5.168) once and making use of equation (5.167), we can obtain the following differential equation:

$$v''''' + k^2 v''' = 0 \tag{5.179}$$

where

$$k = \frac{T}{\sqrt{EI_y EI_z}} \tag{5.180}$$

The general solution to equation (5.179) is

$$v = c_1 \sin kx + c_2 \cos kx + c_3 x^2 + c_4 x + c_5 \tag{5.181}$$

where $c_1, c_2, \ldots, c_5$ are integration constants. Similarly, the general solution for w is

$$w = d_1 \sin kx + d_2 \cos kx + d_3 x^2 + d_4 x + d_5 \tag{5.182}$$

where $d_1, d_2, \ldots, d_5$ are integration constants. It should be noted that in the reduction of the two-variable system [equations (5.167) and (5.168)] to the single-variable system [equation (5.179)], two extraneous coefficients have been introduced. This problem can be resolved by substituting the solution for v and w back into the original equations and making sure that they are identically satisfied. For instance, by substituting the solutions for v and w into equation (5.167), we can find that

$$d_1 = \sqrt{\frac{I_z}{I_y}} c_2, \quad d_2 = -\sqrt{\frac{I_z}{I_y}} c_1 \tag{5.183a,b}$$

which can be substituted back into equation (5.182) to yield

$$w = \sqrt{\frac{I_z}{I_y}} c_2 \sin kx - \sqrt{\frac{I_z}{I_y}} c_1 \cos kx + d_3 x^2 + d_4 x + d_5 \tag{5.184}$$

The solutions for v and w must satisfy all the boundary conditions given in (5.173)–(5.178). First, by substituting the expressions for v and w into the conditions given in (5.175) and (5.176), we obtain $c_3 = d_3 = 0$. Thus,

$$v = c_1 \sin kx + c_2 \cos kx + c_4 x + c_5 \tag{5.185}$$

$$w = \sqrt{\frac{I_z}{I_y}} c_2 \sin kx - \sqrt{\frac{I_z}{I_y}} c_1 \cos kx + d_4 x + d_5 \tag{5.186}$$

which can then be substituted into the remaining boundary conditions to result in the following homogeneous equations:

$$\begin{bmatrix} 0 & 1 & 0 & 1 & 0 & 0 \\ k & 0 & 1 & 0 & 0 & 0 \\ -(I_z/I_y)^{1/2} & 0 & 0 & 0 & 0 & 1 \\ 0 & k(I_z/I_y)^{1/2} & 0 & 0 & 1 & 0 \\ -k\cos kL & k\sin kL & 1 & 0 & 0 & 0 \\ -k^2\sin kL & -k^2\cos kL & 0 & 0 & T/(EI_z) & 0 \end{bmatrix} \begin{Bmatrix} c_1 \\ c_2 \\ c_4 \\ c_5 \\ d_4 \\ d_5 \end{Bmatrix} = \begin{Bmatrix} 0 \\ 0 \\ 0 \\ 0 \\ 0 \\ 0 \end{Bmatrix} \tag{5.187}$$

For nontrivial solution of the integration constants, the determinant of the coefficient matrix of the preceding homogeneous equations must be equal to zero. From this condition, the following *characteristic equation* can be obtained:

$$k^3(1 + \cos kL) = 0 \tag{5.188}$$

from which the critical load can be solved,

$$T_{cr} = \pm \frac{\pi\sqrt{EI_y EI_z}}{L} \tag{5.189}$$

The "±" sign indicates that the torque can be applied either about the positive or the negative axis of the column. For the special case of a circular column, i.e., for cross-sections with $I_y = I_z = I$, the present solution reduces to that given by Ziegler (1977):

$$T_{cr} = \pm \frac{\pi EI}{L} \tag{5.190}$$

The solution procedure presented in this section for solving the critical load of a torsionally loaded column can be applied to solving columns with other boundary conditions and subjected to other types of torque. No attempt will be made here to carry out such operations, however, as they are in principle a repetition of the present procedure. For the special case of a circular column, the results for columns with various boundary conditions under the action of various torques are available in Ziegler (1977).

Applicability of the incremental theories is far beyond that illustrated above for a single compressed or twisted member. From the analytical side of achieving closed-form solutions, they can be applied to the buckling of single members subjected to other types of loadings or the buckling of planar frames restricted to deform in the plane of the frame. As the solutions for problems of this sort are abundant in the literature, we do not feel it necessary to present any other examples in this chapter.

At this point, we feel the most urgent matter is to derive the stiffness matrices for the space frame element in Chapter 6 to follow, such that readers can use them for general wider applications. For this reason, more analytical solutions for the buckling of two-member frames involving the torsional deformations have to be deferred to Chapter 9, using the same set of equations derived in this chapter. In solving these problems, focus will be placed on the establishment of the equilibrium conditions for structural joints connecting non-collinear members in the buckling configuration, using the derived natural boundary conditions.

References

Argyris, J.H., Dunne, P.C., Malejannakis, G.A., and Scharpf, D.W. (1978). On large displacement-small strain analysis of structures with rotational degrees of freedom, *Comp. Meth. Appl. Mech. Eng.*, 14, 401–151, and 15, 99–135.

Argyris, J.H., Hilbert, O., Malejannakis, G.A., and Scharpf, D.W. (1979). On the geometrical stiffness of a beam in space-A consistent VW approach, *Comp. Meth. Appl. Mech. Eng.*, 20, 105–131.

Bleich, F. (1952). *The Buckling Strength of Metal Structures*, McGraw-Hill, New York, N.Y.

Chajes, A. (1974). *Principles of Structural Stability Theory*, Prentice Hall, Englewood Cliffs, N.J.

Chen, W.F. and Lui, E.M. (1987). *Structural Stability-Theory and Implementation*, Elsevier, New York, N.Y.

Gjelsvik, A. (1981). *The Theory of Thin Walled Bars*, John Wiley, New York, N.Y.

Simitses, G.J. (1976). *An Introduction to the Elastic Stability of Structures*, Prentice Hall, Englewood Cliffs, N.J.

Timoshenko, S.P. and Gere, J.M. (1961). *Theory of Elastic Stability*, 2nd edn, McGraw-Hill, New York, N.Y.

Vlasov, V.Z. (1961). *Thin-Waled Elastic Beams*, Israel Program for Scientific Translation, Jerusalem, Israel.

Washizu, K. (1982). *Variational Methods in Elasticity and Plasticity*, 3rd edn, Pergamon Press, Oxford, England.

Yang, Y.B. and Kuo, S.R. (1991a). Out-of-plane buckling of angled frames, *Int. J. Mech. Sci.*, 33(1), 55–67.

Yang, Y.B. and Kuo, S.R. (1991b). Buckling of frames under various torsional loadings, *J. Eng. Mech.*, ASCE, 117(8), 1681–1697.

Yang, Y.B., Kuo, S.R., and Yau, J.D. (1991). Use of straight-beam approach to study buckling of curved beams, *J. Struct. Eng.*, ASCE, 117(7), 1963–1978.

Yang, Y.B. and Kuo, S.R. (1992). Frame buckling analysis with full consideration of joint compatibilities, *J. Eng. Mech.*, ASCE, 118(5), 871–889.

Yang, Y.B. and Kuo, S.R. (1994). *Theory of Nonlinear Framed Structures*, Prentice Hall, Englewood Cliffs, New Jersey.

Yang, Y.B. and McGuire, W. (1986a). Stiffness matrix for geometric nonlinear analysis, *J. Struct. Eng.*, ASCE, 112(4), 853–877.

Yang, Y.B. and McGuire, W. (1986b). Joint rotation and geometric nonlinear analysis, *J. Struct. Eng.*, ASCE, 112(4), 879–905.

Ziegler, H. (1977). *Principles of Structural Stability*, 2nd edn, Birkhäuser, Stuttgart, Germany.

Chapter 6

Stiffness Matrices of Space Frame Element

6.1 Introduction

In order not to distract the readers from the formulation of the space frame element, we decide to defer the typical presentation of analytical solutions for some benchmark problems concerning the lateral buckling of some plane frames in Chapter 9. One key point in this regard has been the rigorous incorporation of conditions of equilibrium for structural joints where members lying along different directions are connected in the deformed C_2 configuration. The benchmark solutions in Chapter 9 will be useful for checking the validity of the solutions obtained by the finite element method or other methods.

In this chapter, we shall proceed one step further to derive a general three-dimensional beam element that will be suitable for the buckling analysis of space frames. It seems that a good starting point here is the principle of virtual displacements based on the updated Lagrangian formulation that has been used exclusively in the previous chapters. Moreover, a basic requirement along these lines is the capability of the general finite element to take into account all the factors addressed in the preceding chapter. Among the factors that have been considered, the conditions of equilibrium for structural joints of non-aligned members appear to be the most important. One feature of these conditions is that they are irrelevant to any elastic deformations. In a finite element formulation employing the principle of virtual displacements, they can only be included through a rigorous account of the virtual work done by the nodal forces, as will be demonstrated in the following sections of the chapter.

6.2 Principle of Virtual Displacements

The principle of virtual displacements has been developed for a three-dimensional beam in Section 5.4. For the reason of self-completeness, some basic issues will be repeated herein. As was stated previously, the nonlinear deformation behavior of a solid beam can be described by three typical configurations in an incremental analysis: the initial undeformed configuration C_0, the last calculated configuration C_1, and the current unknown configuration C_2. For the cases where iterations are performed at each incremental step of a nonlinear analysis, the last calculated configuration C_1 is assumed to be known and the element is in equilibrium at this configuration. With the updated Lagrangian formulation, the equilibrium of a solid at the C_2 configuration can be expressed with reference to the C_1 configuration as in equation (1.123), i.e.,

$$\int_{{}^1V} {}^2_1S_{ij}\delta^2_1\epsilon_{ij}\,{}^1dV = {}^2_1R \tag{6.1}$$

By neglecting the effect of body forces, the external virtual work ${}^2_1R$ in equation (6.1) is

$${}^2_1R = \int_{{}^1S} {}^2_1t_i\delta u_i\,{}^1dS \tag{6.2}$$

In the preceding two equations, δ denotes the variational operator, ${}^2_1S_{ij}$ is the *updated Kirchhoff stresses*, ${}^2_1\epsilon_{ij}$ the *Green–Lagrange strains*, ${}^2_1t_i$ the surface tractions, u_i the displacement components, and ${}^1V$ and ${}^1S$ the volume and surface area, respectively, of the solid element at C_1.

Following Chapter 5, the *engineering approach* will be adopted in the formulation of this chapter, in that three components of strains and stresses, i.e., the axial strain and cross-sectional shear strains and their corresponding stresses, will be adopted in the formulation of the three-dimensional beam element.

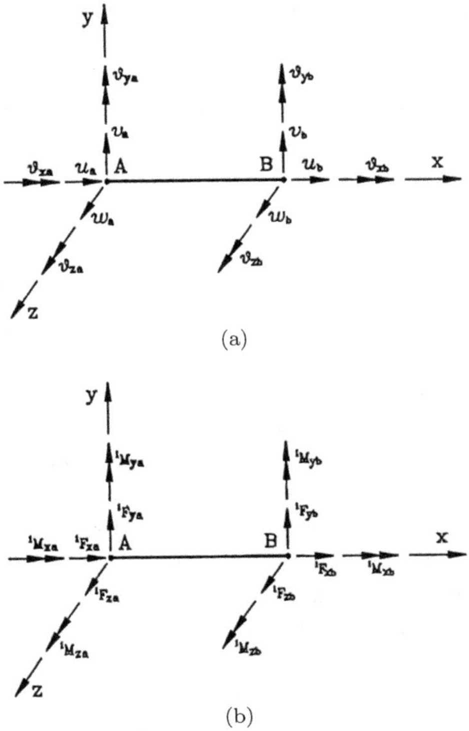

Fig. 6.1. Space frame element: (a) nodal degrees of freecom and (b) initial nodal forces.

Let x denote the centroidal axis and (y, z) the two *principal axes* of the cross-section of the solid beam, shown in Figure 6.1. for the space frame element. As detailed in the preceding chapter, only three components of strains and stresses are included in the potential energy of the three-dimensional solid beam. Following the procedure of Section 5.4, we can transform the nonlinear virtual work equation (6.1) into a linearized virtual work equation for the three-dimensional solid beam as follows:

$$\begin{aligned}&\int_V (Ee_{xx}\delta e_{xx} + 4Ge_{xy}\delta e_{xy} + 4Ge_{xz}\delta e_{xz})dV\\&\qquad + \int_V ({}^1\tau_{xx}\delta\eta_{xx} + 2{}^1\tau_{xy}\delta\eta_{xy} + 2{}^1\tau_{xz}\delta\eta_{xz})dV\\&= {}^2R - {}^1R \qquad (6.3)\end{aligned}$$

in which the subscript "1" has been omitted for clarity; E and G denote the moduli of elasticity and rigidity, respectively; ${}^1\tau_{ij}$ is the Cauchy stresses existing at C_1; and e_{ij} and η_{ij} the linear and nonlinear components of the strain increments. For the beam in equilibrium at C_1, the virtual work ${}^1R$ done by the surface tractions ${}^1t_i$ at C_1 over the surface area S is

$$ {}^1R = \int_S {}^1t_i \delta u_i dS \tag{6.4} $$

In the preceding two equations, all variables should be recognized as the quantities measured with respect to the last calculated configuration C_1.

The first and second integrals on the left-hand side of equation (6.3) represent the variation in strain energy and potential energy, respectively, associated with the deformation of the beam from C_1 to C_2. The terms on the right-hand side, i.e., ${}^2R - {}^1R$, represent the increment in external virtual work done by the surface tractions, resulting in the deformation of the beam from C_1 to C_2. In Section 5.5, it has been demonstrated that if consistent static and kinematic relations have been adopted for the three-dimensional beam, we can derive from equation (6.3) the following virtual work equation of equilibrium in incremental form, as reproduced from equation (5.103):

$$
\begin{aligned}
&\int_0^L (EAu'\delta u' + EI_y w''\delta w'' + EI_z v''\delta v'' + GJ\theta'_x \delta\theta'_x) dx \\
&\quad + \frac{1}{2}\int_0^L \left[{}^1F_x \delta(u'^2 + v'^2 + w'^2) \right. \\
&\quad \left. + {}^1F_x \left(\frac{I_y}{A}\delta w''^2 + \frac{I_z}{A}\delta v''^2 \right) + \bar{K}\delta\theta_x'^2 \right] dx \\
&\quad + \int_0^L [-{}^1M_z \delta(w'\theta'_x) - {}^1M_y \delta(v'\theta'_x) \\
&\quad - {}^1M_y \delta(u'w'') + {}^1M_z \delta(u'v'')] dx \\
&\quad + \int_0^L [{}^1F_y \delta(w'\theta_x - u'v') - {}^1F_z \delta(v'\theta_x + u'w')] dx
\end{aligned}
$$

$$+\frac{1}{2}\int_0^L [{}^1M_x\delta(v''w') - {}^1M_x\delta(v'w'')]dx$$

$$= \langle\delta u\rangle(\{{}^2f\} - \{{}^1f\}) + \left[\begin{matrix} \left({}^1M_z\theta_x - \frac{1}{2}{}^1M_x\theta_z\right)\cdot\delta\theta_y \\ + \left(-{}^1M_y\theta_x + \frac{1}{2}{}^1M_x\theta_y\right)\cdot\delta\theta_z \end{matrix}\right]\Bigg|_0^L \quad (6.5)$$

in which (u, v, w) should be recognized as the displacement increments of the centroid of the cross-section, θ_x denotes the angle of twist, A the area of the cross-section, L the length of the beam, J the torsion constant, I_y and I_z the moments of inertia, and $\bar{K}$ the Wagner coefficient (see Section 5.4 for the definition of each of these parameters). Particularly, it should be noted that $\theta_z = v'$ and $\theta_y = -w'$ have been utilized. In equation (6.5), the *semitangential* property of the torque ${}^1M_x$ and the *quasitangential* property of the bending moments ${}^1M_y$ and ${}^1M_z$ can be observed from the boundary (bracketed) terms on the right-hand side. They have been derived as a natural result of the formulation.

6.3 Finite Element Formulation

In this section, the incremental virtual work equation (6.5) will be used as the basis for the derivation of the three-dimensional beam element. In the following formulation, a row vector is denoted by $\langle\ \ \rangle$, a column vector by $\{\ \ \}$, and a matrix by $[\ \ ]$. As shown in Figure 6.1(a), the displacement vector $\{u\}$ for the three-dimensional beam is defined as

$$\{u\}^T = \langle u_a \quad v_a \quad w_a \quad \theta_{xa} \quad \theta_{ya} \quad \theta_{za} \quad u_b \quad v_b \quad w_b \quad \theta_{xb} \quad \theta_{yb} \quad \theta_{zb}\rangle \quad (6.6)$$

which consists of three translations and three rotations at each node. Correspondingly, the nodal forces $\{{}^1f\}$ acting on the element at C_1 are those depicted in Figure 6.1(b), which can be written as follows:

$$\{{}^1f\}^T = \langle {}^1F_{xa} \quad {}^1F_{ya} \quad {}^1F_{za} \quad {}^1M_{xa} \quad {}^1M_{ya} \quad {}^1M_{za} \quad {}^1F_{xb} \quad {}^1F_{yb} \quad {}^1F_{zb} \quad {}^1M_{xb} \quad {}^1M_{yb} \quad {}^1M_{zb}\rangle \quad (6.7)$$

The nodal forces $\{^2f\}$ acting on the element at C_2 can be defined in a similar way by switching all the superscripts from "1" to "2", i.e.,

$$\{^2f\}^T = \langle ^2F_{xa} \quad ^2F_{ya} \quad ^2F_{za} \quad ^2M_{xa} \quad ^2M_{ya} \quad ^2M_{za} \quad ^2F_{xb} \quad ^2F_{yb} \quad ^2F_{zb} \quad ^2M_{xb} \quad ^2M_{yb} \quad ^2M_{zb}\rangle \tag{6.8}$$

In arriving at equation (6.5), the following assumptions have been made. First, the *Bernoulli-Euler hypothesis* of plane section applies, that is, the displacements of each cross-section can be described by those given in equations (5.7)–(5.9). Second, the initial forces $^1F_x, {}^1F_y \ldots$, and ^1M_z have been defined as the stress resultants over the cross-section of the beam, as in equations (5.20)–(5.25). Finally, the axes x, y, and z have been adopted as the *principal centroidal axes* of the beam, which satisfy the orthogonality conditions given in equations (5.31) and (5.90) for bi-symmetrical cross-sections.

For convenience, we shall denote the first integral (strain energy) on the left-hand side of the equal sign of equation (6.5) by δU and the remaining integrals (potential energy) on the same side by δV. What we will be interested in the following is to derive the *elastic stiffness matrix* $[k_e]$ and *geometric stiffness matrix* $[k_g]$ from the two terms δU and δV, respectively. In a displacement-based finite element formulation, the displacements at any point within a finite element should be related to those at the element nodes by *interpolation functions.* For a three-dimensional beam, the axial displacement u and the angle of twist θ_x can be interpolated by linear functions, and the transverse displacements v and w by cubic functions,

$$u = \{n_1\}^T\{\bar{u}\} \tag{6.9}$$

$$v = \{n_3\}^T\{\bar{v}\} \tag{6.10}$$

$$w = \{n_3\}^T\{\bar{w}\} \tag{6.11}$$

$$\theta_x = \{n_1\}^T\{\overline{\theta}_x\} \tag{6.12}$$

By letting $i = x/L$, the linear and cubic interpolation functions $\{n_1\}$ and $\{n_3\}$ can be expressed as

$$\langle n_1\rangle = \langle (1-i) \quad i\rangle \tag{6.13a}$$

$$\langle n_3\rangle = \langle (1-3i^2+2i^3) \quad (i-2i^2+i^3) \quad (3i^2-2i^3) \quad (i^3-i^2)\rangle \tag{6.13b}$$

and the displacement vectors as

$$\langle \bar{u} \rangle = \langle u_a \quad u_b \rangle \tag{6.14a}$$

$$\langle \bar{v} \rangle = \langle v_a \quad L\theta_{za} \quad v_b \quad L\theta_{zb} \rangle \tag{6.14b}$$

$$\langle \bar{w} \rangle = \langle w_a \quad - L\theta_{ya} \quad w_b \quad -L\theta_{yb} \rangle \tag{6.14c}$$

$$\langle \overline{\theta}_x \rangle = \langle \overline{\theta}_{xa} \quad \overline{\theta}_{xb} \rangle \tag{6.14d}$$

where each nodal degree of freedom has been defined in Figure 6.1(a). As was stated in Section 2.2.2, the above interpolation functions remain an exact solution to the linear differential equations of the three-dimensional solid beam free of any distributed loads.

6.3.1 *Elastic stiffness matrix*

As was noted previously, the first integral on the left-hand side of equation (6.5) represents the variation in strain energy (δU) of the beam occurring during the incremental deformation from C_1 to C_2. This term is identical in form to the one given on the left-hand side of equation (2.72) for the space frame element based on the linear theory. By substitution of equations (6.9)–(6.12) for the displacements, the first integral in (6.5) can be transformed into the following:

$$\begin{aligned}\delta U = {} & \{\delta \bar{u}\}^T \int_0^1 \frac{EA}{L} \{n_1'\}\{n_1'\}^T di \{\bar{u}\} \\ & + \{\delta \bar{v}\}^T \int_0^1 \frac{EI_z}{L^3} \{n_3''\}\{n_3''\}^T di \{\bar{v}\} \\ & + \{\delta \bar{w}\}^T \int_0^1 \frac{EI_y}{L^3} \{n_3''\}\{n_3''\}^T di \{\bar{w}\} \\ & \times \{\delta \overline{\theta}_x\}^T \int_0^1 \frac{GJ}{L} \{n_1'\}\{n_1'\}^T di \{\overline{\theta}_x\}\end{aligned} \tag{6.15}$$

or as follows:

$$\begin{aligned}\delta U = {} & \{\delta \bar{u}\}^T \frac{EA}{L} [K_{11}^{110}]\{\bar{u}\} + \{\delta \bar{v}\}^T \frac{EI_z}{L^3} [K_{33}^{220}]\{\bar{v}\} \\ & + \{\delta \bar{w}\}^T \frac{EI_y}{L^3} [K_{33}^{220}]\{\bar{w}\} + \{\delta \overline{\theta}_x\}^T \frac{GJ}{L} [K_{11}^{110}]\{\overline{\theta}_x\}\end{aligned} \tag{6.16}$$

using the notation $[K_{gh}^{stv}]$ of equation (2.44). Here, each of the integral matrix $[K_{gh}^{stv}]$ has been made available in Appendix A.

With the element displacement vector $\{u\}$ defined in equation (6.6) and the component displacement vectors $\{\bar{u}\}$, $\{\bar{v}\}$, $\{\bar{w}\}$, and $\{\overline{\theta}_x\}$ in equation (6.14), we can further combine the terms in (6.15) into a matrix product as follows:

$$\delta U = \{\delta u\}^T [k_e] \{u\} \tag{6.17}$$

where $[k_e]$ is the *elastic stiffness matrix* for the space frame element. The elastic stiffness matrix $[k_e]$, which has a dimension of 12×12, has been identified as the $[k]$ matrix and given explicitly in (2.89) based on the linear theory, which is reproduced below for easy reference:

$$[k_e] = \begin{bmatrix} [k_1] & [k_2] \\ [k_2]^T & [k_3] \end{bmatrix} \tag{6.18}$$

where the submatrices are

$$[k_1] = \begin{bmatrix} \frac{EA}{L} & 0 & 0 & 0 & 0 & 0 \\ & \frac{12EI_z}{L^3} & 0 & 0 & 0 & \frac{6EI_z}{L^2} \\ & & \frac{12EI_y}{L^3} & 0 & -\frac{12EI_y}{L^2} & 0 \\ & & & \frac{GJ}{L} & 0 & 0 \\ & \text{Symm.} & & & \frac{4EI_y}{L} & 0 \\ & & & & & \frac{4EI_z}{L} \end{bmatrix} \tag{6.19a}$$

$$[k_2] = \begin{bmatrix} -\frac{EA}{L} & 0 & 0 & 0 & 0 & 0 \\ 0 & -\frac{12EI_z}{L^3} & 0 & 0 & 0 & \frac{6EI_z}{L^2} \\ 0 & 0 & -\frac{12EI_y}{L^3} & 0 & -\frac{6EI_y}{L^2} & 0 \\ 0 & 0 & 0 & -\frac{GJ}{L} & 0 & 0 \\ 0 & 0 & \frac{6EI_y}{L^2} & 0 & \frac{2EI_y}{L} & 0 \\ 0 & -\frac{6EI_z}{L^2} & 0 & 0 & 0 & \frac{2EI_z}{L} \end{bmatrix} \tag{6.19b}$$

$$[k_3] = \begin{bmatrix} \frac{EA}{L} & 0 & 0 & 0 & 0 & 0 \\ & \frac{12EI_z}{L^3} & 0 & 0 & 0 & -\frac{6EI_z}{L^2} \\ & & \frac{12EI_y}{L^3} & 0 & \frac{6EI_y}{L^2} & 0 \\ & & & \frac{GJ}{L} & 0 & 0 \\ & \text{Symm.} & & & \frac{4EI_y}{L} & 0 \\ & & & & & \frac{4EI_z}{L} \end{bmatrix} \tag{6.19c}$$

By multiplying the elastic stiffness matrix $[k_e]$ by the element displacement vector $\{u\}$, we can obtain the nodal actions caused by the elastic deformations of the beam element during the incremental step.

6.3.2 *Geometric stiffness matrix*

In order to derive the *geometric stiffness matrix* $[k_g]$ for the space frame element, we shall relate all the member forces, i.e., ${}^1F_x, {}^1F_y, \ldots,$ and ${}^1M_z$, initially acting on section x of the member at C_1 to those at the two ends, which was assumed to be in a state of equilibrium. Based on the conditions of equilibrium for the member at C_1, such relations can be expressed as follows:

$${}^1F_x = {}^1F_{xb} \tag{6.20}$$

$${}^1F_y = -\frac{{}^1M_{za} + {}^1M_{zb}}{L} \tag{6.21}$$

$${}^1F_z = \frac{{}^1M_{ya} + {}^1M_{yb}}{L} \tag{6.22}$$

$${}^1M_x = {}^1M_{xb} \tag{6.23}$$

$${}^1M_y = -{}^1M_{ya}(1 - i) + {}^1M_{yb}i \tag{6.24}$$

$${}^1M_z = -{}^1M_{za}(1 - i) + {}^1M_{zb}i \tag{6.25}$$

where each nodal force has been added a subscript "a" or "b" to indicate the associated end A or B of the beam [Figure 6.1(b)].

The terms with the initial forces ${}^1F_x, {}^1F_y, \ldots,$ and ${}^1M_z$ on the left-hand side of equation (6.5) for the three-dimensional beam element correspond to the terms with the initial stresses ${}^1\tau_{xx}$, ${}^1\tau_{xy}$, ${}^1\tau_{xz}$, ${}^1\tau_{yy}$, ${}^1\tau_{yz}$ and ${}^1\tau_{zz}$ on the left-hand side of the virtual work equation (6.3). They represent in effect, the variation in *potential energy* (δV) or the *effect of instability* of the three-dimensional beam during the deformation process from C_1 to C_2. With the substitution of the interpolation functions (6.9)–(6.12) for the cross-sectional displacements and (6.20)–(6.25) for the cross-sectional forces, each of the terms with the initial forces in equation (6.5) can be expressed in terms of the nodal forces and nodal displacements. For instance, the following operations can be carried out for the term with the bending moment ${}^1M_z$:

$$\begin{aligned}
&-\int_0^L {}^1M_z\delta(w'\theta_x')dx \\
&= -\int_0^L {}^1M_z(\theta_x'\delta w')dx - \int_0^L {}^1M_z(w'\delta\theta_x')dx \\
&= \{\delta\bar{w}\}^T \int_0^1 \frac{{}^1M_{za}(1-i) - {}^1M_{zb}i}{L^2}\{n_3'\}\{n_1'\}^T di\{\bar{\theta}_x\} \\
&\quad + \{\delta\bar{\theta}_x\}^T \int_0^1 \frac{{}^1M_{za}(1-i) - {}^1M_{zb}i}{L^2}\{n_1'\}\{n_3'\}^T di\{\bar{w}\} \\
&= \{\delta\bar{w}\}^T \left(\frac{{}^1M_{za}}{L^2}([K_{31}^{110}] - [K_{31}^{111}]) - \frac{{}^1M_{zb}}{L^2}[K_{31}^{111}]\right)\{\bar{\theta}_x\} \\
&\quad + \{\delta\bar{\theta}_x\}^T \left(\frac{{}^1M_{za}}{L^2}([K_{13}^{110}] - [K_{13}^{111}]) - \frac{{}^1M_{zb}}{L^2}[K_{13}^{111}]\right)\{\bar{w}\} \qquad (6.26)
\end{aligned}$$

where it should be noted that $di = dx/L$, the primes denote differentiation with respect to the nondimensional coordinate i and the integral matrix $[K_{gh}^{stv}]$ as defined in equation (2.44) for the integrals has been utilized.

By the same procedure, the terms containing the initial forces on the left-hand side of equation (6.5) can be rearranged in the following form:

$$\begin{aligned}
\delta V &= \{\delta\bar{u}\}^T \frac{{}^1F_{xb}}{L}[K_{11}^{110}]\{\bar{u}\} + \{\delta\bar{v}\}^T \frac{{}^1F_{xb}}{L}[K_{33}^{110}]\{\bar{v}\} \\
&\quad + \{\delta\bar{w}\}^T \frac{{}^1F_{xb}}{L}[K_{33}^{110}]\{\bar{w}\} + \{\delta\bar{w}\}^T \frac{{}^1F_{xb}I_y}{AL^3}[K_{33}^{220}]\{\bar{w}\}
\end{aligned}$$

$$+\{\delta\bar{v}\}^T\frac{{}^1F_{xb}I_z}{AL^3}[K_{33}^{220}]\{\bar{v}\}+\{\delta\overline{\theta}_x\}^T\,{}^1F_{xb}\frac{I_y+I_z}{AL}[K_{11}^{110}]\{\overline{\theta}_x\}$$

$$-\{\delta\bar{w}\}^T\frac{{}^1M_{za}+{}^1M_{zb}}{L}[K_{31}^{100}]\{\overline{\theta}_x\}$$

$$-\{\delta\overline{\theta}_x\}^T\frac{{}^1M_{za}+{}^1M_{zb}}{L}[K_{13}^{010}]\{\bar{w}\}$$

$$+\{\delta\bar{w}\}^T\left(\frac{{}^1M_{za}}{L^2}[K_{31}^{110}]-[K_{31}^{111}]-\frac{{}^1M_{zb}}{L^2}[K_{31}^{111}]\right)\{\overline{\theta}_x\}$$

$$+\{\delta\overline{\theta}_x\}^T\left(\frac{{}^1M_{za}}{L^2}[K_{13}^{110}]-[K_{13}^{111}]-\frac{{}^1M_{zb}}{L^2}[K_{13}^{111}]\right)\{\bar{w}\}$$

$$-\{\delta\bar{v}\}^T\frac{{}^1M_{ya}+{}^1M_{yb}}{L}[K_{31}^{100}]\{\overline{\theta}_x\}$$

$$-\{\delta\overline{\theta}_x\}^T\frac{{}^1M_{ya}+{}^1M_{yb}}{L}[K_{13}^{010}]\{\bar{v}\}$$

$$+\{\delta\bar{v}\}^T\left(\frac{{}^1M_{ya}}{L^2}([K_{31}^{110}]-[K_{31}^{111}])-\frac{{}^1M_{yb}}{L^2}[K_{31}^{111}]\right)\{\overline{\theta}_x\}$$

$$+\{\delta\overline{\theta}_x\}^T\left(\frac{{}^1M_{ya}}{L^2}([K_{13}^{110}]-[K_{13}^{111}])-\frac{{}^1M_{yb}}{L^2}[K_{13}^{111}]\right)\{\bar{v}\}$$

$$+\{\delta\bar{v}\}^T\frac{{}^1M_{za}+{}^1M_{zb}}{L^2}[K_{31}^{110}]\{\bar{u}\}+\{\delta\bar{u}\}^T\frac{{}^1M_{za}+{}^1M_{zb}}{L^2}[K_{13}^{110}]\{\bar{v}\}$$

$$-\{\delta\bar{w}\}^T\frac{{}^1M_{ya}+{}^1M_{yb}}{L^2}[K_{31}^{110}]\{\bar{u}\}-\{\delta\bar{u}\}^T\frac{{}^1M_{ya}+{}^1M_{yb}}{L^2}[K_{13}^{110}]\{\bar{w}\}$$

$$-\frac{1}{2}\{\delta\bar{v}\}^T\frac{{}^1M_{xb}}{L^2}([K_{33}^{120}]-[K_{33}^{210}])\{\bar{w}\}$$

$$-\frac{1}{2}\{\delta\bar{w}\}^T\frac{{}^1M_{xb}}{L^2}([K_{33}^{210}]-[K_{33}^{120}])\{\bar{v}\}$$

$$+\{\delta\bar{w}\}^T\left(\frac{{}^1M_{ya}}{L^2}([K_{31}^{210}]-[K_{31}^{211}])-\frac{{}^1M_{yb}}{L^2}[K_{31}^{211}]\right)\{\bar{u}\}$$

$$+\{\delta\bar{u}\}^T\left(\frac{{}^1M_{ya}}{L^2}([K_{13}^{120}]-[K_{13}^{121}])-\frac{{}^1M_{yb}}{L^2}[K_{13}^{121}]\right)\{\bar{w}\}$$

$$+\{\delta\bar{v}\}^T\left(-\frac{{}^1M_{za}}{L^2}([K_{31}^{210}]-[K_{31}^{211}])+\frac{{}^1M_{zb}}{L^2}[K_{31}^{211}]\right)\{\bar{u}\}$$

$$+\{\delta\bar{u}\}^T\left(-\frac{{}^1M_{za}}{L^2}([K_{13}^{120}]-[K_{13}^{121}])+\frac{{}^1M_{zb}}{L^2}[K_{13}^{121}]\right)\{\bar{v}\} \quad (6.27)$$

In arriving at the preceding equation, the coefficient $\bar{K}$ has been taken as ${}^1F_{xb}(I_y+I_z)/A$, according to equation (5.89).

Based on the fact that the component displacement vectors $\{\bar{u}\}$, $\{\bar{v}\}$, $\{\bar{w}\}$ and $\{\overline{\theta}_x\}$ are related to the displacement vector $\{u\}$ [see equations (6.9)–(6.12)], all the submatrices in equation (6.27) can be augmented to a 12×12 matrix corresponding to the 12 degree-of-freedom displacement vector $\{u\}$ and combined to yield the following matrix product:

$$\delta V=\{u\}^T[k_g]\{u\} \quad (6.28)$$

The *geometric stiffness matrix* $[k_g]$ as derived above is valid for a three-dimensional solid beam based on the *engineering approach*, which considers three components of strains and stresses, i.e., the axial strain and shearing strains and the corresponding stresses. By letting

$$a=\frac{{}^1F_{xb}}{L},\quad b=\frac{6\,{}^1F_{xb}}{5L}+\frac{12\,{}^1F_{xb}I_z}{AL^3},\quad c=\frac{6\,{}^1F_{xb}}{5L}+\frac{12\,{}^1F_{xb}I_y}{AL^3}$$

$$d=\frac{{}^1M_{ya}}{L},\quad e=\frac{{}^1M_{za}}{L},\quad f=\frac{{}^1F_{xb}J}{AL},\quad g=\frac{{}^1M_{xb}}{L}$$

$$h=\frac{{}^1F_{xb}}{10}+\frac{6\,{}^1F_{xb}I_y}{AL^2},\quad i=\frac{{}^1M_{za}+{}^1M_{zb}}{6},\quad j=\frac{2\,{}^1F_{xb}L}{15}+\frac{4\,{}^1F_{xb}I_y}{AL}$$

$$k=\frac{{}^1F_{xb}}{10}+\frac{6\,{}^1F_{xb}I_z}{AL^2},\quad l=-\frac{{}^1M_{ya}+{}^1M_{yb}}{6},$$

$$m=\frac{2\,{}^1F_{xb}L}{15}+\frac{4\,{}^1F_{xb}I_z}{AL}$$

$$n=\frac{{}^1M_{yb}}{L},\quad o=\frac{{}^1M_{zb}}{L},\quad p=-\frac{{}^1F_{xb}L}{30}+\frac{2\,{}^1F_{xb}I_y}{AL}$$

$$q=-\frac{{}^1M_{xb}}{2},\quad r=-\frac{{}^1F_{xb}L}{30}+\frac{2\,{}^1F_{xb}I_z}{AL} \quad (6.29\text{a-r})$$

The geometric stiffness matrix $[k_g]$ can be explicitly given as follows:

$$[k_g] = \begin{bmatrix} a & 0 & 0 & 0 & -d & -e & -a & 0 & 0 & 0 & -n & -o \\ & b & 0 & d & g & k & 0 & -b & 0 & n & -g & k \\ & & c & e & h & g & 0 & 0 & -c & o & -h & -g \\ & & & f & i & l & 0 & -d & -e & -f & -i & -l \\ & & & & j & 0 & d & -g & h & -i & p & -q \\ & & & & & m & e & -k & -g & -l & q & r \\ & & & & & & a & 0 & 0 & 0 & n & o \\ & & & & & & & b & 0 & -n & g & -k \\ & & & & & & & & c & -o & h & g \\ & & & & & & & & & f & i & l \\ & \text{Symm.} & & & & & & & & & j & 0 \\ & & & & & & & & & & & m \end{bmatrix} \quad (6.30)$$

The geometric stiffness matrix $[k_g]$ as derived above for the three-dimensional solid beam is featured by the fact that all the member actions, i.e., ${}^1F_x$, ${}^1F_y, \ldots,$ and ${}^1M_z$, and their instability effects on the beam have all been taken into account. For this reason, the space frame element developed herein is most suitable for general applications to the buckling and nonlinear analyses of framed structures.

Of interest is that the bending moments and torsional moments have been defined as the stress resultants of the cross-section of the member at the *deformed* C_2 configuration in equations (5.40)–(5.42). By following the rigorous derivation, the bending moments were *naturally* shown to be of the *quasitangential* type and the torsional moments of the *semitangential* type, as indicated by the boundary terms on the right-hand side of the virtual work equation (5.103). It is essential that such boundary terms be included in the formulation of the element stiffness matrices to guarantee that the element can pass the rigid body test. In Section 6.3.3 to follow, the so-called *induced moment matrix* will be derived from the boundary terms mentioned above. In addition, the rigid body test will be conducted for the element in Section 6.4.

6.3.3 *Induced moment matrix*

For a three-dimensional beam, the virtual work done by the surface tractions at the C_1 and C_2 configurations have been denoted by ${}^1R$ and ${}^2R$, respectively. The general definitions for ${}^1R$ and ${}^2R$ excluding the effects of body forces have been given in equations (6.4) and (6.2). In Section 5.4, it has been demonstrated that if the static and kinematic expressions given in Section 5.2 were strictly adopted for all the variables involved, then from the external virtual work increment $({}^2R-{}^1R)$, we are able to show the rotational properties of the bending moments and torsional moments in the three-dimensional space. Such a fact can be observed from the expression on the right-hand side of equation (5.103) or (6.5), i.e.,

$$V.W.I = \langle \delta u \rangle (\{{}^2f\} - \{{}^1f\}) + \left[\begin{array}{c} \left({}^1M_z\theta_x - \frac{1}{2}{}^1M_x\theta_z\right) \cdot \delta\theta_y \\ + \left(-{}^1M_y\theta_x + \frac{1}{2}{}^1M_x\theta_y\right) \cdot \delta\theta_z \end{array} \right] \Bigg|_0^L \tag{6.31}$$

Here, we like to reiterate that the bracketed term represents the virtual work done by the moments induced by the torque ${}^1M_x$ and bending moments ${}^1M_y$ and ${}^1M_z$ undergoing rotations in the three-dimensional space. It is with this term that we interpret the torque ${}^1M_x$ as *semitangential*, and the bending moments ${}^1M_y$ and ${}^1M_z$ as *quasitangential* (see Table 5.1 for the moments induced by various types of moments undergoing rotations).

In recognition of the fact that the moments induced by the initial nodal moments upon rotations relate only to the rotational degrees of freedom of the beam element, equation (6.31) can be rewritten as

$$V.W.I = \langle \delta u \rangle (\{{}^2f\} - \{{}^1f\}) + \{\delta u\}^T [k_i] \{u\} \tag{6.32}$$

where the displacement vector $\{u\}$ has been given in equation (6.6) and the *induced moment matrix* $[k_i]$ is

$$[k_i] = \begin{bmatrix} [0] & & & \\ & [k_i]_a & & \\ & & [0] & \\ & & & [k_i]_b \end{bmatrix} \tag{6.33}$$

with the submatrices $[k_i]_a$ and $[k_i]_b$ defined as follows:

$$[k_i]_a = \begin{bmatrix} 0 & 0 & 0 \\ -{}^1M_{za} & 0 & \frac{{}^1M_{xa}}{2} \\ {}^1M_{ya} & -\frac{{}^1M_{xa}}{2} & 0 \end{bmatrix} \tag{6.34}$$

$$[k_i]_b = \begin{bmatrix} 0 & 0 & 0 \\ -{}^1M_{zb} & 0 & \frac{{}^1M_{xb}}{2} \\ {}^1M_{yb} & -\frac{{}^1M_{xb}}{2} & 0 \end{bmatrix} \tag{6.35}$$

Evidently, the induced moment effects as represented by the $[k_i]$ matrix are of the same order as that of the geometric stiffness matrix $[k_g]$. The inclusion of such an effect in the formulation of the three-dimensional beam element is essential for the element to pass the rigid body test, as will be demonstrated in the section to follow. Unfortunately, such an effect has not received enough attention from researchers previously working on the stability of space frames. It should be noted that the induced moment matrix $[k_i]$ derived above for a single, *discrete* element is asymmetric. Such a property can be attributed to the lack of *conjugateness* between the bending moments and nodal rotational degrees defined as the displacement derivatives (Reissner, 1973). The symmetry of the stiffness matrix will be restored when the element is connected to the other elements, that is, when the conditions of equilibrium for structural joints are taken into account in the element assembly process.

6.3.4 *Element stiffness equation*

As was stated previously, at each incremental step in nonlinear analysis, we shall increase the surface tractions acting on the beam from ${}^1t_i$ to ${}^2t_i$. This will result in an increase in external virtual work from ${}^1R$ to ${}^2R$. According to the incremental virtual work equation (6.5), such an increase in the external virtual work, denoted as *V.W.I.* and represented by the right-hand side of the equal sign of the equation, will be transformed into the *strain energy* (δU) and *potential energy* (δV) of the beam, as represented by the terms on the left-hand side of the equation. Namely,

$$\delta U + \delta V = V.W.I \tag{6.36}$$

Using the relations given in equations (6.17), (6.28), and (6.32) for the corresponding terms in the preceding equation and taking the arbitrary nature of the virtual displacement $\{\delta u\}$, we can obtain the following stiffness equation for the *discrete* space frame element:

$$[k_e]\{u\} + ([k_g] + [k_i])\{u\} = \{^2f\} - \{^1f\} \tag{6.37}$$

where the induced moment matrix $[k_i]$ has been relocated to the left-hand side. In the following section, we shall demonstrate that it is only with the inclusion of the induced moment matrix $[k_i]$ that the stiffness equation derived for the space frame element can pass the rigid body test. Moreover, in calculating the *element forces* in an incremental-iterative analysis, it is essential that the equation of the form in (6.37) be duly used, preferably as follows:

$$\{^2f\} = [k_e]\{u\} + ([k_g] + [k_i])\{u\} + \{^1f\} \tag{6.38}$$

where the role of the initial forces $\{^1f\}$ can be clearly observed.

In the literature, the incremental stiffness equation for the three-dimensional beam element has often been *oversimplified* as follows:

$$[k_e]\{u\} + [k_g]\{u\} = \{\Delta f\} \tag{6.39}$$

This expression has two drawbacks. One is the omission of the induced moment matrix $[k_i]$ mentioned above, which is essential for the beam element to pass the rigid body test.

The other is the incomplete inclusion of the initial forces $\{^1f\}$. It is known that the geometric stiffness matrix $[k_g]$ is a function of the initial forces $\{^1f\}$. But this is only part of the effect. When given in the form of equation (6.39), we fail to give the initial forces $\{^1f\}$ its own identity as in equation (6.38). This is particularly crucial to the recovery of element forces in an incremental-iterative analysis, in which case the initial forces play a role no less than the incremental forces, as will be elaborated in Chapter 7.

6.4 Rigid Body Test

The only assumption with the stiffness equations (6.37) or (6.38) derived in the preceding section for the space frame element is that the displacements occurring within the incremental step from the C_1 to the C_2 configuration is small. Beyond this, there is virtually no restriction on how the beam element moves from C_1 to C_2. One useful test with the stiffness equation is to see if it can really cope with the rigid body motion of the element. For a rigid displacement vector $\{u\}_r$, we can rewrite equation (6.38) as follows:

$$\{^2 f\} = [k_e]\{u\}_r + ([k_g] + [k_i])\{u\}_r + \{^1 f\} \tag{6.40}$$

Here, it should be noted that the initial nodal forces $\{^1 f\}$ acting on the beam element are in equilibrium at C_1, and that the *effects of initial forces* are considered not only through the element force vector $\{^1 f\}$, but also through the geometric stiffness matrix $[k_g]$ and the induced moment matrix $[k_i]$.

According to the rigid body rule, for an initially stressed element subjected to a rigid rotation, the initial forces acting on the element will be directed along the rotated axes after the rigid rotation at C_2, while their magnitudes remain unchanged (see Section 3.5). Consequently, the element forces $\{^2 f\}$ acting on the element after the rigid rotation can be related to the initial forces $\{^1 f\}$ with reference to the C_1 coordinates as follows:

$$\{^2 f\} = [T]\{^1 f\} \tag{6.41}$$

where the *transformation matrix* $[T]$ serves to transform the coordinate axes of the beam from the C_2 to the C_1 configuration (see equation (3.48) for the two-dimensional case):

$$[T] = \begin{bmatrix} [R] & & & \\ & [R] & & \\ & & [R] & \\ & & & [R] \end{bmatrix} \tag{6.42}$$

and $[R]$ is the *rotation matrix*.

Since the strain energy of the beam will not be affected by the rigid body motion, it is easy to verify that the following is valid for the truss and beam elements of concern in this book:

$$[k_e]\{u\}_r = \{0\} \tag{6.43}$$

Conventionally, the preceding equation has been used for testing the legitimacy of the linear stiffness matrix. By substituting equation (6.41), along with equation (6.42), into equation (6.40), it is required that the following relation be valid for rigid body motion $\{u\}_r$ of any type:

$$[k_g]\{u\}_r + [k_i]\{u\}_r = ([T] - [I])\{^1 f\} \tag{6.44}$$

where $[I]$ is a unity matrix. In the following, we shall show how this equation can be used to test the rationality of the geometric stiffness matrix $[k_g]$ and the induced moment matrix $[k_i]$ derived.

Consider, for example, a space frame element subjected to a small rigid rotation θ_r about the x-axis of the beam. For this case, we have

$$\theta_{xa} = \theta_r, \quad \theta_{xb} = \theta_r \tag{6.45}$$

In accordance, the rigid displacement vector $\{u\}_r$ is

$$\{u\}_r = \langle 0 \quad 0 \quad 0 \quad \theta_r \quad 0 \quad 0 \quad 0 \quad 0 \quad 0 \quad \theta_r \quad 0 \quad 0\rangle^T \tag{6.46}$$

and the rotation matrix $[R]$ is

$$[R] = \begin{bmatrix} 1 & 0 & 0 \\ 0 & \cos\theta_r & -\sin\theta_r \\ 0 & \sin\theta_r & \cos\theta_r \end{bmatrix} \approx \begin{bmatrix} 1 & 0 & 0 \\ 0 & 1 & -\theta_r \\ 0 & \theta_r & 1 \end{bmatrix} \tag{6.47}$$

based on the assumption that the angle of rotation θ_r is small.

Firstly, by the use of equations (6.30) and (6.46), the following can be obtained:

$$[k_g]\{u\}_r = \begin{Bmatrix} \{f_a\} \\ \{f_b\} \end{Bmatrix}_g \cdot \theta_r \tag{6.48}$$

where

$$\{f_a\}_g = \langle 0, \quad \frac{^1M_{ya}+^1M_{yb}}{L}, \quad \frac{^1M_{za}+^1M_{zb}}{L}, \quad 0, \quad 0, \quad 0\rangle^T \tag{6.49}$$

$$\{f_b\}_g = \langle 0, \quad -\frac{^1M_{ya}+^1M_{yb}}{L}, \quad -\frac{^1M_{za}+^1M_{zb}}{L}, \quad 0, \quad 0, \quad 0\rangle^T \tag{6.50}$$

Next, with the aid of equations (6.33)–(6.35), and (6.46), it is possible to show that

$$[k_i]\{u\}_r = \begin{Bmatrix} \{f_a\} \\ \{f_b\} \end{Bmatrix}_i \cdot \theta_r \tag{6.51}$$

where

$$\{f_a\}_i = \langle 0, \quad 0, \quad 0, \quad 0, \quad -{}^1M_{za}, \quad {}^1M_{ya} \rangle^T \tag{6.52}$$

$$\{f_a\}_i = \langle 0, \quad 0, \quad 0, \quad 0, \quad -{}^1M_{zb}, \quad {}^1M_{zb} \rangle^T \tag{6.53}$$

On the other hand, using equations (6.42) and (6.47), we can derive the following:

$$([T] - [I])\{{}^1f\} = \begin{Bmatrix} \{f_a\} \\ \{f_b\} \end{Bmatrix}_r \cdot \theta_r \tag{6.54}$$

where

$$\{f_a\}_r = \langle 0, \quad -{}^1F_{za}, \quad {}^1F_{ya}, \quad 0, \quad -{}^1M_{za}, \quad {}^1M_{ya} \rangle^T \tag{6.55}$$

$$\{f_b\}_r = \langle 0, \quad -{}^1F_{zb}, \quad {}^1F_{yb}, \quad 0, \quad -{}^1M_{zb}, \quad {}^1M_{yb} \rangle^T \tag{6.56}$$

For a beam in equilibrium at C_1, the shear forces acting on the two ends of the beam can be related to the bending moments as follows:

$${}^1F_{yb} = -{}^1F_{ya} = -\frac{{}^1M_{za} + {}^1M_{zb}}{L} \tag{6.57}$$

$${}^1F_{zb} = -{}^1F_{za} = \frac{{}^1M_{ya} + {}^1M_{yb}}{L} \tag{6.58}$$

With the preceding conditions and the expressions given in (6.48), (6.51), and (6.54), we can verify that equation (6.44) is exactly satisfied. Hence, we have shown that for this particular mode of rigid rotation, the stiffness equation derived for the element in equation (6.37) can pass the rigid body test. The capabilities of the derived element in dealing with other rigid rotational and translational modes can be verified using the same procedure.

Before closing this section, we like to note that if the induced moment matrix $[k_i]$ were omitted from the formulation, then the element derived will not be able to accommodate the rigid body motion. It should be emphasized that an element not qualified by the rigid body motion test may produce forces that do not form an equilibrium set at the C_2 configuration. One lesson from this section is that the rigid body motion test, which appears to be natural and simple, can be employed as an effective tool for testing the quality of sophisticated finite elements, including the space frame element.

6.5 Joint Equilibrium Conditions for Space Frames

The element stiffness equation (6.37) as derived in Section 6.3 is valid for a *discrete* element in that the conditions of equilibrium for structural joints, where members lying along different directions are connected, have not been taken into account. Once such conditions are considered, a matrix known as the *joint moment matrix* $[k_j]$ can be derived. In our previous works (Yang and Kuo, 1991a, 1992), an element consisting of the joint moment matrix $[k_j]$ has been referred to as a *connected* element, since it is ready for connection to other elements.

In this section, a general procedure for deriving the joint moment matrix $[k_j]$ of the space frame element will be presented. For the present purposes, only end B of the element will be dealt with. It is realized, however, that the procedure can be easily extended to the other end A of the element.

Consider end B of the space frame element. The induced moment matrix $[k_i]_b$ as derived in equation (6.35) for this end can be decomposed into a symmetric part $[s]_b$ and an antisymmetric part $[a]_b$ as

$$[k_i]_b = [s]_b + [a]_b \tag{6.59}$$

in which

$$[s]_b = \frac{1}{2}\begin{bmatrix} 0 & -{}^1M_{zb} & {}^1M_{yb} \\ -{}^1M_{zb} & 0 & 0 \\ {}^1M_{yb} & 0 & 0 \end{bmatrix} \tag{6.60}$$

$$[a]_b = \frac{1}{2}\begin{bmatrix} 0 & {}^1M_{zb} & -{}^1M_{yb} \\ -{}^1M_{zb} & 0 & {}^1M_{xb} \\ {}^1M_{yb} & -{}^1M_{xb} & 0 \end{bmatrix} \tag{6.61}$$

To simplify the present derivation using the tensorial parameters, the subscript "b" that relates a quantity to end B of the element will be temporarily dropped. By making use of the permutation symbol e_{ijk} (Malvern, 1969), the antisymmetric matrix $[a]$ in equation (6.61) can be expressed as follows:

$$a_{ij} = \frac{1}{2} e_{ijk} {}^1M_k \quad (i, j, k = 1, 2, 3) \tag{6.62}$$

where it is implied that ${}^1M_1 = {}^1M_x$, ${}^1M_2 = {}^1M_y$, etc.

Let $[\Phi]$ denote the *transformation matrix* for the element from the local coordinates to the global coordinates. Then,

$$[\Phi]^T[\Phi] = [I] \tag{6.63}$$

For the identity matrix $[I]$,

$$\Phi_{pi}\Phi_{pj} = \delta_{ij} \quad (i, j, p = 1, 2, 3) \tag{6.64}$$

where δ_{ij} is the Kronecker delta, which equals one for $i = j$, and zero otherwise. Based on the tensorial operations, the following relation can be shown to be valid:

$$e_{ijk}\Phi_{pi}\Phi_{qj}\Phi_{rk} = e_{pqr} \quad (i, j, k, p, q, r = 1, 2, 3) \tag{6.65}$$

The nodal moments $\{M\}$ in the element coordinates can be transformed to the structure coordinates as

$$\{\bar{M}\} = [\Phi]\{M\} \tag{6.66}$$

or

$${}^1\bar{M}_p = \Phi_{pi} {}^1M_i \quad (i, p = 1, 2, 3) \tag{6.67}$$

Similarly, the antisymmetric matrix $[a]$ can be transformed to the global coordinates as

$$[\bar{a}] = [\Phi][a][\Phi]^T \tag{6.68}$$

or

$$\bar{a}_{pq} = \Phi_{pi}a_{ij}\Phi_{qj} \quad (i, j, p, q = 1, 2, 3) \tag{6.69}$$

Substituting equation (6.62) into (6.69) yields

$$\bar{a}_{pq} = \Phi_{pi}\Phi_{qj}a_{ij} = \frac{1}{2}e_{ijk}\Phi_{pi}\Phi_{qj}{}^1M_k \quad (i,j,k,p,q,r = 1,2,3) \quad (6.70)$$

By the following relations:

$$e_{ijk}\Phi_{pi}\Phi_{qj} = e_{ijt}\Phi_{pi}\Phi_{qj}\delta_{tk} = e_{ijt}\Phi_{pi}\Phi_{qj}\Phi_{rt}\Phi_{rk} = e_{pqr}\Phi_{rk} \quad (6.71)$$

the preceding equation reduces to

$$\bar{a}_{pq} = \frac{1}{2}e_{pqr}\Phi_{rk}{}^1M_k \quad (k,p,q,r = 1,2,3) \quad (6.72)$$

In view of equation (6.67),

$$\bar{a}_{pq} = \frac{1}{2}e_{pqr}{}^1\bar{M}_r \quad (p,q,r = 1,2,3) \quad (6.73)$$

Next, consider a structural joint of the space frame where n members lying along different directions are connected. For the joint to be in equilibrium at the deformed configuration, the sum of moments exerted by all the members connected to the joint along the three global axes must be equal to zero,

$$\sum_{s=1}^{n} {}^1\bar{M}_r^{(s)} = 0 \quad (r = 1,2,3) \quad (6.74)$$

Consequently, we can derive the following:

$$\sum_{s=1}^{n} {}^1\bar{a}_{pq}^{(s)} = \sum_{s=1}^{n} \frac{1}{2}e_{pqr}{}^1\bar{M}_r^{(s)} = \frac{1}{2}e_{pqr}\sum_{s=1}^{n} {}^1\bar{M}_r^{(s)} = 0 \quad (6.75)$$

where $p,q,r = 1,2,3$. The implication here is that when the anti-symmetric matrices $[\bar{a}]$, which are referred to the global coordinates, are summed over all the elements that share a common joint, the resulting asymmetric stiffness associated with that joint is nothing but zero. Thus, by enforcing the conditions of equilibrium to be satisfied by all elements connected to the same joint, only the symmetric portion of the induced moment matrix, i.e., the $[s]_b$ matrix in (6.60), associated with that joint will be retained after the element assembly. Such a symmetric portion of the matrix in (6.60) has been referred to as the *joint moment matrix* $\lceil k_j \rceil$ (Yang and Kuo, 1991a, 1992).

The above derivation has been made for end B. The same procedure can be repeated for end A to derive another joint moment matrix $[s]_a$ for end A from the conditions of equilibrium for all members meeting at that joint. In combination, the joint moment matrix $\lceil k_j \rceil$ for an element that is *connected* to other elements at both ends is

$$\lceil k_j \rceil \equiv [s] = \begin{bmatrix} [0] & & & \\ & [s]_a & & \\ & & [0] & \\ & & & [s]_b \end{bmatrix} \tag{6.76}$$

where [0] denotes a zero matrix and the submatrix $[s]_a$ can be given similarly as

$$[s]_a = \begin{bmatrix} 0 & -\frac{{}^1M_{za}}{2} & \frac{{}^1M_{ya}}{2} \\ -\frac{{}^1M_{za}}{2} & 0 & 0 \\ \frac{{}^1M_{ya}}{2} & 0 & 0 \end{bmatrix} \tag{6.77}$$

In contrast with the original stiffness equation (6.37) for *discrete elements*, i.e., for elements that are not yet connected to adjacent elements, an element containing such a joint moment matrix $[s] \equiv [k_j]$ is said to be ready for connection to the other elements to form a structure, as it satisfies all conditions of equilibrium for the two ends of the element in the deformed configuration. Namely, the stiffness equation for the *connected elements* (i.e., for elements that are ready to be connected) is

$$[k_e]\{u\} + ([k_g] + [k_j])\{u\} = \{{}^2f\} - \{{}^1f\} \tag{6.78}$$

where the original induced moment matrix $[k_i]$ has been replaced by the joint moment matrix $[k_j] \equiv [s]$ for its compatibility with neighboring elements. Unlike the equation in (6.37) that is strictly valid for a single element, this equation is the one that should be used for the structure, since the joint compatibility has been taken into account.

In the literature, the symmetric induced moment matrix $[s]$ has also been referred to as the *correction matrix*, generated by enforcing the nodal bending moments to behave in a semitangential way, rather than in their original quasitangential way, for the sake of joint equilibrium in the deformed configuration (Argyris *et al.*, 1979; Yang and McGuire, 1986).

6.6 Externally Applied Moments

Moments that are not generated as cross-sectional resultants of a beam but by external physical devices will be discussed in this section. Consider an external moment generated by a couple of direct force ${}^1P$ with a lever arm of length h at node B (free end or unconstrained node) of the beam in the C_1 configuration, as shown in Figure 6.2. Let (α, β, γ) denote a set of arbitrarily oriented, orthogonal coordinates at section B, and $({}^1\vec{e}_\alpha, {}^1\vec{e}_\beta, {}^1\vec{e}_\gamma)$ the corresponding base vectors of the system. The lever arm (lying along the β-axis) and direct forces (lying along the α-axis) constituting the couple in the C_1 configuration can be represented by vectors such as

$$ {}^1\vec{r}_{CD} = h\,{}^1\vec{e}_\beta \tag{6.79} $$

$$ \vec{F}_D = -{}^1P\,{}^1\vec{e}_\alpha \tag{6.80} $$

where ${}^1P$ denotes the magnitude of the direct forces. Consequently, the external moment vector ${}^1\vec{M}$ at the C_1 configuration is

$$ {}^1\vec{M} = {}^1\vec{r}_{CD} \times \vec{F}_D = {}^1M_0\,{}^1\vec{e}_\gamma \tag{6.81} $$

in which ${}^1M_0 = {}^1P \cdot h$ is the moment couple.

When the beam deforms during the incremental step from C_1 to C_2, the end section B will undergo rotations $(\theta_\alpha, \theta_\beta, \theta_\gamma)$ about the

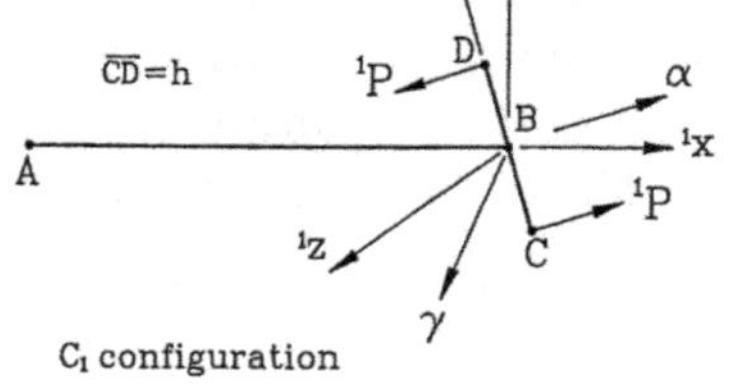

Fig. 6.2. Externally applied moment.

local axes (α, β, γ), which can be expressed as follows:

$$\vec{\theta} = \theta_\alpha \vec{e}_\alpha + \theta_\beta \vec{e}_\beta + \theta_\gamma \vec{e}_\gamma \tag{6.82}$$

As a result, the lever arm of the couple at the C_2 configuration becomes

$${}^2\vec{r}_{CD} = {}^1\vec{r}_{CD} + \vec{\theta} \times {}^1\vec{r}_{CD} \tag{6.83}$$

or

$${}^2\vec{r}_{CD} = -h\theta_\gamma {}^1\vec{e}_\alpha + h\, {}^1\vec{e}_\beta + h\theta_\alpha {}^1\vec{e}_\gamma \tag{6.84}$$

Assuming that a couple of forces are fixed in their directions upon rotations (as a kind of conservative forces), the external moment ${}^2\vec{M}$ at the C_2 configuration can be written as follows:

$${}^2\vec{M} = {}^2\vec{r}_{CD} \times \vec{F}_D \tag{6.85}$$

or

$${}^2\vec{M} = -{}^1M_0\theta_\alpha {}^1\vec{e}_\beta + {}^1M_0 {}^1\vec{e}_\gamma \tag{6.86}$$

In view of equations (6.81) and (6.82), the virtual potential of the external moment ${}^1\vec{M}$ at C_1 is

$$-\delta\vec{\theta} \cdot {}^1\vec{M} = -{}^1M_0\delta\theta_\gamma \tag{6.87}$$

Similarly, by the use of equations (6.86) and (6.82), the virtual potential of the moment ${}^2\vec{M}$ at C_2 is

$$-\delta\vec{\theta} \cdot {}^2\vec{M} = -{}^1M_0\delta\theta_\gamma + {}^1M_0\theta_\alpha\delta\theta_\beta \tag{6.88}$$

Subtracting equation (6.87) from equation (6.88) yields the virtual potential increment generated by the initial moment ${}^1M_0$ upon rotations:

$${}^1M_0\theta_\alpha\delta\theta_\beta = \langle \delta\theta_\alpha \quad \delta\theta_\beta \quad \delta\theta_\gamma \rangle \begin{bmatrix} 0 & 0 & 0 \\ {}^1M_0 & 0 & 0 \\ 0 & 0 & 0 \end{bmatrix} \begin{Bmatrix} \theta_\alpha \\ \theta_\beta \\ \theta_\gamma \end{Bmatrix} \tag{6.89}$$

Here, the square matrix on the right-hand side has been referred to as the *applied moment matrix* $[k_m]$ (Yang and Kuo, 1994), which plays the same role as that of the induced moment matrix $[k_i]_b$ in equation (6.33). Again, it is possible to decompose the applied moment matrix $[k_m]$ into a symmetric and an antisymmetric part,

$$\begin{bmatrix} 0 & 0 & 0 \\ {}^1M_0 & 0 & 0 \\ 0 & 0 & 0 \end{bmatrix} = \frac{1}{2}\begin{bmatrix} 0 & {}^1M_0 & 0 \\ {}^1M_0 & 0 & 0 \\ 0 & 0 & 0 \end{bmatrix} + \frac{1}{2}\begin{bmatrix} 0 & -{}^1M_0 & 0 \\ {}^1M_0 & 0 & 0 \\ 0 & 0 & 0 \end{bmatrix} \tag{6.90}$$

The antisymmetric matrix on the right-hand side of equation (6.90) has the same transformation characteristic as the asymmetric part of the joint moment matrix $[k_j]$ in equation (6.61). Following the same lines of derivation as those of the preceding section, we can prove that this antisymmetric matrix will cancel out the corresponding terms of elements connected to the same joint upon which the applied moment is acting. As a result, only the symmetric portion of the applied moment matrix on the right-hand side of equation (6.90) needs to be retained in the element assembly process for the buckling analysis.

For simplicity in the following discussions, we choose to mention only the *joint moment matrix*, but not joint and symmetric part of applied moment matrices, since it is understood that the symmetric part of the applied moment matrix should be included as well for the case with externally applied moments.

6.7 Buckling Analysis

Corresponding to the stiffness equation in (6.78) for the connected elements, the stiffness equation can be written for a framed structure (with connected elements) during the incremental step from C_1 or C_2 as

$$[K_e]\{U\} + ([K_g] + [K_j])\{U\} = \{^2P\} - \{^1P\} \tag{6.91}$$

where all the variables in capital are those for the structure assembled from the element variables in lower case. Namely, $[K_e]$ is the linear stiffness matrix, $[K_g]$ the geometric stiffness matrix, $[K_j]$ the joint moment matrix, $\{U\}$ the structural displacements, and $\{^1P\}$ and $\{^2P\}$ are the external loads acting on the structure at C_1 and C_2, respectively.

In case there is a free end or an unconstrained end subjected to an applied moment, the applied moment matrix $[k_m]$ in equation (6.90) should be included in the assembly of the joint moment matrix $[K_j]$ in equation (6.91) as well, in a way the same as the assembly of the joint moment matrix $[k_j]$ in equation (6.76) for all the connected elements. We shall not mention this point again in the remaining parts of the book.

As was stated in Sections 5.2.1 and 5.2.2, the behavior of a structure that is susceptible to *buckling* may be divided into two stages as the prebuckling stage and buckling stage. In the prebuckling stage, the structure is characterized by the application of the external loadings from zero at the C_0 configuration to a load $\{^1P\}$ smaller than the critical value at the C_1 configuration. Since the deformations of the structure at this stage are generally small, the effects of change in the geometry of the structure can be neglected. It becomes immaterial whether the configuration C_0 or C_1 is selected as the reference, as is typical in a linear analysis.

In the buckling stage, the structure deforms from C_1 or C_2 in a direction different from the loading direction by an amount that is generally not small, but with basically no change on the magnitude of the applied loads, i.e., $\{^2P\} = \{^1P\}$. In other words, for the buckling analysis, the eigenvalue equation can be written as follows:

$$[K_e]\{U\} + [F(\lambda\{P\}_{\text{ref}})]\{U\} = \{0\} \tag{6.92}$$

where the stiffness matrix $[F(\lambda\{P\}_{\text{ref}})] = ([K_g] + [K_j])$ is a function of the *reference loads* $\{P\}_{\text{ref}}$, defined to be proportional to the initial loads $\{^1P\}$, and λ is a load parameter. Using proper eigenvalue solution methods, the *critical load factor* λ_{cr} that results in the vanishing of the determinant $|[K_e]+[K_g]+[K_j]|$ can be determined. With this, the *buckling load* is simply equal to $\lambda_{cr}\{P\}_{\text{ref}}$, and the corresponding eigenvector $\{U\}$ is the *buckling mode shape*.

Before proceeding to the buckling analysis, some remarks should be given here. In the study by Argyris *et al.* (1979), the joint matrix $[k_j]$ (for elements) or $[K_j]$ (for structure) was arrived at by enforcing the nodal bending moments to behave in a semitangential manner, same as the torsional moments, and they called this matrix the *correction matrix*. Aside from this, most previous researchers fail to recognize the semi- and quasi-tangential properties of nodal bending moments and torques, or the existence of the joint matrix $[k_j]$ and applied moment matrix $[k_m]$ (for elements) or $[K_j]$ (for structure). As this is the case, they are essentially using a buckling equation of the form:

$$[K_e]\{U\} + [K_g]\{U\} = \{^2P\} - \{^1P\} \tag{6.93}$$

Such an expression is incorrect and will be referred to as the *conventional approach* in the numerical studies.

6.8 Numerical Examples

In this section, the effects of both the joint moment matrix $[k_j]$ (for connected joints) and applied moment matrix $[k_m]$ (for free or unrestrained ends) that have been derived as a result of enforcement of the equilibrium conditions for structural joints at the C_2 configuration will be numerically evaluated in the buckling analysis of Examples (6.8.1)–(6.8.7). As was stated, the approach that has no knowledge of joint equilibrium conditions and rotational properties of nodal moments, as implied by equation (6.93), will be referred to as the *conventional approach*. It is noteworthy that the joint moment matrix $[k_j]$ was referred to as the correction matrix and included in the buckling analysis by Argyris *et al.* (1997).

In the first five examples, each member of the frame will be represented by four elements with the following properties: $E = 71{,}240\,\text{N/mm}^2$, $G = 27{,}190\,\text{N/mm}^2$, $J = 2.16\,\text{mm}^4$, $I_z = 0.54\,\text{mm}^4$, and $L = 240\,\text{mm}$. The critical load solutions will be obtained for each of the following examples by the present approach using the eigenvalue equation in (6.92) and by the conventional approach in (6.93).

6.8.1 *Symmetric frame 1-Simply supported out-of-plane*

The symmetric angled frame shown in Figure 6.3 is simply supported both in and out of its own plane, and is under the action

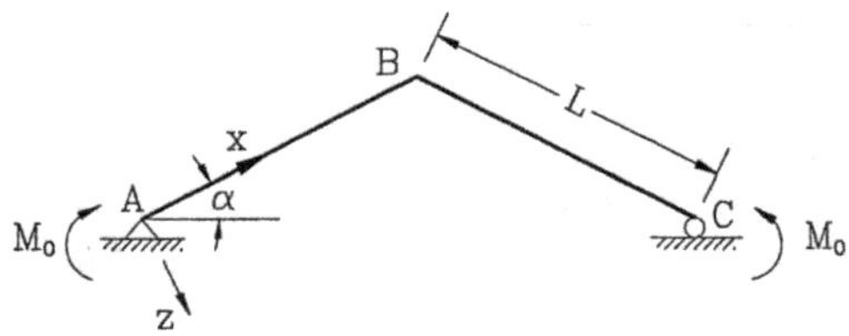

Fig. 6.3. Symmetric frame under uniform bending.

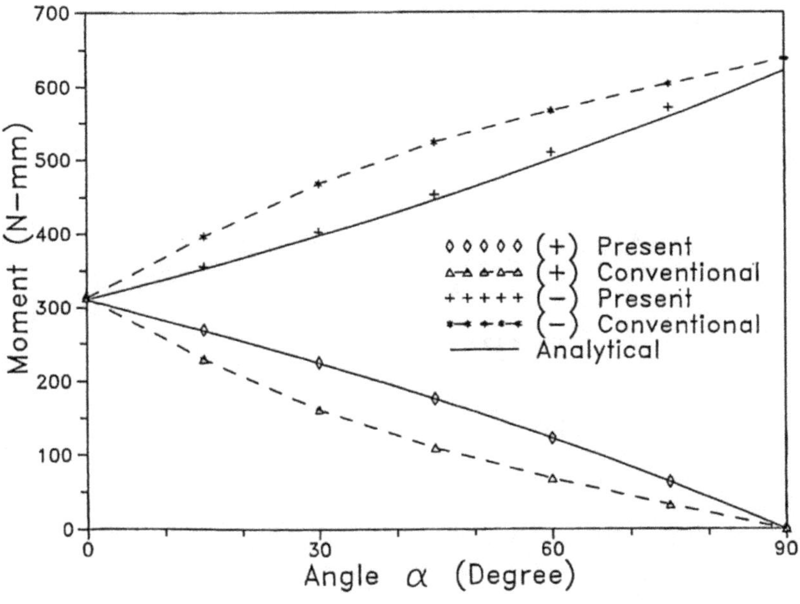

Fig. 6.4. Critical moments for frame 1-symmetric mode.

of uniform bending moment M_0. The present solutions, considering the $[k_j]$ matrix, have been compared with those neglecting the $[k_j]$ matrix (conventional approach) with respect to the inclined angle α in Figure 6.4. Both positive (concave upward) and negative (concave downward) bending cases have been studied. As can be seen from the figure, the conventional approach tends to underestimate the positive critical moment, while overestimating the negative critical moment. For this particular frame, no distinction can be made among the QT-1, QT-2, and ST moments, because of the restraint effect exerted by the two supports. It should be noted that the present

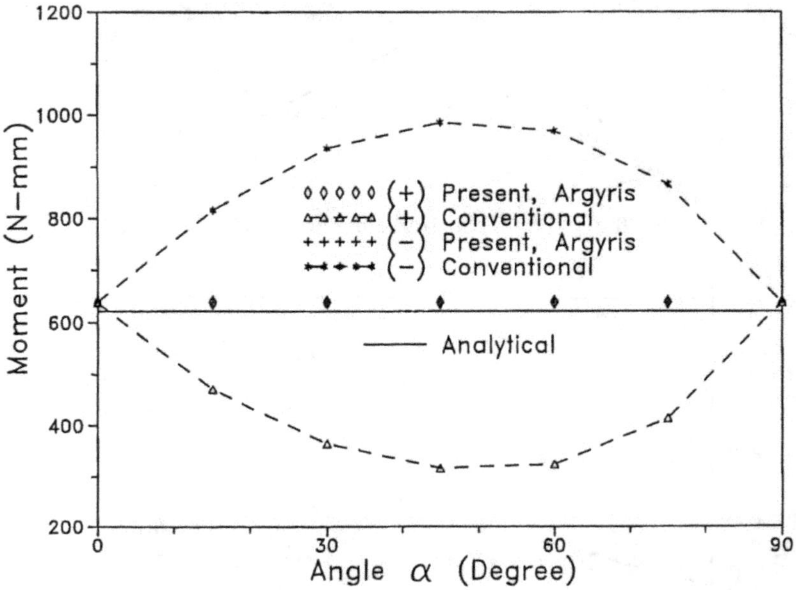

Fig. 6.5. Critical moments for frame 2-symmetric mode.

solutions are in excellent with the analytical solutions (see Chapter 9 or Yang and Kuo, 1991b).

6.8.2 *Symmetric frame 2-Fixed out-of-plane*

The frame is identical to the one in Figure 6.3, except that the out-of-plane rotations are restrained at the two supports. By comparing the present solutions with the conventional ones in Figure 6.5, the effect of joint moment matrix $[k_j]$ on the critical loads of the frame has been demonstrated to be extremely large. One observation from the figure is that the beam has equal resistance against lateral or out-of-plane buckling under the action of either positive or negative moments. Again, it should be added that the rotational effect of the applied moment M_0 has been suppressed by the supports, and that agreement has been made either with previous finite element solutions (Argyris *et al.*, 1979) or analytical solutions (see Chapter 9 or Yang and Kuo, 1991b).

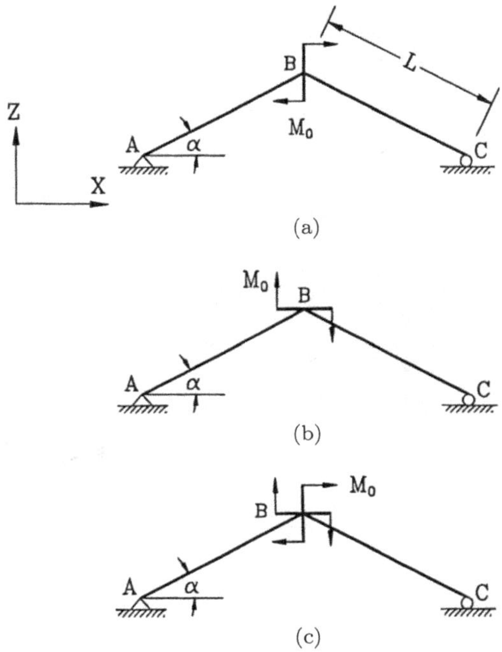

Fig. 6.6. Angled frame with tip moment: (a) QT-1; (b) QT-2; and (c) ST.

6.8.3 *Symmetric frame 1 with tip moment*

This example differs from the first one in that the moment M_0 is applied at the tip of the frame, as shown in Figure 6.6. Since the tip is free to rotate about the three directions, this example allows us to investigate the effect of applied moments through the applied moment matrix $[k_m]$. As can be seen from the results shown in Figure 6.7, a significant difference exists among the critical values for buckling of the frame under the action of QT-1, QT-2, and ST moments (see Table 5.1 for definitions of various moments), which generally cannot be neglected in practice. Of interest to note that the solution obtained by the conventional approach does not coincide with any of the three solutions obtained for the QT-1, QT-2, and ST moments. It varies from the high side of the ST and QT-1 solutions to the low side of the QT-2 solution as the angle of inclination α increases.

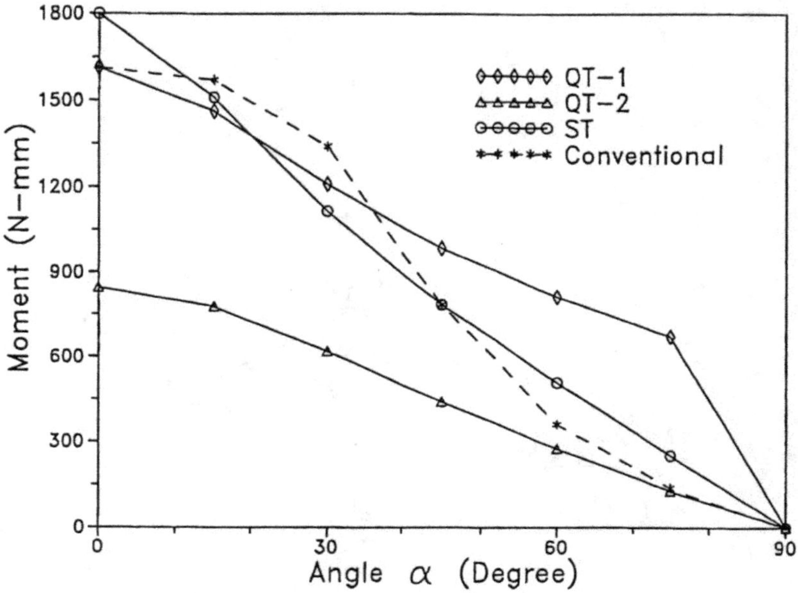

Fig. 6.7. Critical moments for angled frame with tip moment.

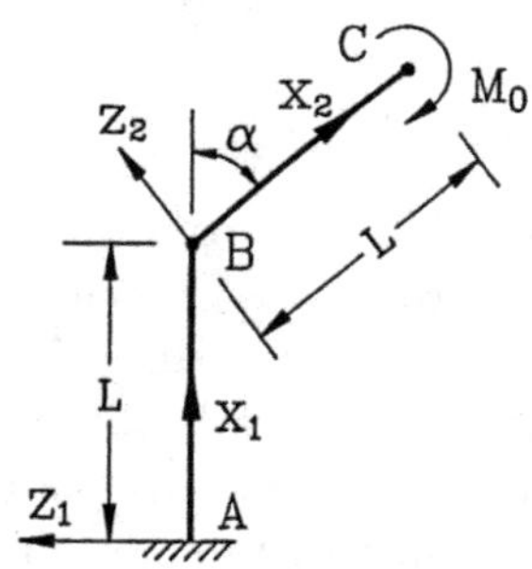

Fig. 6.8. Fixed angled frame with moment load.

6.8.4 *Fixed angled frame with moment load*

Figure 6.8 shows a frame fixed at its base, composed of two members of equal length, and subjected to a bending moment M_0 at the free end. The solutions obtained for the QT-1, QT-2, and ST moments using the present approach have been plotted in Figure 6.9, which are in good agreement with the analytical solutions presented

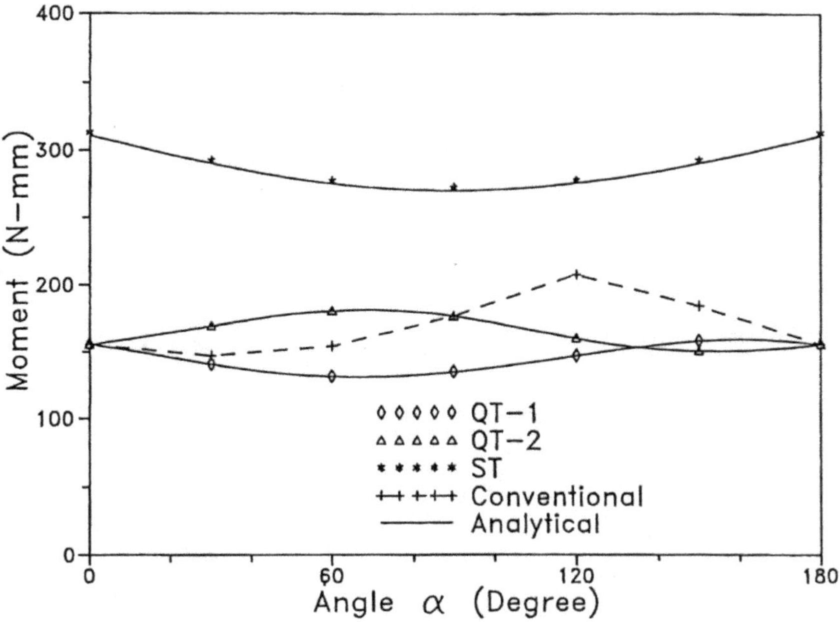

Fig. 6.9. Critical loads for fixed angled frame with moment load.

in Yang and Kuo (1991b) or Chapter 9. Of particular interest in Figure 6.9 is that for the applied moment of the ST type, the conventional approach underestimates dramatically the critical load, while for applied moments of the QT-1 and QT-2 types, it produces solutions significantly deviating from the present ones.

6.8.5 *Fixed angled frame with moment load*

As shown in Figure 6.10, the angled frame is subjected to an in-plane shear force F_z at the tip. The results obtained by the present approach have been plotted in Figure 6.11, along with those by the conventional approach, for the positive and negative shear loads F_z. Again, the difference between the present solutions and the conventional ones is so large that it cannot be neglected in practice. No analytical solutions are available for this example. For the special case of $\alpha = 90°$, the present solutions appear to coincide with those given by Argyris *et al.* (1979).

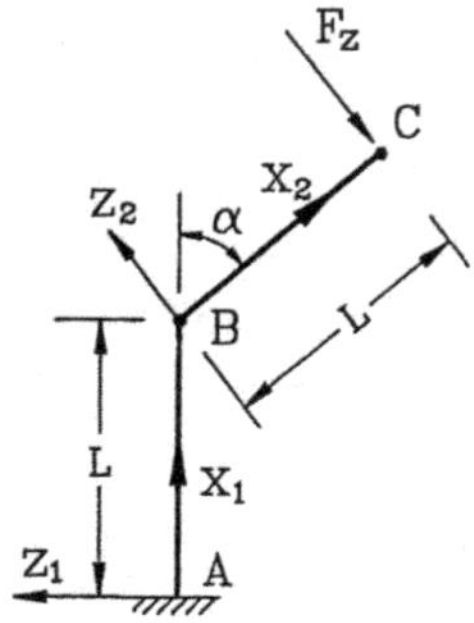

Fig. 6.10. Fixed angled frame subjected to in-plane shear.

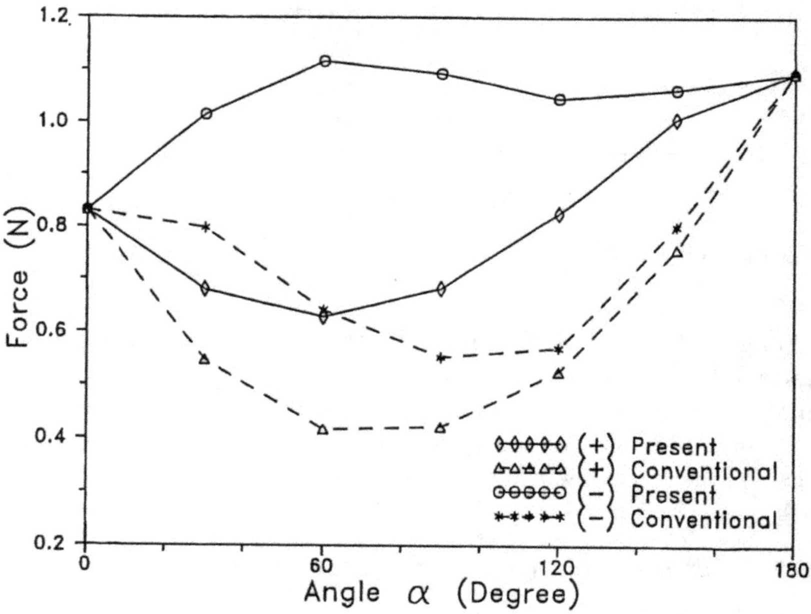

Fig. 6.11. Critical loads for fixed angled frame with in-plane shear.

6.8.6 *Fixed angled frame with lateral shear load*

The angled frame in Figure 6.12 is subjected to a lateral shear load F_y at the free end C. In analysis, each member of the frame is represented by four elements with the following properties: $E = 29{,}000\,\text{ksi}$, $G = 11{,}200\,\text{ksi}$, $A = 3.14\,\text{in.}^2$, $J = 1.57\,\text{in.}^4$, $I_y = I_z = 0.785\,\text{in.}^4$, and

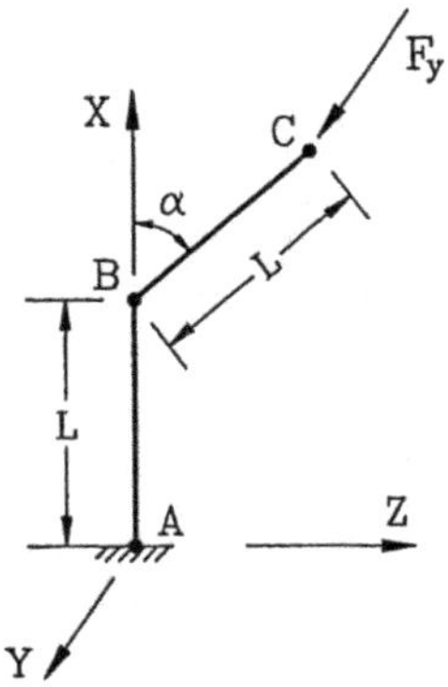

Fig. 6.12. Fixed angled frame subjected lateral shear.

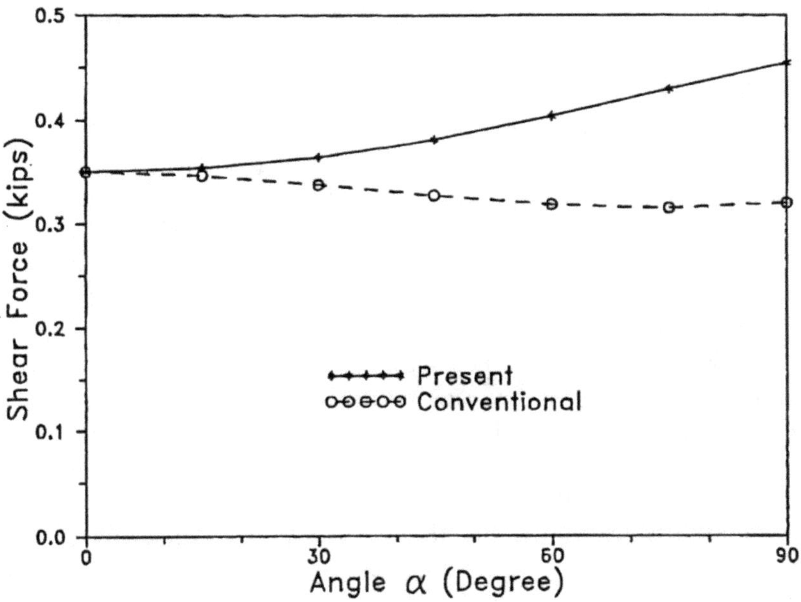

Fig. 6.13. Critical values for fixed angled frame with lateral shear.

$L = 240\,\text{in}$. From the results plotted in Figure 6.13, we observe that for the special case with $\theta = 90°$, the critical load predicted by the conventional approach appears to be 30% lower than that of the present approach.

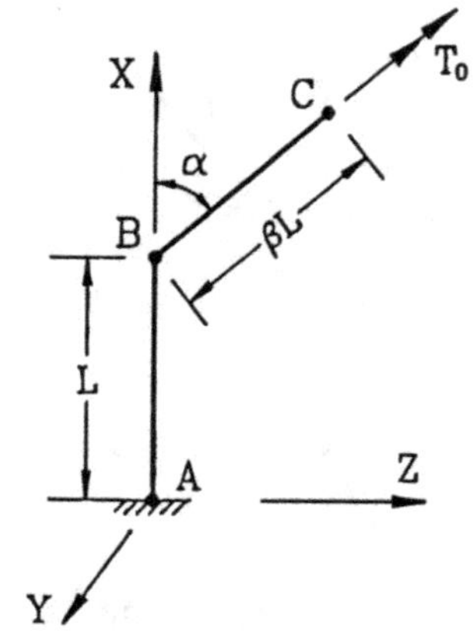

Fig. 6.14. Fixed angled frame subjected to torsional load.

6.8.7 *Fixed angled frame with torsional load*

Consider the two-member fixed angled frame subjected to a torque T_0 at the free end in Figure 6.14. This problem was analytically studied by Yang and Kuo (1991c), which will also be presented in Chapter 9. The torque T_0 can be generated by either of the three mechanisms shown in Figures 6.15(a)–6.15(c). The elastic properties adopted for the two members of the frame are $E = 71{,}240\,\text{N/mm}^2$ and $G = 27{,}190\,\text{N/mm}^2$, and each member has a length of $L = 240\,\text{mm}$.

Three types of cross-sections are considered for the two members, as shown in Figure 6.16. For case 1 with $I_y \gg I_z$, the following properties are adopted for members 1 and 2: $I_y = 50\,\text{mm}^4$, $I_z = 0.18\,\text{mm}^4$, and $J = 0.72\,\text{mm}^4$. For case 2 with $I_y \ll I_z$, the values for I_y and I_z are reversed. And for case 3 with $I_y = I_z$, it is assumed that $I_y = I_z = 12.56\,\text{mm}^4$ and $J = 25.12\,\text{mm}^4$.

The critical loads obtained by the present approach for the case with $\beta = 0$ (i.e., without member 2) have been plotted in Figures 6.17(a)–6.17(c). As can be seen, the present finite element solutions and the previous analytical solutions are in good agreement, while the solutions obtained by the conventional approach that excludes the joint moment matrix $[k_j]$ and applied moment matrix $[k_m]$ are in general not acceptable. Noteworthy is that the buckling

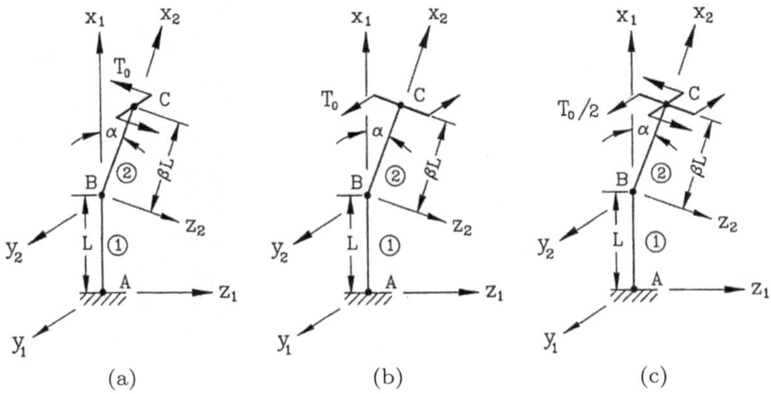

Fig. 6.15. Two-member angled frames subjected to end torques: (a) QT-1; (b) QT-2; and (c) ST.

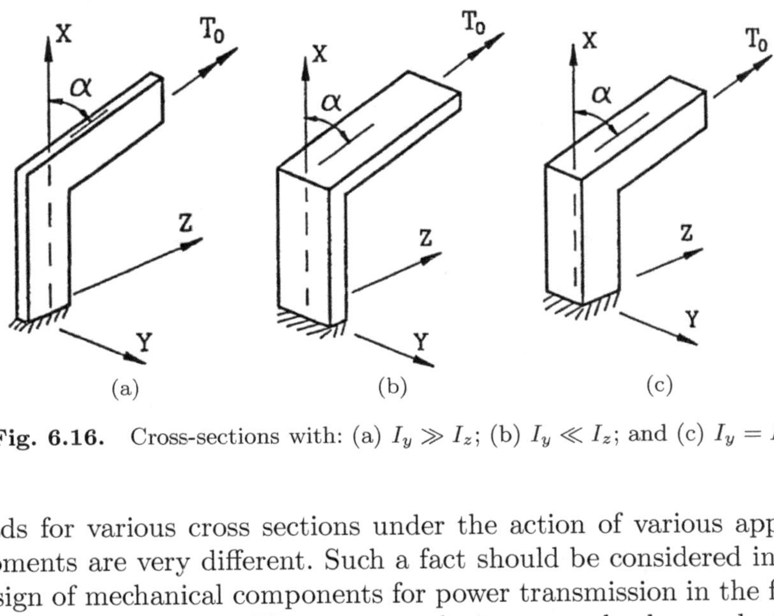

Fig. 6.16. Cross-sections with: (a) $I_y \gg I_z$; (b) $I_y \ll I_z$; and (c) $I_y = I_z$.

loads for various cross sections under the action of various applied moments are very different. Such a fact should be considered in the design of mechanical components for power transmission in the form of torsional moments. The same conclusions can also be reached for the case with $\beta = 1$ (i.e., with identical length for members 1 and 2) in Figures (6.18)(a)–(6.18)(c).

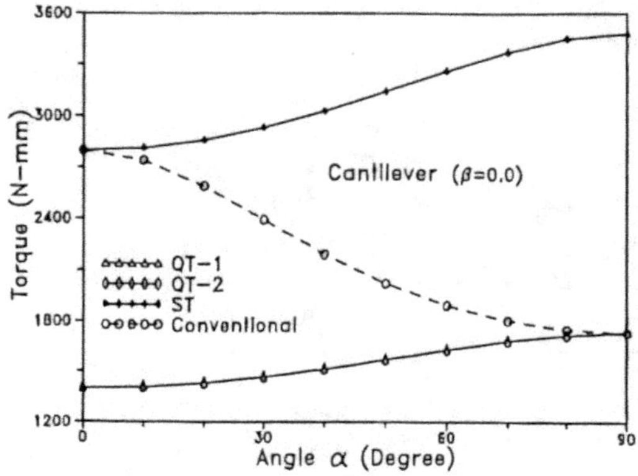

(a)

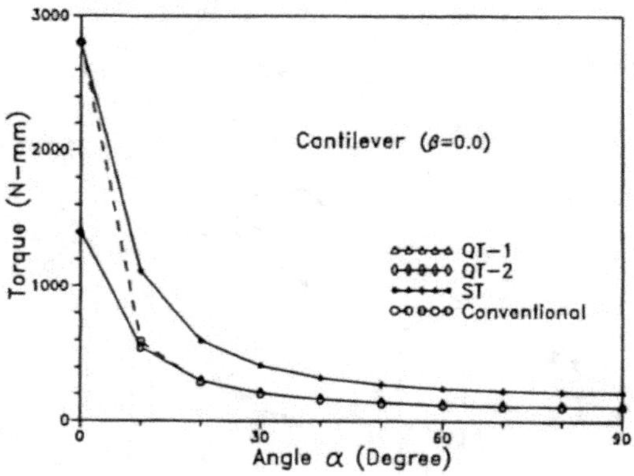

(b)

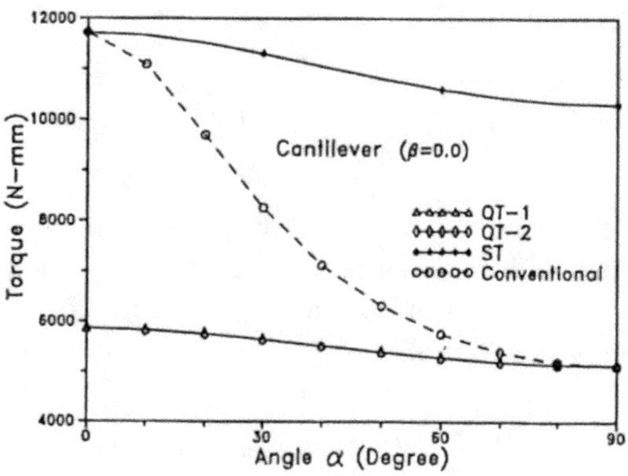

(c)

Fig. 6.17. Critical torques with $\beta = 0$: (a) $I_y \gg I_z$; (b) $I_y \ll I_z$; and (c) $I_y = I_z$.

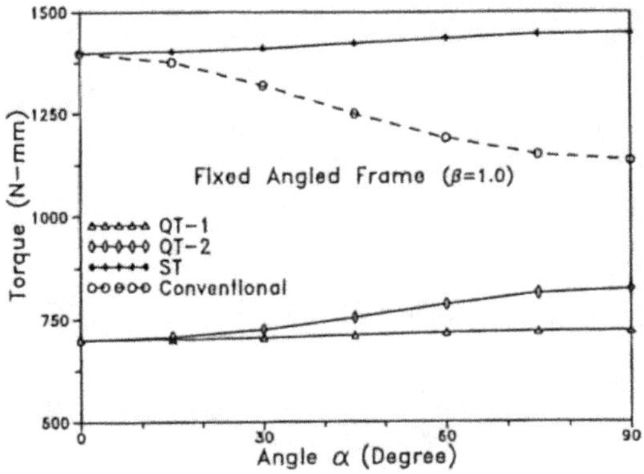

(a)

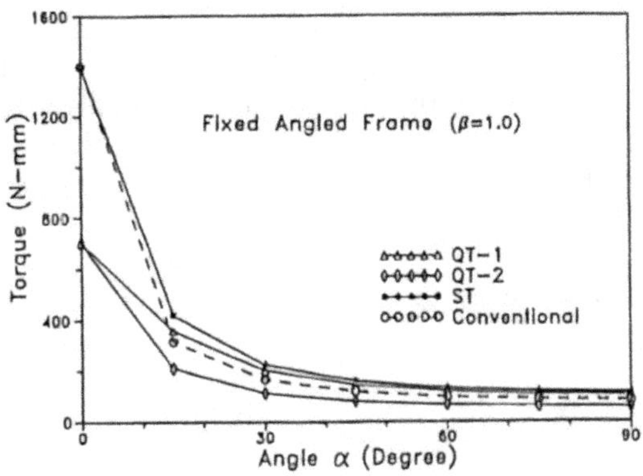

(b)

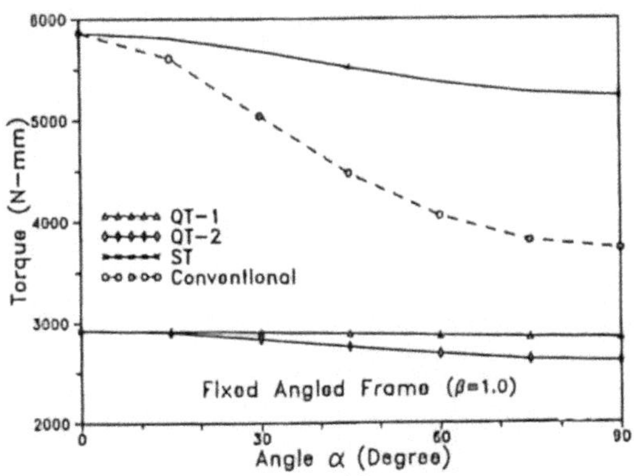

(c)

Fig. 6.18. Critical torques with $\beta = 1$: (a) $I_y \gg I_z$; (b) $I_y \ll I_z$; and (c) $I_y = I_z$.

References

Argyris, J.H., Hilbert, O., Malejannakis, G.A., and Scharpf, D.W. (1979). On the geometric stiffness of a beam in space-a consistent V.W. approach, *Comp. Meth. Appl. Mech. Eng.*, 20, 105–131.

Malvern, L.E. (1969). *Introduction to the Mechanics of a Continuous Medium*, Prentice-Hall, Englewood Cliffs, N.J.

Reissner, E. (1973). One-dimensional large-displacement finite-strain beam theory, *Studies Appl. Math.*, LII(2), The Massachusetts Institute of Technology.

Yang, Y.B. and Kuo, S.R. (1991a). Consistent frame buckling analysis by finite element method, *J. Struct. Eng.*, ASCE, 117(4), 1053–1069.

Yang, Y.B. and Kuo, S.R. (1991b). Out-of-plane buckling of angled frames, *Int. J. Mech. Sci.*, 33(1), 55–67.

Yang, Y.B. and Kuo, S.R. (1991c). Buckling of frames under various torsional loadings, *J. Eng. Mech.*, ASCE, 117(8), 1681–1697.

Yang, Y.B. and Kuo, S.R. (1992). Frame buckling analysis with full consideration of joint compatibilities, *J. Struct. Eng.*, ASCE, 118(5), 871–889.

Yang, Y.B., and Kuo, S.R. (1994). *Theory and Analysis of Nonlinear Framed Structures*, Prentice Hall, Englewood Cliffs, New Jersey.

Yang, Y.B. and McGuire, W. (1986). Joint rotation and geometric nonlinear analysis, *J. Struct. Eng.*, ASCE, 112(4), 879–905.

Chapter 7

Incremental-Iterative Nonlinear Analysis

7.1 Introduction

In an incremental-iterative analysis, we need to include the feature of iterations at each incremental step. To this end, we shall use the *right superscript* of a variable to indicate an *incremental step* and the *right subscript* an *iterative step*. Both the right superscript and right subscript may be placed either outside or inside the symbols [ ] (for matrices) and { } (for vectors) merely to avoid confusion with other indexes. In addition, the symbol Δ will be used to indicate an incremental quantity for clarity. This notation is different from those used in the previous chapters in the formulation of the stiffness matrices for various elements.

Conceptually, three configurations will be used to describe the incremental motion of a nonlinear structure, i.e., the initial undeformed configuration C^0, the last known configuration C^{i-1}, and the current unknown configuration C^i. For the structure at the ith incremental step, it is assumed that all the information of the structure from C^0 to C^{i-1} is known, including the history of loadings, the structural deformations, and element forces, and that we are interested in the behavior of the structure during the ith incremental step from C^{i-1} to C^i, in response to the small increase in external loadings. Although the deformations of the structure during each incremental step, as typified by the process from C^{i-1} to C^i, are small in magnitude, the total deformations accumulated from all the incremental steps prior to C^{i-1} can be arbitrarily large. This is certainly one advantage with the incremental nonlinear analysis. Namely, by dividing a nonlinear problem involving large deformations into a number of incremental steps, we need only deal with the sub-problem of small deformations at each incremental step, previously known as the *incrementally small-deformation theory* (Yang *et al.*, 2002). Based on the updated Lagrangian formulation, the last calculated configuration C^{i-1} is selected as the reference for describing the motion of the structure during the incremental step from C^{i-1} to C^i.

In the previous chapters, we have devoted most of the effort to derive the incremental element equations for the truss and beam elements, either two-dimensional or three-dimensional, to characterize the behavior of the element from C^{i-1} to C^i. Using the present notation, the stiffness equation derived for the *discrete element* is the one given in equation (6.37), which can be recast in a general form as

$$([k_e] + [k_g] + [k_i])^{i-1}\{\Delta u\} = \{f^i\} - \{f^{i-1}\} \tag{7.1}$$

where the elastic stiffness matrix $[k_e]$, geometric stiffness matrix $[k_g]$ and induced moment matrix $[k_i]$ are those available for the element at C^{i-1}, $\{\Delta u\}$ are the incremental displacements from C^{i-1} to C^i, $\{f^i\}$ the nodal loads acting at C^i and $\{f^{i-1}\}$ the nodal loads acting at C^{i-1}.

The *induced moment matrix* $[k_i]$ is asymmetric and needed only for the *discrete* space frame element prior to its connection with the other elements. It will reduce to the symmetric *joint moment matrix* $[k_j]$ when *connected* to other elements. For the case of applied moments at free or unrestrained ends of a space frame, another matrix that is the symmetric part of the *applied moment matrix* $[k_m]$ should be included as part of the joint moment matrix. See Sections 6.5 and 6.6 for the details. For simplicity in further discussions concerning the structure equations (with joint compatibility considered), we choose to mention only the joint moment matrix $[k_j]$, but not joint and (symmetric part of) applied moment matrices, since it is understood that the effect of the applied moment matrix should be included as well for the case with externally applied moments.

To reflect the joint compatibility of *connected* elements for the structure, we shall modify equation (7.1) as follows:

$$([k_e] + [k_g] + [k_j])^{i-1}\{\Delta u\} = \{f^i\} - \{f^{i-1}\} \tag{7.2}$$

which can be assembled over all elements of a structure to form the incremental equations that characterize the behavior of the structure from C^{i-1} to C^i:

$$([K_e] + [K_g] + [K_j])^{i-1}\{\Delta U\} = \{P^i\} - \{P^{i-1}\} \tag{7.3}$$

where for the structure, $[K] = [K_e] + [K_g] + [K_j]$, the elastic stiffness matrix $[K_e]$, geometric stiffness matrix $[K_g]$, and (symmetric) joint moment matrix $[K_j]$ are made available only for the preceding configuration C^{i-1}, $\{\Delta U\}$ are the incremental displacements from C^{i-1} to C^i, $\{P^i\}$ the applied nodal loads acting at C^i, and $\{P^{i-1}\}$ the applied nodal loads acting at C^{i-1}. To simplify the present discussion, only *proportional loadings* are considered herein, implying that the applied loads $\{P^i\}$ at C^i are proportional to the applied loads $\{P^{i-1}\}$ at C^{i-1}.

The equations as given in (7.3) are nonlinear, since the stiffness matrices $[K_g]+[K_j]$ are functions of the initial applied loads $\{P^{i-1}\}$ of the structure, or the initial internal forces $\{f^{i-1}\}$ of each element of the structure. Such a problem cannot be directly solved but needs to be solved by methods that are of incremental-iterative nature, i.e., by incorporating iterations at each incremental step of the analysis. In most cases, the nonlinear solution of a structure so obtained is presented in the form of load-deflection curves for the degree(s) of freedom of interest. In fact, an incremental-iterative analysis is a combined and repetitive use of the *structural equation* (7.3) and *element equation* (7.2) presented above in the iterative loop in searching for the solutions, as will be discussed in detail in the sections to follow.

7.2 Mechanism of Iterations for Each Incremental Step

Central to the incremental-iterative analysis of nonlinear analysis is the adoption of the *rigid body rule* in both the predictor and corrector. Because of this, the procedure of iterations has been referred to as the *physics-based computational method.* The mechanism of iterations for each incremental step involving the predictor and corrector

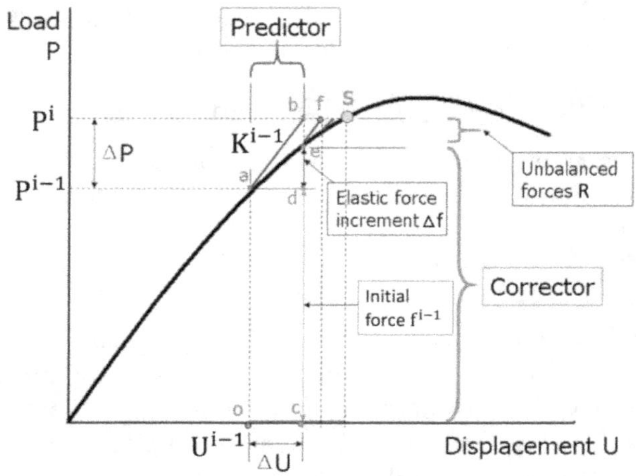

Fig. 7.1. Mechanism of iterations performed for each incremental step.

phases has been conceptually presented in Figure 7.1, in which the symbols such as { } for vectors and [ ] for matrices were dropped for clarity. This figure is not new to most researchers, but it contains messages beyond the common understanding. For convenience in interpretation, let us use the simplified notation in Figure 7.1 and follow the procedural steps for the structure to deform from C^{i-1} to C^i in response to a load increment from P^{i-1} to P^i:

1. Structure at the initial configuration C^{i-1}: Let us start with point a for the C^{i-1} configuration of the structure as typified by the displacement U^{i-1} and load P^{i-1}.
2. Incremental step in loading: Consider the incremental step with the load P^{i-1} increased by a small amount of ΔP to P^i.
3. *Predictor phase*: Use the tangent stiffness K^{i-1} to solve for the structural displacement increment ΔU based on equation (7.3). Now, we can obtain the trial (approximate) solution at point b. Clearly, the tangent stiffness K^{i-1} affects only the speed of convergence from point b to f and to s. It can never be exact, but only approximate. As a minimum requirement, the tangent stiffness K^{i-1} should be accurate to the level that the direction of iterations was not misguided (Kuo and Yang, 1995). This can be easily met by duly considering rigid-rotation characteristics in the

geometric stiffness matrices, or by using geometric stiffness matrices that can pass the rigid boy test.

4. *Corrector phase*: This phase is concerned with the calculation of the element forces as indicated by point e in Figure 7.1. Three sub-steps are involved, as explained in the following:

   (i) From points a to d: The refers to *updating of the geometry* of each element at point d. This can be done by updating the nodal coordinates using the solved structural displacements ΔU, and by computing the element displacements Δu using the connectivity between the structure and the constituting elements. More details on this will be given in Section 7.3.1 for trusses, plane frames, and space frames (with small rotations) and in Appendix B for space frames (with large rotations).

   (ii) *Updating of the initial element forces* at point d. By the rigid body rule, the initial element forces acting at point d can be simply taken as the initial forces f^{i-1} acting at point a (corresponding to P^{i-1}), which is also meant by the figure.

   (iii) The *force increment* Δf (segment de) can be calculated by multiplying the displacement increment Δu (element level) by the tangent stiffness k^{i-1} (using the stiffness $[k_e]+[k_g]+[k_i]$ in equation (7.1) with or without updating) for each element, since the deformations Δu are known to be small.

   Consequently, the total element forces at point e are $f^i = f^{i-1} + \Delta f$. By summing over all elements of the structure, the total internal forces F^i of the structure can be obtained.

5. *Error-checking phase*: Once the total internal forces F^i of the structure are obtained, the *unbalanced forces* R can be calculated as the difference between the total applied loads P^i and the internal forces F^i, i.e., $R = P^i - F^i$, in the updated configuration.
6. *Next iteration*: By treating the unbalanced forces R as the applied loads, another iteration to solve for the structural displacements and element forces can be conducted. It does not mean much whether the same stiffness matrix K^{i-1} or an updated stiffness is used. This only affects the number of iterations or the speed of convergence. The iteration process will be terminated when some preset tolerance criteria, in terms of force increments and/or displacement increments, are met.

In the above procedure, it is the *corrector phase* that governs the accuracy of the entire nonlinear solution. Aside from updating of the structural and element geometry, this phase deals with the calculation of the internal forces F^i and further the unbalanced forces R. The reason is quite simple. If the unbalanced forces R are not correctly calculated, then the iterations to follow will be wrongly directed and converge to incorrect solutions. In contrast, the *preditor phase*, based mainly on equation (7.3), is approximate in nature, which at its best can affect the number of iterations or the speed of convergence. The only requirement for the predictor phase or the stiffness K^{i-1} used is that it should not misguide the direction of iterations in the postbuckling response (Kuo and Yang, 1995).

Unfortunately, concerning the nonlinear and postbuckling analysis of structures in the literature, most studies have come to the point of presenting only the governing stiffness equations similar to or much simpler than equation (7.3) for the problem considered, which represents merely the predictor phase. Only very few works have proceeded further to describe the corrector phase noted here, concerning the three points (i)–(iii) mentioned above, which governs essentially the accuracy of the solution. The purpose of this chapter is to remove the ambiguity of incremental-iterative analysis by looking into the various phases involved.

7.3 Corrector Phase that Governs the Accuracy of Nonlinear Solutions

As was mentioned above, the corrector phase is the one that governs the accuracy of incremental nonlinear analysis. However, this phase has not been clearly unveiled in most previous researches. In many cases, the improper treatment of the corrector phase has been the main cause for non-convergent or erroneous iterations. As was stated, the corrector phase can be divided into three sub-steps, as will be discussed in the following.

7.3.1 *Updating of geometry*

Since the accuracy of an incremental-iterative analysis is governed primarily by the corrector phase, it is essential to have this phase correctly performed in the incremental-iterative process. First of all, for a typical incremental step (or for the first iteration of the incremental step), the structural displacement increments $\{\Delta U\}$ can be solved from the structural equations (7.3) in the predictor phase. This means that the displacement increments $\{\Delta u\}$ can be easily made available for each element of the structure through the connectivity between the structure and constituting elements.

For plane and space trusses and plane frames, the updating of the geometry of the structure from C^{i-1} to C^i is quite straightforward, due to the fact that all the displacement increments are additive. The same is also true for the *in-plane rotation increments* of plane frames.

A *truss* is known to be a *two-force member*. It contains only the translational, but no rotational degrees of freedom at the two nodes. As can be seen from Chapter 4, the element stiffness matrices are functions of the elastic modulus E, cross-sectional area A, element length L, *relative nodal displacements* Δu, Δv, Δw, and element forces F_x. For most trusses, the cross-sectional areas A are assumed to remain unchanged in the deformation process. The mechanical properties of a truss element can be determined totally by its two nodal points.

For *plane trusses*, the *element axis* (x) is simply the one that connects the two nodes of the element. It is easy to choose another axis (y) lying on the plane and normal to the element axis (x), to form an orthogonal set of coordinates for formulating the stiffness matrices. The coordinates of the two nodes determine the length L of the element. The updating of geometry for each element starts with the updating of the nodal coordinates for the two ends. Namely, by adding the nodal displacement increments generated during the incremental step from C^{i-1} to C^i (obtained in the predictor phase) to the original nodal coordinates, the updated nodal coordinates can be obtained. Then, the updated coordinates of the two nodes can be used to update the element axes xy, element length L, and relative nodal displacements Δu, Δv, Δw. The element force F_x has yet to be updated by the force recovery procedure to be presented in the following two sub-sections. Once the element axes and stiffness matrices are updated. All the element stiffness matrices can be assembled to form the global stiffness matrix, based on which a new iteration (or incremental step) can be performed.

The analysis of a *space truss* is no different from that of a plane truss since no rotational degrees of freedom are involved at the two nodes of each element. Aside from the element axis x that is determined exclusively by the two nodal points, we can arbitrarily choose two additional axes y and z to form an orthogonal set of coordinates at the beginning of the analysis. Once the coordinates are updated for the two nodes of an element, the updated element axis x is determined. We can take the cross-product of the updated x-axis and the original y-axis to form the updated z-axis, and the cross-product of the updated z-axis and the x-axis to form the updated y-axis. With the updated coordinate system established for each element, the space truss can be analyzed in the same iterative way as the plane truss. Again, the element force F_x has yet to be updated by the procedure to be presented in the following two sub-sections.

A *plane frame* is featured by the fact that *in-plane rotations* are considered for the two nodal points of each element. For the plane frame element, the element axis x is determined from the coordinates of the two end nodes. For each cross-sections of the element, we can choose the *major principal axis* y to be lying on the plane and the other minor principal axis z normal to the plane. Again, it is assumed that the cross-sectional properties of the plane frame, e.g., E, A, I_z, remain unchanged when in loading. Similar to the truss element, the nodal coordinates can be updated by adding the nodal displacement increments to the original nodal coordinates. Then, we can determine the updated element axis x and element length L. One thing that makes the analysis of plane frames different from that of plane trusses is that each frame element will be deformed and become curved in the loading process.

The above problem can be resolved by adopting the *convected coordinates* that rotate but do not deform with the element (Belytschko and Hsieh, 1974), which is good for problems with small displacements and rotations at each incremental step. Using the convected coordinate system (see Figure 7.2), the updated element axis x and updated element length L are determined from the updated coordinates of the two end nodes. The updated y-axis is taken to be normal to the x-axis and lying on the same plane. The element forces F_x and M_z involved in the element stiffness matrices in Chapter 3 have yet to be determined by the force recovery procedure to be presented in the following two sub-sections. With the geometry updated

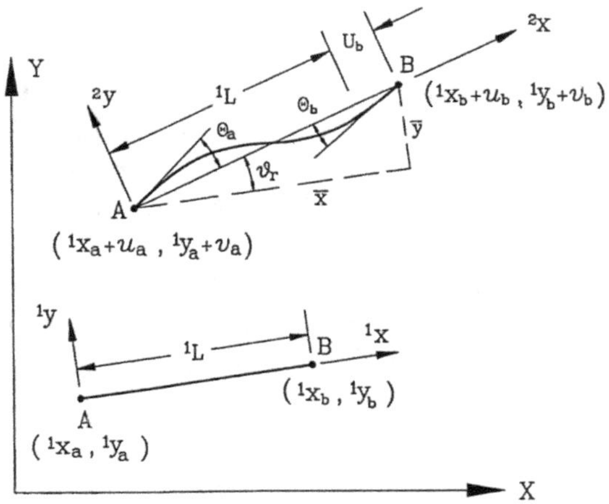

Fig. 7.2. Convected coordinates $(^2x, {}^2y)$.

for each element, another iteration can be performed for the plane frame.

In contrast, the analysis of *space frames* is much more complicated than any of the structures mentioned above due to the involvement of three rotational degrees of freedom at each end of the frame element. Conventionally, the end rotations of each element of the space frame are assumed to be small (Connor *et al.*, 1968; Tezcan and Mahapatra, 1969; Oran, 1973; Remseth, 1979; Chan and Kitipornchai, 1987). Since the *law of commutativity* remains valid for small rotations about the three axes, the end rotations of each element can be calculated simply by adding the rotation increments generated at each incremental step to those accumulated up to the previous step. Thus, based on the assumption of *small rotations*, the procedure for updating the end orientations is not that difficult, if the concept of convected coordinates is adopted.

For an incremental step with small rotations, the first task is to update the coordinates and rotations (in element axes) of the two nodal points. Then, the (convected) element axis x and element length L can be determined. In general, the two ends will be differently twisted, meaning that the two major axes for the cross sections of the two ends will not be on the same plane. Based on the concept of convected coordinates, we can take the average of the two major principal axes of the two ends as the updated major principal axis y of the element, and the cross product of the x and y-axes as the updated minor principal axis z of the element. The element forces F_x, M_y, and M_z have yet to be determined from the force recovery procedure to be presented in the following two sub-sections. With all the above data, another iteration can be conducted for the structure.

However, for the cases where the incremental end rotations are of finite magnitudes, the law of *commutativity* does not apply. The procedure for updating the end orientations of each frame element in the three-dimensional sense has to be developed based on the theory of *finite rotations*, as was presented in Appendix B.

Up to now, it is assumed that the *predictor phase* of the analysis has been completed. Namely, the structural displacement increments $\{\Delta U\}$ have been made available by solving the structural equations in (7.3) for the load increment, as typified by the movement from point a to b, along with the movement of displacement from point o to point c or from point a to point d in Figure 7.1. Also, the geometry of each element of the structure has been updated. In other words, the nodal coordinates of the two ends of the element are known, and so are the element axes and cross-sectional axes at the two ends. Hence, we are ready to update the initial forces and calculate the force increments for each element at the end of the incremental step, as will be briefed below.

7.3.2 *Updating of initial element forces*

The element displacement increment $\{\Delta u\}$ can be obtained from $\{\Delta U\}$ by using the connectivity between the elements and the structure. We will use the element equation (7.1) to compute the total element force $\{f^i\}$ for the end of the ith increment, as typified by point c to e in Figure 7.1. To this end, the element equation (7.1) can be recast as follows:

$$\{f^i\} = \{f^{i-1}\} + \left([k_e] + [k_g] + [k_i]\right)^{i-1}\{\Delta u\} \tag{7.4}$$

which obviously is composed of two parts, the initial force $\{f^{i-1}\}$ and the force increment (second term on the right-hand side). We use the *induced moment matrix* $[k_i]$ here, since it is a rigid-body qualified for discrete elements prior to their connection with other elements to form the structure. In contrast, the joint moment matrix $[k_j]$ that considers the joint compatibility will be used exclusively for the assembly of the structure stiffness matrices, as in the predictor phase.

According to the rigid body rule, the initial nodal forces acting on a finite element will rotate following the rotation of the element with no change in their magnitudes, regardless of how large the rotation is. Hence, *the initial forces* $\{f^{i-1}\}$ *initially acting at* C^{i-1} *with reference to* C^{i-1} *can be directly taken as the forces acting at the updated* C^i *with reference to* C^i. There is no need to perform the transformation for the rotation of axes from C^{i-1} to C^i.

7.3.3 *Calculation of element force increments*

The force increment $\{\Delta f\}$ originally put as the second term on the right-hand side of equation (7.4) can be expressed as follows for all the elements considered in this book:

$$\{\Delta f\} = [k]^{i-1}\{\Delta u\} \tag{7.5}$$

Based on the theoretical derivations in preceding chapters, the stiffness matrix $[k]$ in equation (7.5) should be interpreted as $[k] = [k_e]+[k_g]+[k_i]$ for the (discrete) space frame element (see Chapter 6), as $[k] = [k_e]+[k_g]$ for the plane frame element (see Chapter 3), and as $[k] = [k_e]+[k_g]+[s_1]+[s_2]+[s_3]$ for the truss element (see Chapter 4),

As a matter of fact, for the most common case where the element displacement increment $\{\Delta u\}$ can be regarded as *small* for each incremental step (see Figure 3.7 of Chapter 3 for the illustration), it is generally good to treat $[k]$ as the *elastic stiffness matrix* $[k_e]$. See the proof given for equation (3.55) based on the *natural deformation* approach. Consequently, the force recovery equation becomes

$$\{\Delta f\} = [k_e]^{i-1}\{\Delta u\} \tag{7.6}$$

which is the simplest one existing in the literature. The legitimacy of equation (7.6) for force recovery is also supported by Figure 7.1 in that the length of *de* (force increment) can be precisely calculated by multiplying the length of *ad* (displacement increment) by the tangent stiffness $[k_e]^{i-1}$.

Of interest to note is that the elastic stiffness matrix $[k_e]$ alone has been used in both the *corrector phase* [as in equation (7.6)] and in the *predictor phase* for constructing the structural stiffness matrix for solving displacement increments. It was demonstrated that even with such a simple stiffness matrix, quite a number of nonlinear and postbuckling problems can be solved (Yang *et al.* 2019), if use has been made of the *rigid body rule* and reliable nonlinear solution methods, such as Yang and Shieh's (1990) *generalized displacement control method* (GDC method). In Sections 7.6–7.8, the basic features and logistics of the GDC method will be presented in detail based on Yang and Shieh (1990).

Consequently, based on the rigid body rule for the initial forces and the above reasoning for the force increments, the total internal forces $\{f^i\}$ for each element of the structure at C^i can be updated as follows:

$$\{f^i\} = \{f^{i-1}\} + \{\Delta f\} \tag{7.7}$$

where the initial nodal forces $\{f^{i-1}\}$ are exactly those existing at C^{i-1}, i.e., with no transformation performed for the rotations from C^{i-1} to C^i. With the element forces $\{f^i\}$ made available, the stiffness matrix $[k_g]$ for the plane frame element, plus the induced moment matrix $[k_i]$ for the space frame element, can be calculated. Together with relative nodal displacements Δu, Δv, Δw, the stiffness matrices $[k_g]$, $[s_1]$, $[s_2][s_3]$ for the truss elements can be calculated as well.

With the element forces $\{f^i\}$ made available for each element, we can now loop over all elements of the structure to obtain the *total internal forces* $\{F^i\}$:

$$\{F^i\} = \sum_e^{num} \{f^i\}_e \tag{7.8}$$

where *num* denotes the total number of elements of the structure. The total internal forces $\{F^i\}$ is signified by point *e* in Figure 7.1. Then the *unbalanced forces* for the structure can be computed as follows:

$$\{R^i\} = \{P^i\} - \{F^i\} \tag{7.9}$$

which is signified by the segment *be* in Figure 7.1. With the geometry of the structure updated and the unbalanced forces calculated, we can now proceed to the next iteration or next increment of analysis.

7.4 Pure Incremental Method as a Starting Point

For incremental-iterative analysis, it is essential to distinguish between an incremental step and the iterative steps performed during the incremental step. In the following part of this chapter, we shall use the *right superscript* of a variable to indicate an *incremental step*. For clarity, the symbol Δ will be used to indicate an incremental quantity. In accordance, the symbols and equations used in previous chapters need to be altered slightly.

However, it should be noted that the method presented herein and those in the following for incremental-iterative methods by existing methods deal mainly with the *predictor phase*, concerning the implementation of load increments and the numerical stability around critical points. Basically, no coverage has been made for the *corrector phase*. There is no doubt that the procedure presented in the preceding section for treating the corrector phase is essential to the success of each solution method.

To start, let us consider the *pure incremental method*, i.e., the method with no iterations, as schematically shown in Figure 7.3. Based on the updated Lagrangian formulation and using the present notation, we are saying that for the ith incremental step from C^{i-1} to C^i, the equations of equilibrium can be established for the structure at the desired configuration C^i, but with reference to the last known configuration C^{i-1}. Accordingly, the incremental structural

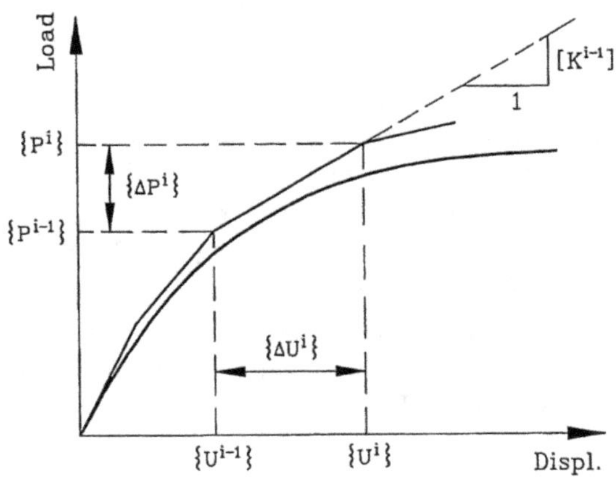

Fig. 7.3. Pure incremental method.

equation (7.3) can be recast as

$$[K^{i-1}]\{\Delta U^i\} = \{\Delta P^i\} \tag{7.10}$$

where the tangent stiffness matrix $[K^{i-1}]$ is evaluated at the beginning of the ith step (or the end of the i-1 step), composed of the contribution of $[K_e]$ and $[K_g]$ for plane frames, plus $[K_j]$ for space frames, and of $[K_e]$, $[K_g]$, $[S_1]$, $[S_2]$, and $[S_3]$ for trusses, as previously presented for C_1. The load increment $\{\Delta P^i\}$ is

$$\{\Delta P^i\} = \{P^i\} - \{P^{i-1}\} \tag{7.11}$$

where $\{P^i\}$ and $\{P^{i-1}\}$ denote the external loads acting on the structure at C^i and C^{i-1}, respectively. In equation (7.10), $\{\Delta U^i\}$ denotes the displacement increments generated during the ith incremental step, in response to the application of the load increments $\{\Delta P^i\}$. The total external loads $\{P^i\}$ acting on the structure at the ith incremental step can be computed as the summation of all the previous load increments, i.e.,

$$\{P^i\} = \sum_{r=1}^{i} \{\Delta P^r\} \tag{7.12}$$

and the total displacements $\{U^i\}$ of the structure as

$$\{U^i\} = \sum_{r=1}^{i} \{\Delta U^r\} \tag{7.13}$$

The above are the types of equations that were often given in the literature, which relate only to the *predictor phase.*

For a nonlinear analysis, we need also to talk about the *corrector phase*, concerning the force recovery for structural members. With the structural displacement increments $\{\Delta U^i\}$ (and therefore the element displacement increments $\{\Delta u^i\}$) made available, the geometry of the structure and elements can be updated as described in Section 7.3.1 for trusses, plane frames, and space frames (with small rotations) and in Appendix B for space frames (with large rotations). Consequently, all the geometry data of the structure are known up to the C^{i-1} configuration, including nodal coordinates, length and axis directions of each element. One thing that was not clearly shown in the above global equations is how the internal forces are calculated for each element of the structure, as they are needed in updating the stiffness matrix $[K^i]$, which is a function of the element forces. In this regard, the procedure described in Sections 7.3.2 and 7.3.3 should be followed.

Using the pure incremental method, the load increments $\{\Delta P^i\}$ are assumed to be such small that the nonlinear behavior of the structure can be approximated by a piecewise linear response. One advantage of this method is simplicity. However, for problems involving large displacements and rotations, the drift-off errors accumulated through each incremental step may not be negligible. Moreover, there is no guarantee that the structure will remain in equilibrium at each of the solution points in Figure 7.1. Hence, recourse should be had to methods that are equipped with iterative means, as will be discussed in the sections to follow.

7.5 Statement of the Problem for Incremental-Iterative Analysis

In Figure 7.4, a typical load-deflection curve that includes the post-buckling behavior of the structure has been schematically shown. Depending on the history of loading, the stiffness of the structure may be softening or stiffening, the equilibrium path may be stable

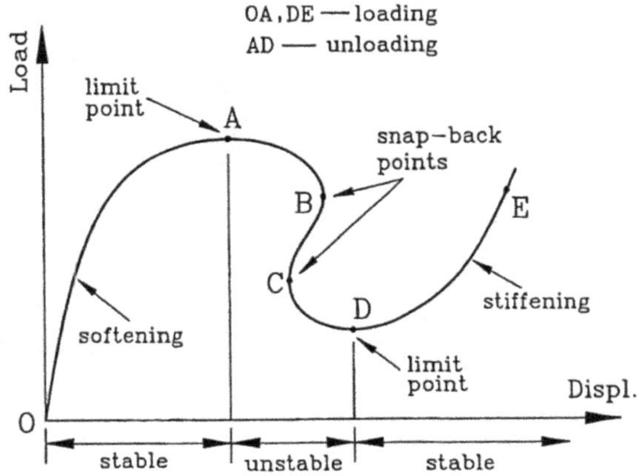

Fig. 7.4. General characteristics of a nonlinear structure.

or unstable, and the structure itself may be on a stage of loading or unloading. All such phenomena are typified by the occurrence of critical points such as the *limit points* and *snap-back points* in the load-deflection curves, which tend to bring computational difficulties of a certain sort in the solution process. There are many points that should be reasonably and precisely dealt with before a nonlinear solution method can be used to successfully trace the postbuckling path of a structure.

Particularly, we shall demonstrate how the iterative procedure can be incorporated at each incremental step to ensure that the conditions of equilibrium are satisfied for the structure via the predictor and corrector while improving the quality of solutions. When talking about the *step-by-step marching techniques*, we are faced with two basic questions: (1) How can the load increment be determined for each incremental step to accommodate softening and stiffening phenomena? (2) What is the guideline for performing iterations within each incremental step to avoid numerical difficulty around the critical points? These two questions relate to two different, but not independent, problems.

For the first question, the appropriateness of a load increment of an incremental-iterative method can be roughly assessed from the number of iterations performed for each incremental step. An ideal incremental-iterative method should allow the load increment to vary according to the degree of nonlinearity as exemplified by the stiffness of the structure considered, to optimize the efficiency of computation.

The second question deals primarily with the characteristics of convergence of each iterative scheme, especially in the neighborhood of various critical points, such as limit points and snap-back points, as indicated in Figure 7.4. Indeed, whether an iterative scheme can successfully trace the load-deflection of structures into the postbuckling range can be judged from its convergent characteristics in the vicinity of *critical points*.

7.5.1 *General equations for incremental-iterative methods*

For incremental-iterative analysis, it is essential to distinguish between an incremental step and the iterative steps performed during the incremental step. As was stated, we shall use the *right superscript* of the variable to indicate an *incremental step* and the *right subscript* an *iterative step*. In addition, the symbol Δ will be used to indicate an incremental quantity for clarity. Consequently, the nonlinear equations of a structure can be written for the jth iteration of the ith incremental step in a general form for use in the predictor phase as

$$[K_{j-1}^{i}]\{\Delta U_{j}^{i}\} = \{P_{j}^{i}\} - \{F_{j-1}^{i}\} \tag{7.14}$$

where for the structure at the ith increment, $\{\Delta U_{j}^{i}\}$ denotes the displacement increments for the jth iteration, and the tangent stiffness $[K_{j-1}^{i}]$ and internal forces $\{F_{j-1}^{i}\}$ for the jth iteration are calculated based on the result of the preceding $(j-1)$th iteration. Using the notation presented in preceding chapters, the tangent stiffness $[K_{j-1}^{i}]$ is composed of the elastic stiffness $[K_e]$ and geometric stiffness $[K_g]$ for plane frames, plus the joint moment stiffness $[K_j]$ for space frames, and of the stiffness matrices $[K_e]$, $[K_g]$, $[S_1]$, $[S_2]$, and $[S_3]$ for trusses, as previously presented for C_1. Particularly, the applied loads $\{P_{j}^{i}\}$ are allowed to *vary* during the jth iteration, as an improvement over the Newton–Raphson method that iterates at *constant* loads, to resolve the divergent problem near the limit point, as will be discussed in Section 7.5.2. The initial conditions to equation (7.14) are

$$[K_{0}^{i}] \equiv [K_{l}^{i-1}], \quad [F_{0}^{i}] \equiv [F_{l}^{i-1}], \quad [U_{0}^{i}] \equiv [U_{l}^{i-1}] \tag{7.15a-c}$$

where the subscript l is used to denote the *last* iteration of the preceding $(i-1)$th increment.

For *proportional loadings* considered exclusively in this book, the external loads $\{P_{j}^{i}\}$ can be decomposed and related to a reference load vector $\{\hat{P}\}$ as follows:

$$\{P_{j}^{i}\} = \{P_{j-1}^{i}\} + \{\Delta P_{j}^{i}\} \tag{7.16}$$

or equivalently as

$$\{P_{j}^{i}\} = \{P_{j-1}^{i}\} + \lambda_{j}^{i}\{\hat{P}\} \tag{7.17}$$

with the initial condition $\{P_{0}^{i}\} \equiv \{P_{l}^{i-1}\}$ is adopted to account for the loads accumulated from the *last* iteration of the preceding $(i-1)$th increment. Using the loading function in equation (7.17), the *load increment factor* λ_{j}^{i} for the jth iteration of the ith increment may be treated as an unknown parameter, in addition to the N unknowns $\{\Delta U_j^i\}$, in the $N+1$ theory to be presented in Section 7.6. With the displacement increments $\{\Delta U_j^i\}$ solved from equation (7.14) for the jth iteration, the total displacements $\{U_j^i\}$ of the structure can be obtained by accumulation:

$$\{U_j^i\} = \{\Delta U_{j-1}^i\} + \{\Delta U_j^i\} \tag{7.18}$$

Conventionally, the difference between the externally applied loads $\{P_{j-1}^i\}$ and the internal forces $\{F_{j-1}^i\}$ (summed over all elements) of the structure, resulting from the preceding $(j-1)$th iteration, has been referred to as the *unbalanced forces* $\{R_{j-1}^i\}$, namely,

$$\{R_{j-1}^i\} = \{P_{j-1}^i\} - \{F_{j-1}^i\} \tag{7.19}$$

Using equation (7.17), the governing equation in (7.14) can be recast as

$$[K_{j-1}^i]\{\Delta U_j^i\} = \lambda_j^i\{\hat{P}\} + \{R_{j-1}^i\} \tag{7.20}$$

which consists of clearly $N+1$ unknowns, i.e., λ_j^i and $\{\Delta U_j^i\}$. For convenience, we can further decompose the preceding equation into the following two separate equations (Batoz and Dhatt, 1979):

$$[K_{j-1}^i]\{\Delta \hat{U}_j^i\} = \{\hat{P}\} \tag{7.21}$$

$$[K_{j-1}^i]\{\Delta \bar{U}_j^i\} = \{R_{j-1}^i\} \tag{7.22}$$

Accordingly, the associated displacement increments $\{\Delta U_j^i\}$ can be broken into two parts as

$$\{\Delta U_j^i\} = \lambda_j^i\{\Delta \hat{U}_j^i\} + \{\Delta \bar{U}_j^i\} \tag{7.23}$$

One merit with the above decomposition of equation (7.20) into two equations is that the two equations can be solved simultaneously by a single call to the equation solver by treating $\{\hat{P}\}$ and $\{R_{j-1}^i\}$ as the parallel applied loads. The other advantage of the above decomposition is that the load increment factor λ_j^i can be treated as an independent unknown, in addition to the N unknowns contained in the displacement vector $\{\Delta U_j^i\}$. Conventionally, to determine the load increment factor λ_j^i requires the imposition of an additional constraint, based on which various solution methods have been proposed.

So far, all the manipulations presented in Section 7.5.1 have been made for the predictor phase concerning the solution of the structural displacements $\{\Delta U_j^i\}$, given the applied loads or unbalanced forces $\{P_j^i\} - \{F_{j-1}^i\}$. Again, we need to supplement the required corrector phase herein. Once the structural displacements $\{\Delta U_j^i\}$ is solved, the element displacements $\{\Delta u_j^i\}$ can be directly made available. Then, the structural geometry (as implied by point d in Figure 7.1) can be updated, and the element forces can be recovered using $\{f^i\} = \{f^{i-1}\} + \{\Delta f\}$ (as implied by segment $c - d - e$ in Figure 7.1). Following the procedures in Sections 7.3.2 and 7.3.3, the total element forces can be computed, so are the total internal forces and unbalanced forces of the structure, which serves as the basis for the next iteration.

7.5.2 *Newton–Raphson method*

This method is probably the oldest iterative method that is still widely used. Most iterative schemes adopted in the solution of nonlinear equations can be regarded as variants of this method. With the standard Newton–Raphson method, the external loads are increased by a constant amount for the first iteration, i.e., for $j = 1$, of each incremental step. For the remaining iterations within the same step, i.e., for $j \geq 2$, the load increment is set equal to zero, namely, the external loads are kept as constant during the iterations. For this reason, the Newton–Raphson method has also been referred to as the *load control method.* In terms of the load increment factor λ_j^i, the constraint condition of the method can be given as follows:

$$\lambda_j^i = \begin{cases} \text{constant} & \text{for } j = 1 \\ 0 & \text{for } j \geq 2 \end{cases} \tag{7.24}$$

A schematic of the iteration procedure for the Newton–Raphson method performed at constant load has been plotted in Figure 7.5.

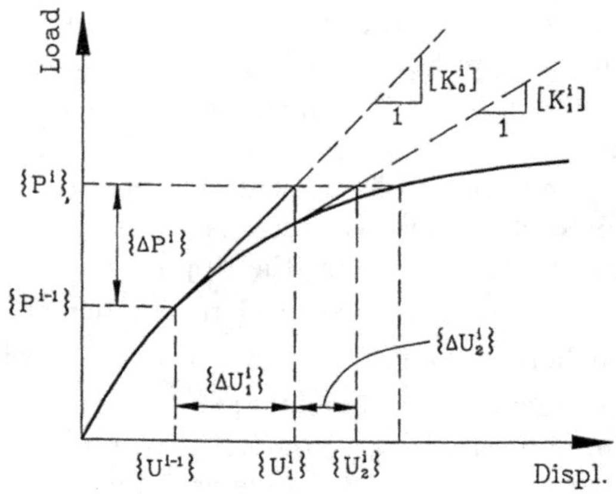

Fig. 7.5. Newton–Raphson method.

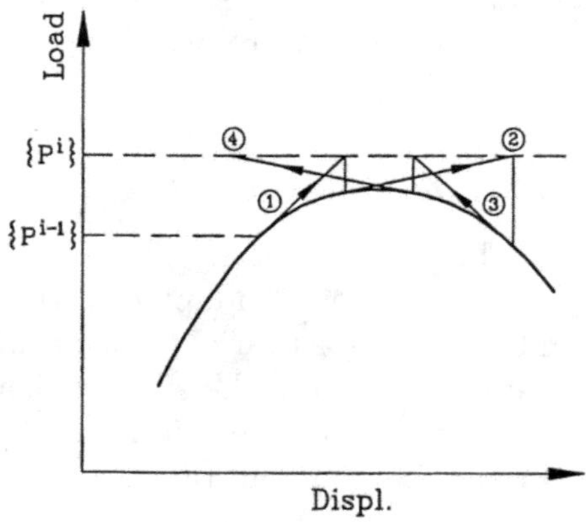

Fig. 7.6. Divergence of Newton–Raphson iterations.

The Newton–Raphson method has been reported to fail in solving problems involving limit points previously (Bergan and Soreide, 1978; Batoz and Dhatt, 1979; Ramm, 1981; Powell and Simons, 1981). This can be attributed to the fact that iterations have been performed at constant loads. As can be seen from Figure 7.6 for the Newton–Raphson method, the direction of iterations is controlled by the horizontal line that represents a preset level of the applied loads. Whenever the applied loads exceed the ultimate loads corresponding to a limit point at a certain step, the horizontal line that guides the direction of iteration will never cross the load-deflection curve. As a result, no convergent point can be found by the iterative process in the neighborhood of the limit point. See Figure 7.6 for an illustration of the divergent iterative process where the number of iterations has been given in circles. This figure explains clearly why the Newton–Raphson method has not been recommended for tracing the response of structures beyond a limit point.

7.5.3 *Displacement control method*

The *displacement control method* has been presented by Argyris (1965), and later modified by others (Pian and Tong, 1971; Zienkiewicz, 1971). It differs from the Newton–Raphson method in that iterations are performed at a constant displacement, rather than a constant load. With the *displacement control method*, we need to select a particular displacement component, say, the qth component, of the structure as the control parameter for guiding the iterations.

Let ΔU_{qj}^i denote the displacement increment of the qth component for the jth iteration of the ith increment. The constraint condition imposed by the displacement control method can be expressed as follows:

$$\Delta U_{qj}^i = \begin{cases} \text{constant} & \text{for } j = 1 \\ 0 & \text{for } j \geq 2 \end{cases} \tag{7.25}$$

According to equation (7.23), the qth component of displacement can be broken down into two parts as

$$\Delta U_{qj}^i = \lambda_j^i \Delta \hat{U}_{qj}^i + \Delta \bar{U}_{qj}^i \tag{7.26}$$

From equation (7.26), the load parameter λ_j^i can be solved,

$$\lambda_j^i = \frac{\Delta U_{qj}^i - \Delta \bar{U}_{qj}^i}{\Delta \hat{U}_{qj}^i} \tag{7.27}$$

For the first iteration of each incremental step, i.e., for $j = 1$, the unbalanced forces $\{R_0^i\}$ are zero. From equation (7.22), it can be seen that the displacements $\Delta \bar{U}_{qj}^i$ also vanish. For the remaining iterative steps ($j \geq 2$), however, the control displacement ΔU_{qj}^i is set to zero according to equation (7.25). It follows that the load parameter λ_j^i of equation (7.27) reduces to

$$\lambda_j^i = \begin{cases} \dfrac{\Delta U_{qj}^i}{\Delta \hat{U}_{qj}^i} & \text{for } j = 1 \\ -\dfrac{\Delta \bar{U}_{qj}^i}{\Delta \hat{U}_{qj}^i} & \text{for } j \geq 2 \end{cases} \tag{7.28}$$

Obviously, the external loads are not kept constant during the iteration process. A schematic of the displacement control method has been given in Figure 7.7.

Just as a method iterating at constant loads fails to pass the limit point, so too, a method iterating at constant displacements may break down when the controlled displacement snaps back from one load level to the other (Ramm, 1981; Crisfield, 1981; Waszczyszyn, 1983). This is certainly one disadvantage of the displacement control method if we realize that the selection of a proper control displacement component for controlling the iterative process is not always obvious for structures consisting of a large number of degrees of freedom. To overcome this drawback, we may adopt procedures that include all of the displacement components, rather than a single component, for controlling the direction of iteration. Both the arc

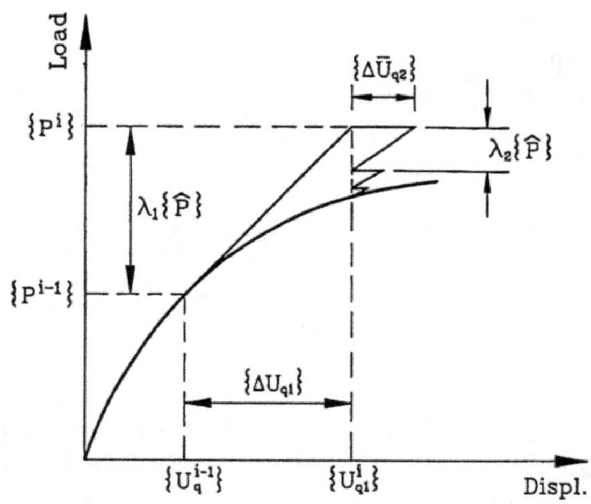

Fig. 7.7. Displacement control method.

length method and work control method to be discussed below can be considered as procedures falling in this category.

7.5.4 *Arc length method*

The *arc length method* in its original form is based on the following constraint condition for determining the load increments and for performing iterations:

$$\{\Delta U_1^i\}^T \Delta U_j^i + \lambda_1^i \lambda_j^i = \Delta S^2 \tag{7.29}$$

where $\{\Delta U_1^i\}$ and ΔU_j^i denote the displacement increments for the first and jth iterations, respectively, of the ith incremental step, and ΔS is the *arc length*,

$$\Delta S = \begin{cases} \text{constant} & \text{for } j = 1 \\ 0 & \text{for } j \geq 2 \end{cases} \tag{7.30}$$

At the beginning of each incremental step, i.e., for $j = 1$, there will be no unbalanced forces, which implies that $\{\Delta \bar{U}_1^i\} = \{0\}$ according to equation (7.22). It follows that the displacements for the first iterative step can be reduced from equation (7.23) to the following:

$$\{\Delta U_1^i\} = \lambda_1^i \{\Delta \hat{U}_1^i\} \tag{7.31}$$

which can be substituted back into equation (7.29) to yield

$$\lambda_1^i = \pm \frac{\Delta S}{\sqrt{\{\Delta \hat{U}_1^i\}^T \{\Delta \hat{U}_1^i\} + 1}} \tag{7.32}$$

This is exactly the load increment to be applied at the beginning of the ith incremental step. One drawback with the arc length method is the lack of information for determining the sign of the load parameter λ_1^i in equation (7.32). It should be noted that in a step-by-step nonlinear analysis, the plus sign refers to the stage of loading, and the minus sign is the stage of unloading for the structure under consideration (see Figure 7.4).

For the remaining iterative steps, i.e., for $j \geq 2$, the iterations are performed such that there will be no increase in the arc length ΔS as indicated by equation (7.30). In this case, the arc length equation (7.29) reduces to an orthogonality condition. By substituting

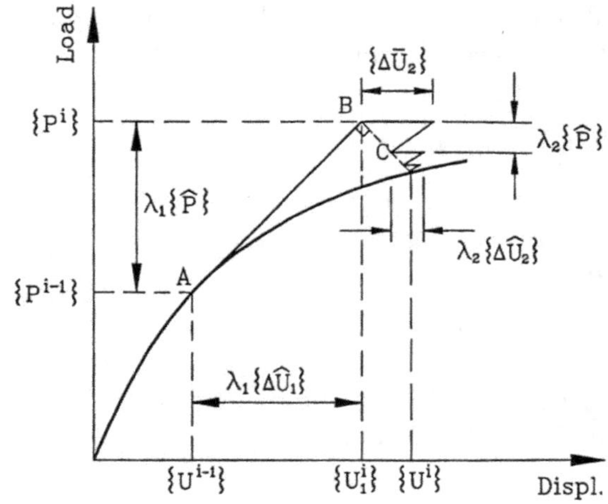

Fig. 7.8. Arc length method.

equation (7.23) for the displacement increments $\{\Delta U_j^i\}$ into equation (7.29) and noting that $\Delta S = 0$ for $j \geq 2$, the load parameter λ_j^i can be solved:

$$\lambda_j^i = -\frac{\{\Delta U_1^i\}^T\{\Delta \bar{U}_j^i\}}{\{\Delta U_1^i\}^T\{\Delta \hat{U}_j^i\} + \lambda_1^i} \quad \text{for } j \geq 2 \tag{7.33}$$

As can be seen, the arc length method iterates neither at a constant load nor at a constant displacement (see Figure 7.8 for a schematic of the iteration procedure). With the load parameter λ_j^i determined, the displacement increments $\{\Delta U_j^i\}$ for the jth iteration can be calculated from equation (7.23).

The arc length method was derived by Wempner (1971) and Riks (1972, 1979), and later improved by Ramm (1981), Crisfield (1981, 1983), and many other investigators. Conceptually, this method can achieve better performance than the methods that employ either constant load or constant displacement iterations. However, the arc length equation (7.29) and the orthogonality condition it implies are not consistent in physical units. For instance, in equations (7.32) and (7.33), the load parameter λ_j^i is by itself a scalar, but the displacement vectors $\{\Delta U_1^i\}$, $\{\Delta \hat{U}_1^i\}$, $\{\Delta \bar{U}_1^i\}$, $\{\Delta \hat{U}_j^i\}$, and $\{\Delta \bar{U}_j^i\}$ contain quantities such as translations and rotations, which are different both in physical units and orders of magnitudes. Numerical difficulties of a certain sort are likely to occur due to such inconsistencies, as will be discussed in Section 7.7.3.

7.5.5 Work control method

The *work control method* proposed by Yang and McGuire (1985) is based on the following constraint equation:

$$\{\Delta U_j^i\}^T(\lambda_j^i\{\hat{P}\}) = \Delta W \tag{7.34}$$

where the work increment ΔW is defined as

$$\Delta W = \begin{cases} \text{constant} & \text{for } j = 1 \\ 0 & \text{for } j \geq 2 \end{cases} \tag{7.35}$$

For the first iteration ($j = 1$), the constraint equation implies that the load parameter λ_1^i is determined on the basis of a constant work increment ΔW. By substituting equation (7.31) for $\{\Delta U_1^i\}$ into equation (7.34), the load parameter λ_1^i can be obtained as

$$\lambda_1^i = \pm \frac{\Delta W}{\sqrt{\{\Delta\hat{U}_1^i\}^T\{\hat{P}\}}} \tag{7.36}$$

where the sign remains to be determined.

For the remaining iterative steps ($j \geq 2$), the load parameter λ_j^i is determined from the condition of zero work increment, i.e., with ΔW set equal to zero. By the use of equation (7.23), we can solve from equation (7.34) the following:

$$\lambda_j^i = \pm \frac{\{\Delta\bar{U}_j^i\}^T\{\hat{P}\}}{\{\Delta\hat{U}_j^i\}^T\{\hat{P}\}} \quad \text{for } j \geq 2 \tag{7.37}$$

Evidently, the physical quantities involved in the preceding two equations are consistent in units, which represents an improvement over the arc length method. A comparison of equation (7.37) with (7.33) for the iterative steps indicates that the work control method differs from the arc length method both in the omission of the correcting term λ_1^i in the denominator and in the use of different weighting vectors for calculating the load parameter λ_j^i.

For problems where the loadings consist of a single concentrated load, equation (7.37) reduces to (7.28) for iterative steps with $j \geq 2$

because the reference load vector $\{\hat{P}\}$ contains only one nonzero entry. For this special case, the work control method reduces to the displacement control method, with the displacement associated with the loading degree of freedom serving as the control parameter. However, for cases where there are more than one concentrated loads, the iteration characteristics of the work control method will be different from those of the displacement control method, in that both the loads and displacements are not kept constant during iteration.

In practice, the load increment λ_1^i can be related to the *current stiffness parameter* (*CSP*) proposed by Bergan (1978, 1980) as

$$\lambda_1^i = \pm\lambda_1^1|CSP|^{1/2} \tag{7.38}$$

where λ_1^1 denotes the load increment for the first incremental step and the *CSP* is defined as follows:

$$CSP = \frac{\lambda_1^i\{\Delta U_1^1\}^T\{\hat{P}\}}{\lambda_1^1\{\Delta U_1^i\}^T\{\hat{P}\}} \tag{7.39}$$

Here $\{\Delta U_1^1\}$ and $\{\Delta U_1^i\}$ denote the displacement increments associated with the first iteration of the first and ith incremental steps, respectively. To verify the relation given in (7.38), we can simply substitute the expression for *CSP* back into the relation to obtain

$$(\lambda_1^i\{\Delta U_1^i\}^T\{\hat{P}\})^{1/2} = (\lambda_1^1\{\Delta U_1^1\}^T\{\hat{P}\})^{1/2} \tag{7.40}$$

Rearranging,

$$\{\Delta U_1^i\}^T(\lambda_1^i\{\hat{P}\}) = \{\Delta U_1^1\}^T(\lambda_1^1\{\hat{P}\}) = \text{constant} \tag{7.41}$$

which is exactly a restatement of the concept of a constant work increment.

The CSP has an initial value of unity for any nonlinear structures. One general property of this parameter is that it tends to increase for structures that are loaded into the stiffening stage and to decrease for structures on the softening stage. For structures reaching the limit points on the load–deflection curves (see Figure 7.4), the CSP becomes exactly equal to zero. Thus, a positive *CSP* refers to a stable region on the load–deflection curves, for which the external loads can be increasingly applied. To comply with this stage of loading, the load parameter λ_1^i in equation (7.38) should be made positive.

On the other hand, a negative *CSP* corresponds to an unstable region on the load–deflection curves, for which the external loads should be decreasingly applied. As a result, the negative sign in equation (7.38) should be selected. One drawback with the CSP is that it approaches infinity in the vicinity of snap-back points, as will be discussed in Section 7.8.1. For this reason, the work control method described herein is generally good for tracing paths with limit points, but with limited success for tracing paths involving snap-back points.

7.6 Nonlinear Structural Theory in $N+1$ Dimensional Space

In a nonlinear analysis, we are basically faced with the problem of solving for the $N+1$ system parameters (i.e., N displacement components $\{\Delta U^i_j\}$ and one load factor λ^i_j) from the $N+1$ system equations (i.e., N equations of equilibrium and one constraint equation). The N equations of equilibrium for the structure that are suitable for the jth of the ith incremental step of a nonlinear analysis have been formulated and given in the form of equation (7.20), as reproduced below:

$$[K^i_{j-1}]\{\Delta U^i_j\} = \lambda^i_j\{\hat{P}\} + \{R^i_{j-1}\} \tag{7.42}$$

where $[K^i_{j-1}]$ is the tangent stiffness of the structure, $\{\Delta U^i_j\}$ the displacement increment vector, $\{\hat{P}\}$ the reference load vector, and $\{R^i_{j-1}\}$ denotes the unbalanced force vector resulting from the last iteration. The reason to include the load parameter $\lambda^i_j x$ as a variable (unknown) is to devise a scheme that can circumvent the limit point problems by performing iterations not at constant loads. To solve the N displacements $\{\Delta U^i_j\}$ and one load parameter λ^i_j, a constraint equation is required in addition to equation (7.42).

For the purpose of incremental-iterative computation, it is convenient to decompose the equation (7.42) into two parts as follows (Batoz and Dhatt, 1979):

$$[K^i_{j-1}]\{\Delta \hat{U}^i_j\} = \{\hat{P}\} \tag{7.43}$$

$$[K^i_{j-1}]\{\Delta \bar{U}^i_j\} = \{R^i_{j-1}\} \tag{7.44}$$

The solutions of the preceding two equations can be combined to yield the structural solution as

$$\{\Delta U_j^i\} = \lambda_j^i\{\Delta\hat{U}_j^i\} + \{\Delta\bar{U}_j^i\} \tag{7.45}$$

where the load parameter λ_j^i has to be determined from the constraint condition.

In the preceding section, we have demonstrated that each of the solution methods discussed differs only in the use of the different control parameters or constraint equations for determining the load increments and for performing iterations. Theoretically, all the $N+1$ system parameters can be possibly selected, whether individually or in combination, as the control parameter. In a general formulation, it is useful to start with the following *constraint equation*:

$$\{C\}^T\{\Delta U_j^i\} + k\lambda_j^i = H_j^i \tag{7.46}$$

in which all the $N+1$ system parameters have been considered. The reliability and effectiveness of a nonlinear solution scheme hinges on the selection of the constants $\{C\}$ and k and the incremental parameter H_j^i for the constraint equation. In the following, we shall demonstrate how the concept of *stability* or *bounded response*, widely used in the study of structural performance can be used as a criterion for evaluating the appropriateness of various constrain equations.

The preceding system equations and constraint equation, i.e., equations (7.43)–(7.46), can be combined into a single matrix equation in the $N+1$ dimensional space as follows:

$$[\hat{K}_{j-1}^i]\begin{Bmatrix}\lambda_j^i\{\Delta\hat{U}_j^i\}\\ \lambda_j^i\end{Bmatrix} + [\bar{K}_{j-1}^i]\begin{Bmatrix}\{\Delta\bar{U}_j^i\}\\ 1\end{Bmatrix} = \begin{Bmatrix}\{0\}\\ H_j^i\end{Bmatrix} \tag{7.47}$$

where the generalized stiffness matrices $[\hat{K}_{j-1}^i]$ and $[\bar{K}_{j-1}^i]$ in the $N+1$ dimensional space are defined as

$$[\hat{K}_{j-1}^i] = \begin{bmatrix}[K_{j-1}^i] & -\{\hat{P}\}\\ \{C\}^T & k\end{bmatrix} \tag{7.48}$$

$$[\bar{K}_{j-1}^i] = \begin{bmatrix}[K_{j-1}^i] & -\{R_{j-1}^i\}\\ \{C\}^T & 0\end{bmatrix} \tag{7.49}$$

For the sake of clarity, the subscript "$j-1$" for each stiffness matrix will be dropped from here on.

The $N \times N$ stiffness matrix $[K^i]$ can be decomposed using the triple-factoring method of Section 2.4 as follows:

$$[K^i] = [L][D][L]^T \tag{7.50}$$

where $[D]$ is a diagonal matrix and $[L]$ a lower triangular matrix with all the diagonal entries set to unity. Accordingly, the determinant of $[K^i]$ is equal to that of $[D]$,

$$\det[K^i] = \det[D] \tag{7.51}$$

By adopting the following definitions for $\{T_1\}$ and $\{T_2\}$:

$$-\{\hat{P}\} = [L][D]\{T_1\} \tag{7.52}$$

$$\{C\} = [L][D]\{T_2\} \tag{7.53}$$

the generalized stiffness matrix $[\hat{K}^i]$ can be decomposed as follows:

$$[\hat{K}^i] = \begin{bmatrix} [L] & 0 \\ \{T_2\}^T & 1 \end{bmatrix} \begin{bmatrix} [D] & 0 \\ 0 & d \end{bmatrix} \begin{bmatrix} [L]^T & \{T_1\} \\ 0 & 1 \end{bmatrix} \tag{7.54}$$

where

$$d = k - \{T_2\}^T[D]\{T_1\} \tag{7.55}$$

Consequently, the determinant of the generalized stiffness matrix $[\hat{K}^i]$ becomes

$$\det[\hat{K}^i] = (k - \{T_2\}^T[D]\{T_1\})\det[K^i] \tag{7.56}$$

On the other hand, from equations (7.52) and (7.53), the following relation can be derived:

$$\{T_2\}^T[D]\{T_1\} = -\{C\}^T[K^i]^{-1}\{\hat{P}\} \tag{7.57}$$

Substituting equation (7.57) into (7.56) yields

$$\det[\hat{K}^i] = (k + \{C\}^T[K^i]^{-1}\{\hat{P}\})\det[K^i] \tag{7.58}$$

which, with the use of (7.43), can also be expressed as

$$\det[\hat{K}^i] = (k + \{C\}^T\{\Delta\hat{U}_j^i\})\det[K^i] \tag{7.59}$$

Rearranging equation (7.47) yields the following form:

$$\begin{Bmatrix} \lambda_j^i\{\Delta\hat{U}_j^i\} \\ \lambda_j^i \end{Bmatrix} = [\hat{K}^i]^{-1}\begin{Bmatrix} \{0\} \\ H_j^i \end{Bmatrix} - [\hat{K}^i]^{-1}[\bar{K}^i]\begin{Bmatrix} \{\Delta\bar{U}_j^i\} \\ 1 \end{Bmatrix} \tag{7.60}$$

This equation indicates clearly that for the parameters λ_j^i and $\{\Delta\hat{U}_j^i\}$ to be bounded, the determinant of the generalized stiffness matrix $[\hat{K}^i]$, rather than the original stiffness matrix $[K^i]$, must not be zero. The inverse of the generalized stiffness matrix $[\hat{K}^i]$ can be represented by

$$[\hat{K}^i]^{-1} = \frac{1}{\det[\hat{K}^i]}\begin{bmatrix} [A] & \{D_1\} \\ \{D_2\}^T & \det[K^i] \end{bmatrix} \tag{7.61}$$

where $[A]$, $\{D_1\}$ and $\{D_2\}$ are matrix and vectors consisting of cofactors of the entries in $[\hat{K}^i]$. Using equations (7.61) and (7.49), we can show that

$$[\hat{K}^i]^{-1}[\bar{K}^i] = \frac{1}{\det[\hat{K}^i]}\begin{bmatrix} [A][K^i] + \{D_1\}\{C\}^T & -[A]\{R_{j-1}^i\} \\ \{D_2\}[K^i] + \det[K^i]\{C\}^T & -\{D_2\}^T\{R_{j-1}^i\} \end{bmatrix} \tag{7.62}$$

Substitution of equations (7.61) and (7.62) into the last row of equation (7.60), along with the use of equations (7.44) and (7.59), yields

$$\lambda_j^i = \frac{1}{\{C\}^T\{\Delta\hat{U}_j^i\} + k}(H_j^i - \{C\}^T\{\Delta\bar{U}_j^i\}) \tag{7.63}$$

Both equations (7.60) and (7.63) serve as a useful basis for evaluating the numerical stability of various incremental-iterative methods for the solution of nonlinear problems. Here, a method is said to be *numerically stable* only when both the load and displacement increments, i.e., λ_j^i and $\{\Delta U_j^i\}$, solved with the method remain *bounded* throughout the entire history of loading. Wherever the load or any of the displacement components ceases to be bounded at certain points, numerical instability or divergence may occur in the step-by-step solution process.

From equations (7.58) or (7.59), it can be seen that once a limit point is encountered, the determinant of the $[K^i]$ matrix becomes equal to zero and so does the $[\hat{K}^i]$ matrix. Evidently, all methods that employ the constraint equation in the form of equation (7.46) will fail right at the limit points. It should be noted, however, that a limit point (or a snap-back point) represents merely an ideal mathematical point in the load-deflection curves, which can hardly be encountered in practice due to the presence of computation errors such as truncation errors and round-off errors. Therefore, when talking about the iterative nature of a solution scheme, we need to be concerned with the general convergence characteristics of the scheme in approaching and in departing from the limit point or other critical points. This is the reason why we prefer to use terms such as the *vicinity* or *neighborhood* of critical points, rather than critical points, themselves.

It should be noted that the load parameter λ_j^i as given in equation (7.63) can also be solved directly from equations (7.45) and (7.46). However, by going through the foregoing procedure, we are able to examine the roles of both the load and displacement parameters in the incremental-iterative process. Moreover, we are able to demonstrate that it is the determinant of the generalized stiffness matrix $[\hat{K}^i]$, rather than the original stiffness matrix $[K^i]$, that determines whether the system parameters λ_j^i and $\{\Delta U_j^i\}$ will remain bounded or not in such a process.

7.7 Comments on Existing Solution Methods

By the concept of bounded response, the incremental-iterative methods presented in Section 7.5 will be reevaluated in this section. It should be noted that the same criterion can be applied to evaluating other methods existing in the literature if constraint equations of the form in equation (7.46) have been adopted.

7.7.1 *Newton–Raphson method*

The original Newton–Raphson method performs iterations at constant load. This is equivalent to the adoption of the following constants for the constraint parameters: $\{C\} = \{0\}$; $k = 1$; $H_1 =$ constant (a prescribed load increment); and $H_j = 0$ for $j \geq 2$. By substituting these values for the parameters into equation (7.63), and noting that $\{R_0\} = \{\Delta \bar{U}_1^i\} = \{0\}$ for the first iteration ($j = 1$), we can easily derive the constraint equation previously presented as in equation (7.24).

With this method, it is true that the load parameter λ_j^i remains bounded, in fact, equal to zero by postulation, for iterative steps with $j \geq 2$. We should not forget, however, that as the limit point is approached, the determinant of the original stiffness matrix $[K^i]$ will approach zero. According to equation (7.59), we observe that the determinant of the generalized stiffness matrix $[\hat{K}^i]$ also approaches zero, since $\{C\} = \{0\}$ and $k = 1$. From equation (7.60), it is ascertained that the displacement components solved here tend to be unbounded. Thus, the Newton–Raphson method is not recommended for solving problems involving limit points.

7.7.2 *Displacement control method*

Let the qth component of displacements be the control parameter. The following are assumed: $\{C\}^T = \{0 \ldots 010 \ldots 0\}^T$, of which all the entries are zero except the qth; $k = 0$; $H_1 = \Delta U_{qj}^i$ (a prescribed displacement increment); and $H_j = 0$ for $j \geq 2$. Substituting these parameters into equation (7.63) yields exactly the equation given in equation (7.28). It can be verified that in the vicinity of limit points, both the generalized stiffness matrix $[\hat{K}^i]$ and $[\bar{K}^i]$ remain nonsingular as the $[K^i]$ matrix is non-singular. Evidently, both the load parameter λ_j^i (see equation (7.63)) and the displacement increments $\{\Delta U_j^i\}$ (see equation (7.60) for $\{\Delta \hat{U}_j^i\}$ and (7.45) for $\{\Delta U_j^i\}$) appear to be of finite amplitudes. This confirms the capability of the displacement method to deal with limit points.

Now, suppose the control displacement parameter, i.e., the qth component of displacements, snaps back at certain loading stages.

As the snap-back point is approached, there will be no increase or decrease in the control displacement, i.e., $\Delta\hat{U}^i_{qj} \to 0$. From equation (7.58), we observe that the determinant of the $[\hat{K}^i]$ matrix approaches zero, although the determinant of the $[K^i]$ matrix remains unequal to zero. As indicated by equation (7.60), numerical instability may occur in regions where the control displacement tends to snap back.

7.7.3 *Arc length method*

Previously, various iterative vectors have been imposed by researchers in applying the arc length method to guide the direction of iteration. For the present purposes, only the arc length method with fixed normal plane in the $N+1$ dimensional space, as was presented in Section 7.5.4 will be discussed here. Assume the constraint parameters as follows: $\{C\} = \{\Delta U^i_1\}(= \lambda^i_1\{\Delta\hat{U}^i_1\})$; $k = \lambda^i_1$; $H_1 = \Delta S^2$, where ΔS is a prescribed arc length; and $H_j = 0$ for $j \geq 2$. For the incremental step (i.e., with $j = 1$), $\{\Delta\bar{U}^i_1\} = \{0\}$, the load parameter λ^i_1 given by equation (7.62) reduces exactly to the one given in (7.32), while for the iterative steps (i.e., with $j \geq 2$), it will reduce to the one in (7.33).

From equation (7.59), it is certain that the determinant of the $[\hat{K}^i]$ matrix will remain unequal to zero in the vicinity of critical points, such as the limit points and snap-back points. However, because of the use of different units for the quantities in equation (7.33), there exists the possibility for the load parameter λ^i_1 (involved in the denominator) to be so large that the sign of λ^i_j depends on the angle between $\{\Delta U^i_1\}$ and $\{\Delta\bar{U}^i_j\}$. For instance, in the vicinity of snap points with sharp gradients, it is likely that improper sign may be obtained for the load parameter λ^i_j which may lead to incorrect iterative directions and eventually numerical divergence.

7.7.4 *Work control method*

Let $\{C\} = \lambda^i_j\{\hat{P}\}$, $k = 0$, $H^i_1 = \Delta W$ (a prescribed work increment), and $H^i_j = 0$ for $j \geq 2$. Substitution of these parameters into equation (7.63) yields exactly the expression for the load parameter λ^i_j in equations (7.36) and (7.37). For structures with a small number of load components, the determinant of the $[\hat{K}^i]$ matrix may approach zero if the degrees of freedom associated with the load components tend to snap back, as can be verified from equation (7.59) by setting $k = 0$ and letting the components of $\{\Delta\hat{U}_j^i\}$ corresponding to the load components approach zero. This has the indication that certain numerical difficulties may occur around the snap-back points.

As an alternative for explaining this phenomenon, we have demonstrated that the load parameter λ_1^i can be related to the *CSP* by equation (7.38). Since the *CSP* tends to approach infinity when any one of the load–deflection curves snaps back (see Section 7.8.1 later on), there exists the possibility for the load parameter λ_1^i to become unbounded in this region. Consequently, the work control method may fail for structures of which the degrees of freedom associated with the load components tend to snap back in the load-deflection curves.

7.8 Generalized Displacement Control Method (GDC Method)

It has been demonstrated in Sections 7.4–7.6 that most existing incremental-iterative techniques are not perfect from the following perspectives: (1) numerical stability in approaching the critical points, such as the limit points and snap-back points, (2) adjustability in load increments to reflect the variation in stiffness of the structure, and (3) self-adaptive capability in determining the direction of loading. In addition, it has been shown that the numerical stability of a solution method hinges on the *bounded* characteristics of the load increment factor λ_j^i and the displacement increments $\{\Delta U_j^i\}$, which, in turn, relies on the selection of proper constraint parameters $\{C\}$, k, and H_j^i, as can be seen from equations (7.59)–(7.63).

Based on the preceding considerations, the constraint parameters have been proposed by Yang and Shieh (1990): $k = 0$, and $\{C\} = \lambda_1^i\{\Delta\hat{U}_1^{i-1}\}$. Consequently, we can obtain from equation (7.63) the following:

$$\lambda_j^i = \frac{H_j^i - \lambda_1^i\{\Delta\hat{U}_1^{i-1}\}^T\{\Delta\bar{U}_j^i\}}{\lambda_1^i\{\Delta\hat{U}_1^{i-1}\}^T\{\Delta\hat{U}_j^i\}} \tag{7.64}$$

where $\{\Delta\hat{U}_1^{i-1}\}$ denotes the displacement increments resulting from the first iteration of the last or $(i-1)$th incremental step, and H_j^i is a generalized displacement (a constant). Since $\{\Delta\bar{U}_j^i\} = \{0\}$ for $j = 1$ and $H_j^i = 0$ for $j \geq 2$, the preceding equation reduces to

$$\lambda_1^i = \sqrt{\frac{H_1^i}{\{\Delta\hat{U}_1^{i-1}\}^T\{\Delta\hat{U}_1^i\}}} \tag{7.65}$$

for the first iterative step, and

$$\lambda_j^i = \frac{\{\Delta\hat{U}_1^{i-1}\}^T\{\Delta\bar{U}_j^i\}}{\{\Delta\hat{U}_1^{i-1}\}^T\{\Delta\hat{U}_j^i\}} \quad \text{for } j \geq 2 \tag{7.66}$$

i.e., for the following iterative steps. This means that the load parameter λ_j^i is not kept constant during the iteration process. As a result, the problem caused by constant load iteration around a limit point is circumvented.

By taking $\{\Delta\hat{U}_1^0\}$ equal to $\{\Delta\hat{U}_1^1\}$, from equation (7.65), we can obtain the generalized displacement H_1^1 as

$$H_1^1 = (\lambda_1^1)^2\{\Delta\hat{U}_1^1\}^T\{\Delta\hat{U}_1^1\} \tag{7.67}$$

which can be substituted back into equation (7.65) to yield

$$\lambda_1^i = \lambda_1^1\sqrt{\frac{\{\Delta\hat{U}_1^1\}^T\{\Delta\hat{U}_1^1\}}{\{\Delta\hat{U}_1^{i-1}\}^T\{\Delta\hat{U}_1^i\}}} \tag{7.68}$$

Further, by introducing the *general stiffness parameter* (GSP) as follows:

$$\text{GSP} = \frac{\{\Delta\hat{U}_1^1\}^T\{\Delta\hat{U}_1^1\}}{\{\Delta\hat{U}_1^{i-1}\}^T\{\Delta\hat{U}_1^i\}} \tag{7.69}$$

the load parameter λ_1^i for the ith step can be computed as

$$\lambda_1^i = \pm\lambda_1^1|\text{GSP}|^{1/2} \tag{7.70}$$

where the load increment λ_1^1 for the first load step is a preset value. In equation (7.69), it is noted that the numerator represents the norm of the displacement increments of the first incremental step under the reference load $\{\hat{P}\}$, and the denominator is approximately the norm of those of the ith incremental step. With this, the *GSP* can be interpreted as an indicator of the stiffness variation of the structure for the ith incremental step with respect to the first incremental step.

Using the present method, there is no need to determine the $\pm$ sign in equation (7.70). In the section to follow, it will be demonstrated that the loading and unloading of the incremental process can be determined by the change in sign the *GSP*. This enables the incremental-iterative process to be conducted automatically to trace the load-deflection curves of structures involving multiple critical points in the postbuckling response.

The method of solution presented herein has been referred to as the GDC method. It is superior to most existing methods for the following reasons: First, numerical stability can always be ensured in regions near the limit points and snap-back points, in the sense that both the load parameter λ_j^i and displacement increments $\{\Delta U_j^i\}$ remain bounded. Second, the variation in nonlinearity of the stiffness of the structure has been taken into account through the *GSP*, as can be seen from equation (7.70). Finally, a change in the sign of *GSP* serves as an indicator for reversing the direction of loading. The reason for this will be given in the following section where the properties of the *GSP* will be explored.

7.8.1 *General stiffness parameter vs. current stiffness parameter*

The *CSP*, as proposed by Bergan (1978, 1980), has often been used to monitor the variation in stiffness of a structure in geometrically nonlinear analysis. Starting with an initial value of unity, the *CSP* will increase as the stiffness of the structure increases, and vice versa. A zero value for the *CSP* implies the occurrence of a limit point. Despite these useful characteristics, the *CSP* is not suitable for application to problems involving multiple critical points. First of all, the *CSP* changes sign at both limit points and snap-back points (see the numerical examples in Section 7.9 for illustration), thereby making it impossible to distinguish between these two different types of critical points. Realizing that the direction of loading needs to be reversed only when passing the limit points, we can see that the *CSP* is not a proper parameter for guiding the direction of loading. Moreover, the *CSP* varies in an abrupt manner whenever any displacement component snaps back. Such a lack of smoothness with the *CSP* in approaching the snap-back points can cause certain numerical problems if we try to link the load parameter λ_j^i to the *CSP*, say, using the formula of the type given in equation (7.38).

The *general stiffness parameter* (*GSP*), as defined in equation (7.69), has several important characteristics:

1. The *GSP* has been defined in (7.69) as the ratio of the norm of the displacement increments at the first incremental step to those (approximately) at the ith (current) incremental step. With such a definition, the *GSP* is representative of the stiffness variation of the structure at the ith incremental step with reference to the first step. One advantage of this parameter is that no jump in numerical value may occur ever in regions near the snap-back points.
2. The *GSP* turns out to be negative only for the incremental steps *immediately after* the limit points, whereas for the other incremental steps, it will always remain positive. This can be attributed to the fact that the sign of *GSP* depends only on the two vectors $\{\Delta\hat{U}_1^{i-1}\}$ and $\{\Delta\hat{U}_1^i\}$, as can be seen from equation (7.69). As illustrated in Figure 7.9, the product of the two vectors $\{\Delta\hat{U}_1^{i-1}\}$ and $\{\Delta\hat{U}_1^i\}$ is negative when they form an obtuse angle or pass

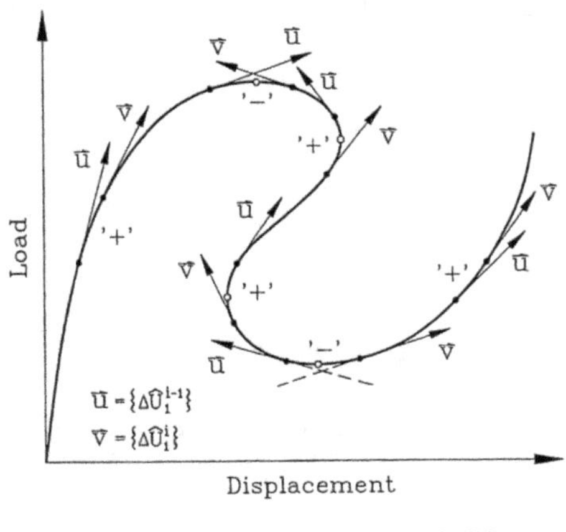

Fig. 7.9. Characteristics of *GSP*.

a limit point. Once a negative *GSP* is detected, it means we are going to pass a limit point and the loading direction should immediately be reversed. Otherwise, the product of two vectors $\{\Delta\hat{U}_1^{i-1}\}$ and $\{\Delta\hat{U}_1^{i}\}$ remains always positive, and no change in loading direction is needed. Since the *GSP* changes signs only when passing a limit point, it serves as a good indicator for guiding the direction of loading.

3. The *GSP*, similar to the *CSP*, starts with the initial value of unity and vanishes at limit points. It tends to increase for structures on the stiffening stage and to decrease on the softening stage. However, the *GSP* is superior to the *CSP* in that it remains bounded throughout the entire history of loading, regardless of the presence of snap-back points. As such, it can be successfully used in formula such as equation (7.70) for adjusting the load increment parameter λ_1^i for each incremental step, as a reflection of the variation in nonlinearity of the structure considered.

7.8.2 *Algorithm for geometric nonlinear analysis*

The method presented in Section 7.8 can be incorporated into a general analysis program for the automatic solution of geometrically nonlinear problems involving the postbuckling response with multiple critical points. For the proportional loadings considered in this book, we can relate the total applied loads $\{P_j^i\}$ at the jth iteration of the ith incremental step to the reference load vector $\{\hat{P}\}$ by the following formula:

$$\{P_j^i\} = \Lambda_j^i\{\hat{P}\} \tag{7.71}$$

where the total load factor Λ_j^i can be related to the load increment parameter λ_j^i by

$$\Lambda_j^i = \Lambda_{j-1}^i + \lambda_j^i \tag{7.72}$$

with the initial condition $\Lambda_0^i = \Lambda_l^{i-1}$ with subscript l indicating the last iteration of the preceding $(i-1)$th incremental step.

The GDC method presented above can be easily incorporated into a general analysis program to solve geometrically nonlinear problems with multi-critical points. Following is a brief description of the procedure for the method:

1. Select a reference load vector $\{\hat{P}\}$ and basic load increment λ_1^1 for starting. Set the initial conditions: $\{P_0^1\} = \{0\}$, $\{U_0^1\} = \{0\}$, $\Lambda_0^1 = 0$, etc.
2. For ith incremental step:
   For the first iteration ($j = 1$):
   (a) Form the structural stiffness matrix $[K_0^i] = [K_l^{i-1}]$.
   (b) Solve the equations of equilibrium (7.21) for the reference load $\{\hat{P}\}$ to obtain $\{\Delta\hat{U}_1^i\}$. For the first iteration ($j = 1$), the unbalanced forces $\{R_0^i\} = \{0\}$. So the displacement increments $\{\Delta\bar{U}_1^i\} = \{0\}$, as determined by equation (7.44).
   For $i = 1$, let *GSP* $= 1$. For $i \geq 2$, use equation (7.69) to determine *GSP*.
   (c) For $i \geq 2$, use the load equation (7.70) to determine λ_1^i.
   (d) Check if *GSP* is negative. If yes, multiply λ_1^i by -1 to reverse the direction of loading.
   (e) Determine the total displacement increments $\{\Delta U_j^i\}$ for the iteration using equation (7.45), noting that $\{\Delta\bar{U}_1^i\} = \{0\}$.
3. For subsequent iterations ($j \geq 2$):
   (a) Determine the unbalanced forces $\{R_{j-1}^i\}$ using equation (7.9).
   (b) Update the stiffness matrix $[K_{j-1}^i]$ (optional).
   (c) Solve equations (7.43) and (7.44) for the two displacement vectors $\{\Delta\hat{U}_j^i\}$ and $\{\Delta\bar{U}_j^i\}$, respectively.
   (d) Use the load equation (7.66) to determine λ_j^i.
   (e) Compute the total displacement increments $\{\Delta U_j^i\}$ for the jth iteration using equation (7.45).
4. Update the total load factor: $\Lambda_j^i = \Lambda_{j-1}^i + \lambda_j^i$, total load vector $\{P_j^i\} = \Lambda_j^i\{\hat{P}\}$, and total structural displacements $\{U_j^i\} = \{U_{j-1}^i\} + \{\Delta U_j^i\}$.
5. Update the coordinates of each node for the structure.

6. Update the geometry of the structure.
7. Loop over each element of the structures:
    (a) Compute the element displacements $\{\Delta u_j^i\}$.
    (b) Calculate the element force increments $\{\Delta f\}$.
    (c) Calculate the total element forces $\{f_j^i\}$ using equation (7.7), considering particularly the initial forces effect.
8. Calculate the total internal forces $\{F_j^i\}$ of the structure using equation (7.8).
9. Repeat steps 3–5 until the desired accuracy is achieved, by using the displacement norm error or the like.
10. If the total load Λ_j^i does not exceed the maximum load allowed, return to step 2 for the next increment. Otherwise, stop the procedure.

In the preceding paragraph, a description has been made of the most essential phases involved in the geometrically nonlinear analysis of framed structures using the GDC method. The procedure presented herein serves merely as a guidance of the conceptual flow for the step-by-step nonlinear analysis. It can never be considered complete without the supplement of related details. For instance, when talking about the formation of a structure stiffness matrix, we have to first establish the element stiffness matrices, and then to rotate all these matrices from the element axes to a common global structural axes, and finally to assemble the matrices to form the structure stiffness matrix, with due account taken of the restrained boundary effect (see Section 7.2 for an example on linear analysis).

7.9 Numerical Examples

This section has the objective of numerically evaluating the applicability and reliability of the procedures presented in this chapter for the nonlinear analysis of framed structures.

7.9.1 *Two-member truss*

The two-member truss under the apex loads P_u and P_v in Figure 7.10 has been used as a benchmark for investigating the adequacy of

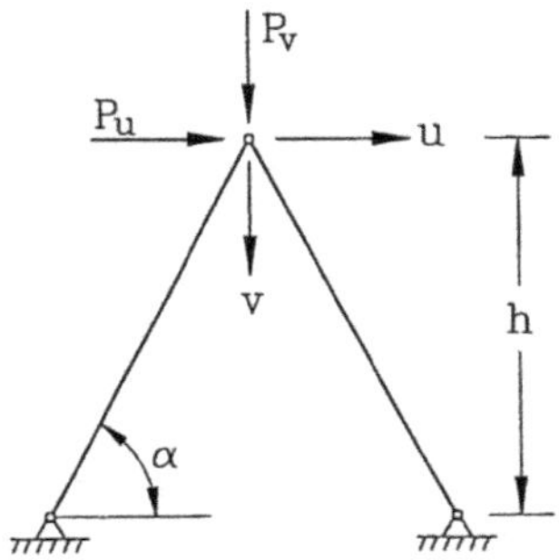

Fig. 7.10. Two-member truss.

various incremental-iterative algorithms. This example is so simple that it can be effectively used to assess the workability of each solution method. Besides, exact analytical solutions are available for the problem. The following data are assumed for the truss: member's axial rigidity $EA = 1{,}884.694\,\text{lb}$ (8,366.88 N), inclined angle $\alpha = 63.4°$, and height of vertex $h = 25.847\,\text{in.}$ (65.651 cm). Two loading cases will be studied.

In the first case, the horizontal load is considered as an imperfection, i.e., by letting $P_u = 0.05P_v$. The solutions automatically solved by the GDC method have been shown in Figures 7.11(a) and (b), which are in good agreement with the exact solutions given by Pecknold *et al.* (1985). As can be seen from Figure 7.11(a) for the vertical load-deflection curve, there are four limit points a, c, d, and f, and two snap-back points b and e. Correspondingly, the horizontal response was shown in Figure 7.11(b). The present analysis has demonstrated the self-adaptive capability of the GDC method in tracing the path involving multi-critical points, in adjusting the load step sizes, and in reversing the direction of loading. The control parameter *GSP*, as shown in Figure 7.11(c), represents a significant improvement over the *CSP* shown in Figure 7.11(d), in that the discontinuities encountered by the *CSP* have been eliminated.

In the second case, the vertical load is treated as an imperfection, i.e., by letting $P_v = 0.05P_u$. The load-deflection curves for the horizontal and vertical responses solved automatically by the GDC method for this case were plotted in Figures 7.12(a) and 7.12(b), respectively, along with the *GSP* and *CSP* curves plotted

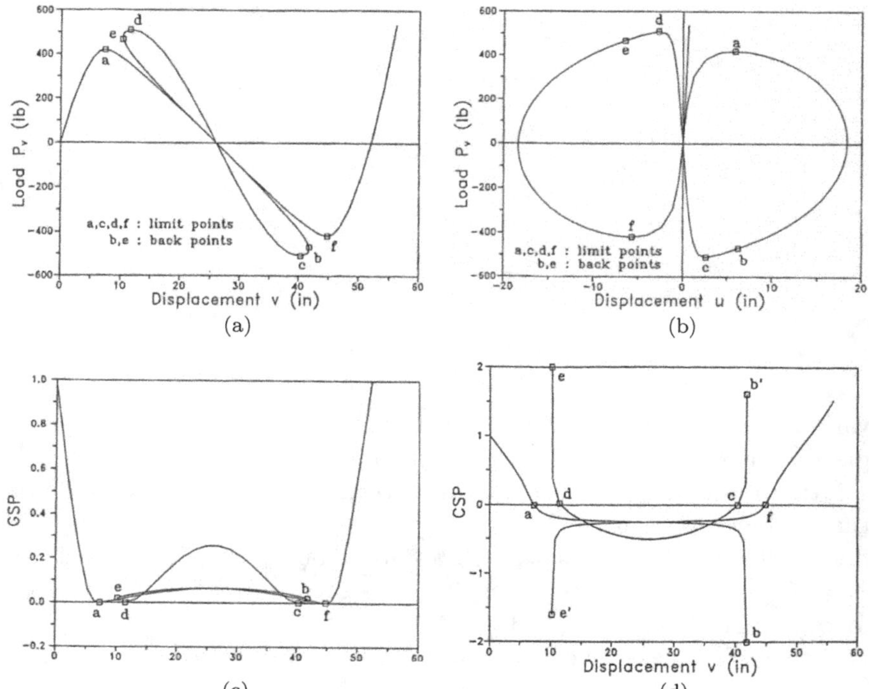

Fig. 7.11. Load case 1 for two-member truss: (a) P_v vs. v; (b) P_v vs. u; (c) GSP vs. v; (d) CSP vs. v.

in Figures 7.12(c) and 7.12(d), respectively, for reference. The observations made for the first case apply here.

7.9.2 *Shallow arch*

The shallow arch shown in Figure 7.13 is hinged at both ends. For this arch, the following data are assumed: span $L = 100$ in. (254 cm), apex height $h = 5$ in. (12.7 cm), elastic modulus $E = 200$ psi (1,378 kPa), moment of inertia $I = 1$ in.^{4} (41.62 cm^4), and cross-sectional area $A = 1$ in.^{2} (6.45 cm^2). In the finite element modeling, the arch was first represented by twenty-five straight beam elements. The portion

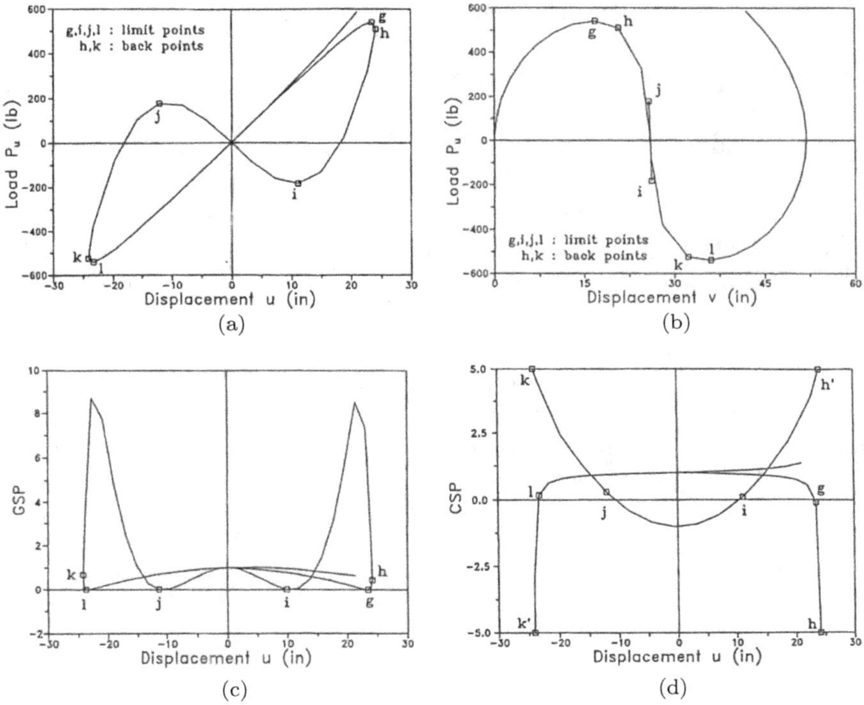

Fig. 7.12. Load case 2 for two-member truss: (a) P_u vs. u; (b) P_u vs. v; (c) GSP vs. u; (d) CSP vs. u.

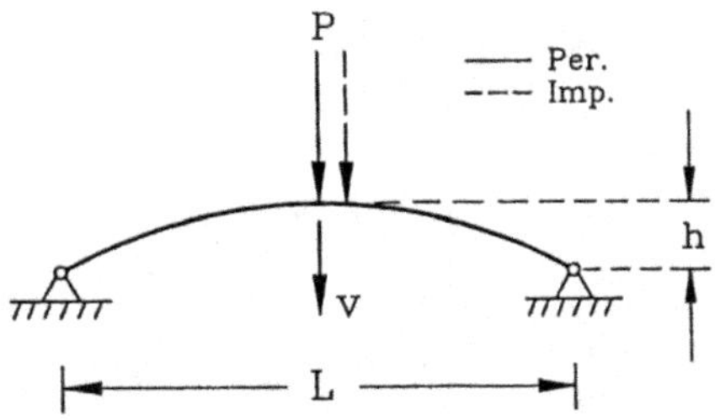

Fig. 7.13. Shallow arch.

of the arch covered by the central element was further divided into two beam elements, in order to provide a node at the apex of the arch for application of the vertical load.

Two loading cases were considered. The first case consists of a vertical load at the central node, which is known to be symmetrical. In the second case, the central load was dislocated and applied at the node nearest to the center of the arch to account for the effect of imperfection in loading. The finite element adopted herein is the plane frame element derived in Chapter 3, while the procedure for recovering and updating the element forces is the one presented in Section 7.3 for incremental-iterative analysis.

The results obtained automatically by the GDC method for both loading cases have been plotted in Figure 7.14, in which the curve for the perfect loading case can be regarded as the primary path, and the looping curves for the imperfect case as the secondary path.

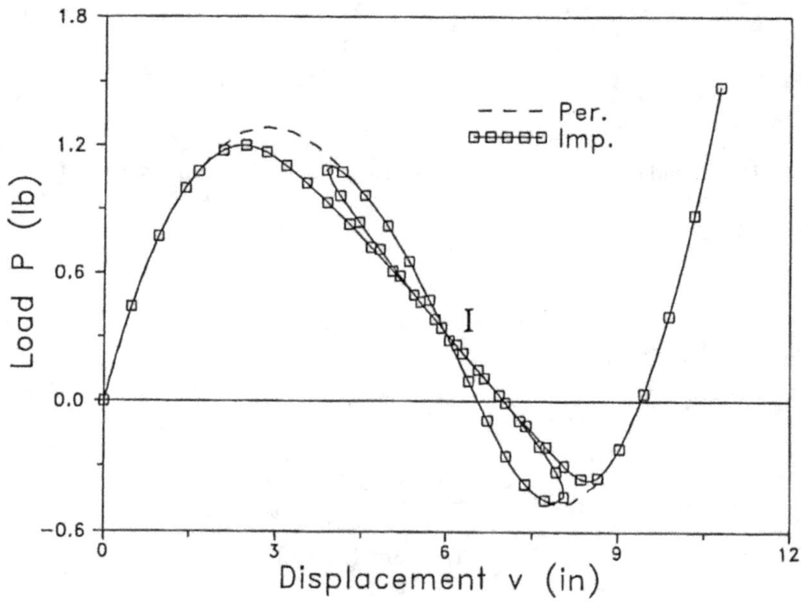

Fig. 7.14. Load–deflection curves for shallow arch.

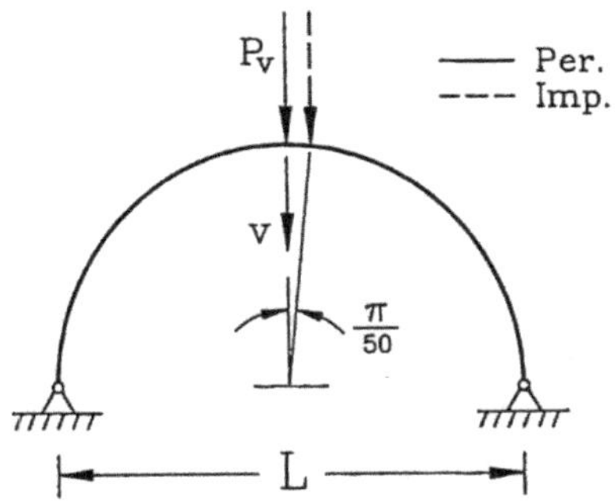

Fig. 7.15. Hinged circular arch.

It is interesting to note that all the secondary paths intersect the primary path at point I, an indication of the accuracy of the present solutions in comparison with the solutions solved by Harrison (1978) using the discrete element method.

7.9.3 *Circular arch with central load*

The hinged circular arch shown in Figure 7.15 has also been studied by Harrison (1978). The following data were assumed for the circular arch: modulus of elasticity $E = 200\,\text{psi}\,(1,378\,\text{kPa})$, diameter $L = 100\,\text{in.}\,(254\,\text{cm})$, moment of inertia $I = 1\,\text{in.}^4\,(41.62\,\text{cm}^4)$, and cross-sectional area $A = 10\,\text{in.}^2\,(64.52\,\text{cm}^2)$. In the present study, the arch was represented by 25 straight beam elements of equal length, with the portion of the arch covered by the central element further divided into two elements, to create the location for the application of the load. Two loading cases are considered. One is the case with load applied symmetrically at the apex, and the other assumes the load to be displaced a small distance to the nearest neighboring node so as to produce the effect of imperfect loading.

The load-deflection curves for the circular arch solved by the GDC method for the symmetric and asymmetric loadings were plotted in Figures 7.16 and 7.17, respectively. Note that for extremely large deformation problems such as the present one, a great number of beam elements should be used to minimize the errors that may result

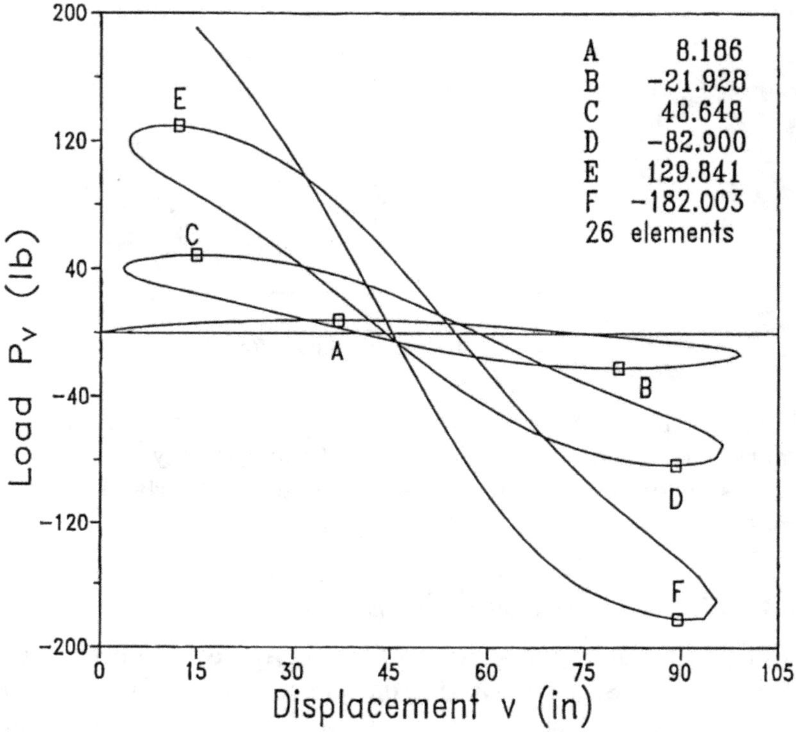

Fig. 7.16. Load–deflection curve for arch under symmetric loading.

from the lack of accuracy in the geometry modeling. With the present 26 elements, however, we are still able to produce solutions that are very close to those obtained by Harrison (1978), who used a finer mesh of 50 elements. Again, the self-adaptive capability of the GDC method in solving problems with complicated postbuckling responses has been clearly demonstrated in this example.

7.9.4 *Angled frame under uniform bending*

In Figure 7.18, an angled frame under uniform bending that is allowed to buckle out-of-plane is considered, which was studied by Argyris *et al.* (1979), Simo and Vu-Quoc (1986), and Yang and Kuo (1992). The three-dimensional characteristics of this frame will be

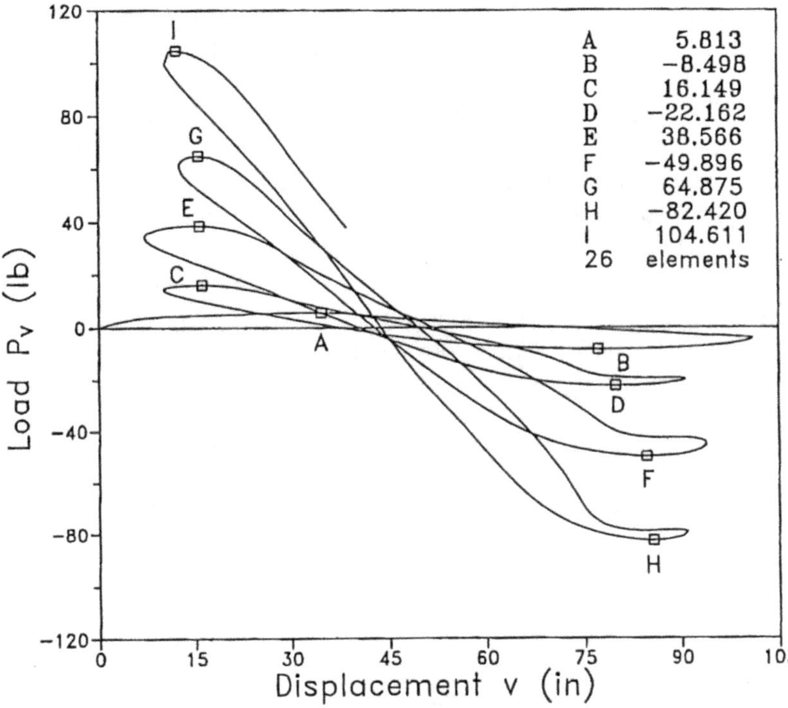

Fig. 7.17. Load–deflection curve for arch under asymmetric loading.

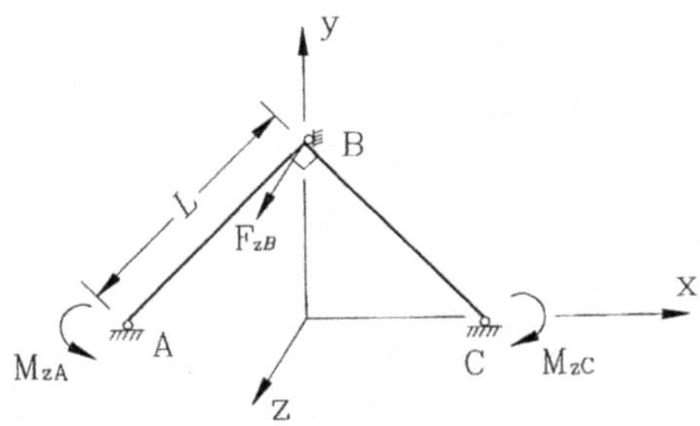

Fig. 7.18. Angled frame under uniform bending.

fully revealed in the postbuckling range since it is not restrained to in-plane deformation.

The following properties are adopted for the angled frame in Figure 7.18: cross-sectional area $A = 18\,\text{mm}^2$, torsional constant

$J = 2.16\,\text{mm}^4$, moment of inertia (minor axis) $I_y = 0.54\,\text{mm}^4$, moment of inertia (major axis) $I_z = 1{,}350\,\text{mm}^4$, length $L = 240\,\text{mm}$, elastic modulus $E = 71{,}240\,\text{N/mm}^2$, and Poisson's ratio $\upsilon = 0.31$. Due to the symmetry of the angled frame, only the left half is analyzed, which is modeled by 10 elements. The member AB is restrained against rotations about the x-and y-axes and translations along the y-and z-axes at node A. The translation along the x-axis at node B is also restrained.

To investigate the effect of positive and negative bending on the angled frame, two different loading cases are analyzed: (1) Applied moment $M_{zA} = -250$ N-mm, perturbation load $F_{zB} = 0.0025$ N; and (2) applied moment $M_{zA} = 150$ N-mm, perturbation load $F_{zB} = 0.0015$ N.

For load case 1, the nodal displacements u_{xA} and u_{zB} have been plotted against the applied moment M_{zA} in Figures 7.19 and 7.20, respectively. The present approach obtained by using

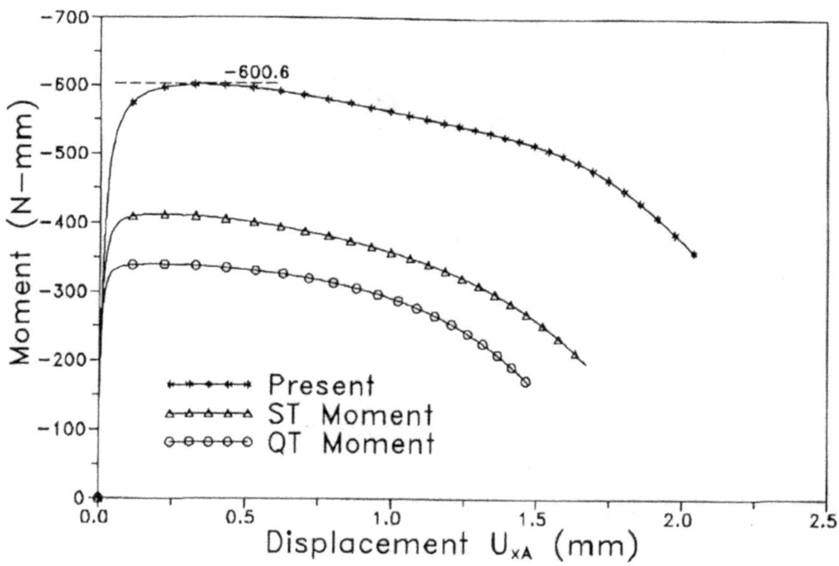

Fig. 7.19. Load case 1: Various solutions for angled frame: u_{xA} vs. M_{zA}.

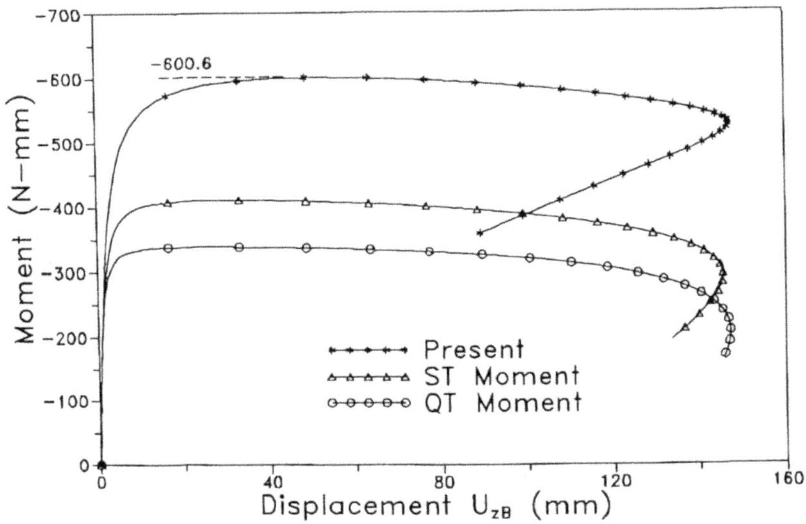

Fig. 7.20. Load case 1: Various solutions for angled frame: u_{zB} vs. M_{zA}.

equation (7.4) for recovering the element forces $\{f^i\}$ predicts a critical moment of -600.6 N-mm, which compares quite well with the theoretical buckling load derived by Yang and Kuo (1991), i.e., $M_{\text{cr}} = \pm(\pi\sqrt{EI_zGJ})/L = \pm 622.2$ N-mm. See also the linearized buckling analysis obtained by using the present beam element in Section 6.8.1. The ST solution in Figures 7.19 and 7.20 are those obtained by treating the nodal moments as semitangential moments, i.e., using element equation of the form in (7.2) that includes the joint moment stiffness [see Yang and Kuo, 1992 or Section 6.5 for explanation). The QT (quasitangential) solution in fact refers to the researches with no knowledge of the rotational properties for nodal moments undergoing out-of-plane rotations, i.e., using only the $[k_e] + [k_g]$ matrix for force increment calculation (see Chapter 6 or Yang and Kuo, 1992).

For load case 2, only the solutions obtained by the present approach for u_{xA} and u_{zB} have been plotted in Figures 7.21 and 7.22. The positive critical moment of 619.6 N-mm obtained herein agrees

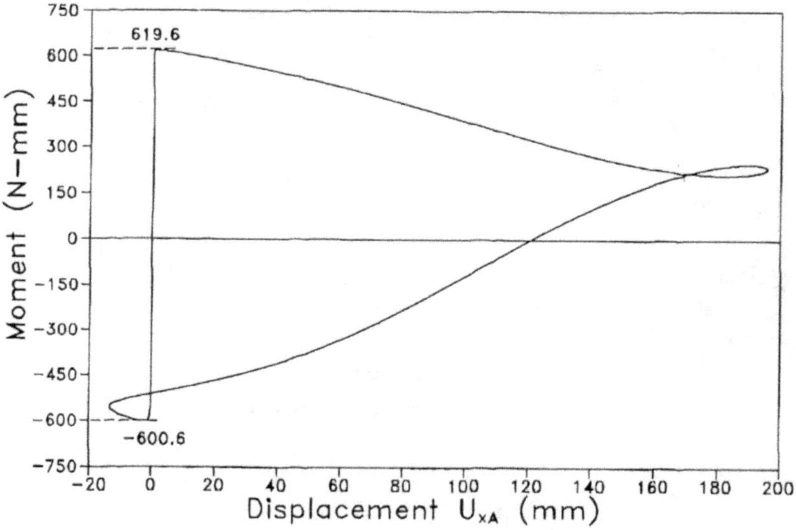

Fig. 7.21. Load case 2: Various solutions for angled frame: u_{xA} vs. M_{zA}.

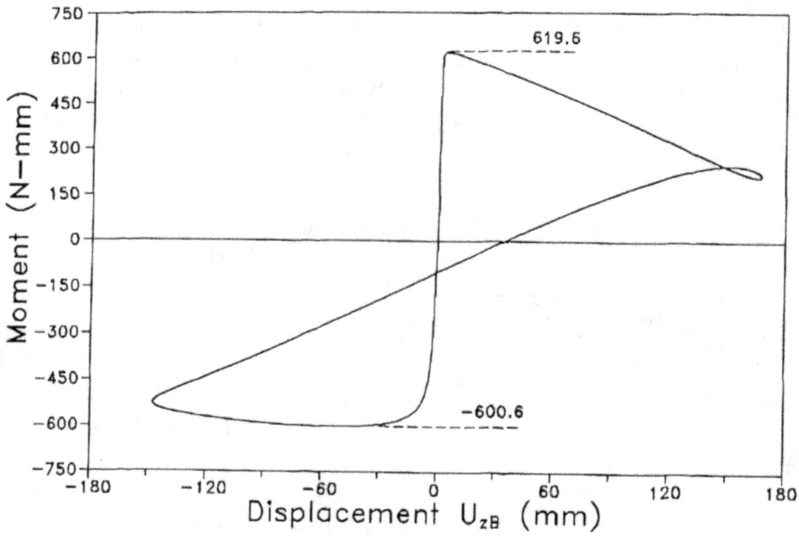

Fig. 7.22. Load case 2: Various solutions for angled frame: u_{zB} vs. M_{zA}.

very well with the linearized buckling value of 622.2 N-mm. For comparison, the critical loads predicted by Argyris *et al.* (1979) and Simo and Vu-Quoc (1986) based on their geometric nonlinear analysis are 624.77 N-mm and 626 N-mm, respectively. From the looping curves of Figures 7.21 and 7.22, we observe that as end A rotates from 0° to 360°, the frame rotates out-of-plane about the axis connecting its supports and returns to its initial configuration. It should be noted that the postbuckling solutions predicted herein are in good agreement with those given by Argyris *et al.* (1979) and Simo and Vu-Quoc (1986).

7.9.5 *Circular arch under uniform bending*

The last problem to be studied here is the out-of-plane buckling of the circular arch under uniform bending, as shown in Figure 7.23, which represents, in effect, the left half of a hinged semi-circular arch under uniform bending. The linearized buckling behavior of this problem was previously studied by Yang and Kuo (1987). The arch is hinged at support A in the sense that rotations about the X- and Y-axes are restrained, i.e., $\Theta_{Xa} = 0$ and $-\Theta_{Ya} = U_{Za} = 0$, and that translations along the Y- and Z-axes are also restrained, i.e., $U_{Ya} = 0$ and $U_{Za} = 0$. Due to the consideration of symmetry, the arch is restrained at end B against rotations about the Y-and Z-axes, i.e., $-\Theta_{Yb} = U'_{Yb} = 0$, and against translation along the X-axis, i.e., $U_{Xb} = 0$. To trigger the out-of-plane buckling mode, a

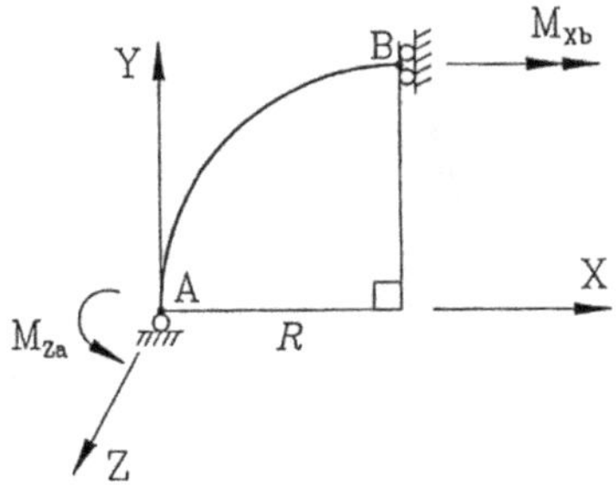

Fig. 7.23. Circular arch under uniform bending.

torsion M_{Xb} with a magnitude equal to 1/500 of the applied moment M_{Za} is applied at end B to serve as the imperfection in loading. The cross-section of the arch is assumed to be rectangular with a height of 30 mm and a width of 0.6 mm. The cross-sectional properties of the arch at end A are as follows: $A = 18\,\text{mm}^2$, torsional constant $J = 2.16\,\text{mm}^4$, moment of inertia $I_z = 1,350\,\text{mm}^4$ and $I_y = 0.54\,\text{mm}^4$. Other related properties of the arch are: $R = 100\,\text{mm}$, modulus of elasticity $E = 71{,}240\,\text{N/mm}^2$ and modulus of rigidity $G = 27{,}190\,\text{N/mm}^2$.

This problem has been selected for the following reasons. First, the analytical solution of the problem is available in Yang and Kuo (1987). Secondly, the problem serves as an effective vehicle for demonstrating the fact that the buckling of curved beams can be accurately predicted by the straight-beam elements if the effects of joint equilibrium have been properly taken into account. Finally, and most importantly, the problem can display the desired phenomenon of finite rotations in the three-dimensional space when loaded into the range of postbuckling.

By approximating the arch by twelve straight-beam elements, the load-deflection curves solved have been plotted for three typical degrees of freedom in Figures 7.24–7.26. As indicated in the figures,

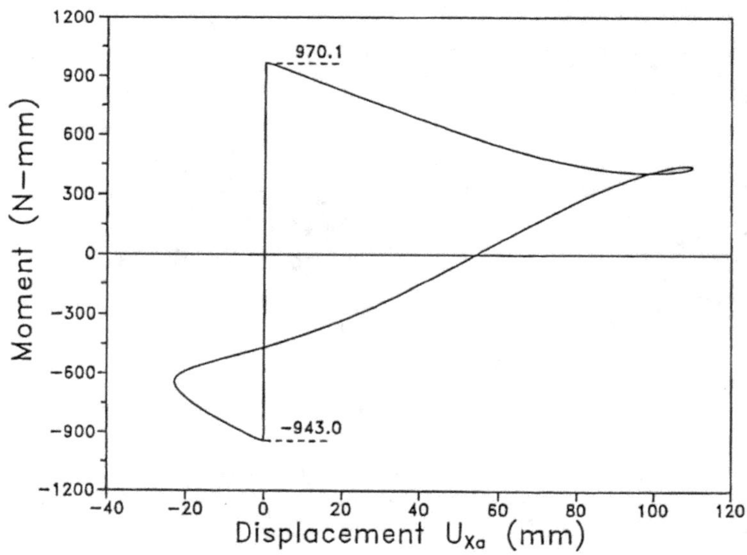

Fig. 7.24. Displacement U_{Xa} vs. bending moment M_{Za}.

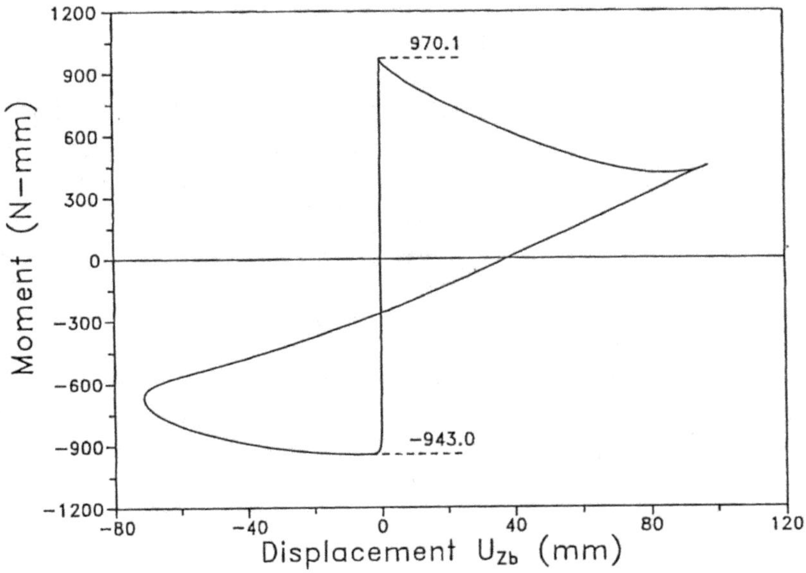

Fig. 7.25. Displacement U_{Xb} vs. bending moment M_{Za}.

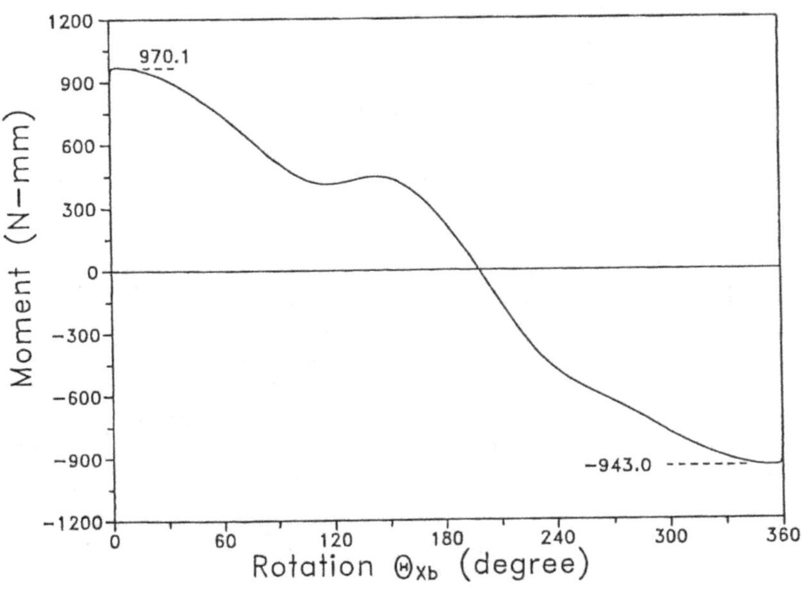

Fig. 7.26. Rotation Θ_{Xb} vs. bending moment M_{Za}.

the buckling loads calculated using the present nonlinear approach for the positive and negative bending cases are 970.1 M-mm and −943.0 N-mm, respectively, which compare quite well with the linearized buckling values of 974.7 N-mm and −947.0 N-mm using the same straight-beam modeling or the theoretical curved-beam buckling solution obtained by the formula of Yang and Kuo (1987).

From the looping curves of Figures 7.24–7.26, we observe that as end B rotates from 0° to 360°, the arch rotates out-of-plane about the horizontal (X) axis and returns to its initial configuration. Such a statement remains true regardless of whether the arch is initially loaded by positive or negative bending moments. As finite rotations have been involved in the postbuckling response of the example, the capability of the procedure described in Appendix B for dealing with the finite-rotation problems in the three-dimensional sense has been clearly demonstrated.

References

Argyris, J.H. (1965). Continua and discontinua, *Proc. Conf. on Matrix Methods in Structural Mechanics*, Wright-Patterson Air Force Base, Ohio, pp. 11–89.

Argyris, J.H., Hilbert, O., Malejannakis, G.A., and Scharph, D.W. (1979). On the geometrical stiffness of a beam in space – A consistent v.w. approach, *Comput. Meth. Appl. Mech. Engrg.*, 20, 105–131.

Batoz, J.L. and Dhatt, G. (1979). Incremental displacement algorithms for nonlinear problems, *Int. J. Numer. Meth. Eng.*, 14, 1262–1266.

Belytschko, T., and Hsieh, B.J. (1974). Nonlinear transient analysis of shells and solids of revolution by convected elements, *AIAA J.*, 12(8), 1031–1035.

Bergan, P.G. (1978). Solution technique for nonlinear finite element problems, *Int. J. Numer. Meth. Eng.*, 12, 1677–1696.

Bergan, P.G. (1980). Solution algorithm for nonlinear structural programs, in *Numerical Methods for Non-linear Problems*, Taylor, C., Hinton, E., and Owen, D.R.J. (eds.), Pineridge, Swansea, pp. 291–305.

Bergan, P.G. and Soreide, T.H. (1978). Solution of large displacement and instability problems using the current stiffness parameter, in *Finite Elements in Nonlinear Mechanics*, Bergan, P.G., *et al.* (eds.), Tapir Publishers, Trondheim, Norway, pp. 647–669.

Chan, S.L. and Kitipornchai, S. (1987). Geometric nonlinear analysis of asymmetric thin-walled beam-columns, *Eng. Struct.*, 9, 243–254.

Connor, J.J., Logcher, D., and Chan, S.C. (1968). Nonlinear analysis of elastic framed structures, *J. Struct. Div*, ASCE, 94(6), 1525–1547.

Crisfield, M.A. (1981). A fast incremental/iterative solution procedure that handles snap-through, *Comput. Struct.*, 13, 55–62.

Crisfield, M.A. (1983). An arc length method including line searches and accelerations, *Int. J. Numer. Meth. Eng.*, 19, 1269–1289.

Harrison, H.B. (1978). Post-buckling behavior of elastic circular arches, *Proc. Inst. Civil Engrs*, 65, Part 2, 283–298.

Kuo, S.R. and Yang, Y.B. (1995). Tracing postbuckling paths of structures containing multi loops, *Int. J. for Numer. Meth. in Eng.*, 38(23), 4053–4075.

Oran, C. (1973). Tangent stiffness in space frames, *J. Struct. Div.*, ASCE, 99(6), 987–1001.

Pecknold, D.A., Ghaboussi, J., and Healey, T.J. (1985). Snap-through and bifurcation in a simple structure, *J. Eng. Mech.*, ASCE, 111(7), 909–922.

Pian, T.H.H. and Tong, P. (1971). Variational formulation of finite displacement analysis, in *Proc. IUTAM Symposium on High Speed Computing for Elastic Structures*, de Veubeke, B.F., (ed.), University of Liege, Liege, Belgium, pp. 43–63.

Powell, G. and Simons, J. (1981). Improved iteration strategy for nonlinear structures, *Int. J. Numer. Meth. Eng.*, 17, 1455–1467.

Ramm, E. (1981). Strategies for tracing the nonlinear response near limit points, in *Nonlinear Finite Element Analysis in Structural Mechanics*, Wunderlich, W., Stein, E. and Bathe, K.J. (eds.), Springer-Verlag, Berlin, FRG, pp. 63–89.

Remseth, S.N. (1979). Nonlinear static and dynamic analysis of framed structures, *Comput. Struct.*, 10, 879–897.

Riks, E. (1972). The application of Newton's methods to the problem of elastic stability, *J. Appl. Mech.*, 39, 1060–1066.

Riks, E. (1979). An incremental approach to the solution of snapping and buckling problems, *Int. J. Solids Struct.*, 15, 529–551.

Simo, J.C. and Vu-Quoc, L. (1986). A three-dimensional finite-strain rod model. Part II: Computational aspects., *Comput. Meth. Appl. Mech. Engrg.*, 58, 79–115.

Tezcan, S.S. and Mahapatra, B.C. (1969). Tangent stiffness matrix for space frame members, *J. Struct. Div.*, ASCE, 95(6), 1257–1270.

Waszczyszyn, Z. (1983). Numerical problems of nonlinear stability analysis of elastic structures, *Comput. Struct.*, 17(1), 13–24.

Wempner, G.A. (1971). Discrete approximation related to nonlinear theories of solids, *Int. J. Solids Struct.*, 7, 1581–1599.

Yang, Y.B., Chen, A., Yan, Y., and Wang, Z.L. (2019). Using only elastic stiffness in nonlinear and postbuckling analysis of structures, *Int. J. Struct. Stability & Dyn.*, 19(10), 1950112.

Yang, Y.B. and Kuo, S.R. (1987). Effect of curvature on stability of curved beams, *J. Struct. Eng.*, ASCE, 113(6), 1185–1202.

Yang, Y.B. and Kuo, S.R. (1991). Out-of-plane buckling of angled frames, *Int. J. Mech. Sci.*, 33(1), 55–67.

Yang, Y.B. and Kuo, S.R. (1992). Frame buckling analysis with full consideration of joint compatibilities, *J. Eng. Mech.*, ASCE, 118(5), 871–889.

Yang, Y.B., Kuo, S.R., and Wu, Y.S. (2002). Incrementally small-deformation theory for nonlinear analysis of structural frames, *Eng. Struct.*, 24(6), 783–798.

Yang, Y.B. and McGuire, W. (1985). A work control method for geometrically nonlinear analysis, in *Proc. 1985 Int. Conf. Numer. Meth. Eng.: Theory and Application,* Middleton, J., Pande, G.N. (eds.), University College, Swansea, UK, pp. 913–921.

Yang, Y.B. and Shieh, M.S. (1990). Solution method for nonlinear problems with multiple critical points, *AIAA J.*, 28(12), 2110–2116.

Zienkiewicz, O.C. (1971). Incremental displacement in non-linear analysis, *Int. J. Numer. Meth. Eng.*, 3, 587–588.

Chapter 8

Rigid Triangular Plate Element for Geometric Nonlinear Analysis of Frames, Plates, and Shells

8.1 Introduction

In the nonlinear analysis of elastic structures, the displacement increments generated at each incremental step can be decomposed into two components as *rigid displacements* and *natural deformations* (see Section 3.9). Based on the updated Lagrangian (UL) formulation, the *geometric stiffness matrix* $[k_g]$ is first derived for a *three-dimensional rigid beam element* from the virtual work equation using a rigid displacement field. Further, by treating the three-node triangular plate element (TPE) as the composition of three rigid beams lying along the three sides, the $[k_g]$ matrix for the TPE can be assembled from those of the rigid beams. The idea for the UL-type incremental-iterative nonlinear analysis is that if the rigid rotation effects are fully taken into account at each stage of analysis, then the remaining effects of natural deformations can be treated using the small-deformation linearized theory. The present approach is featured by the fact that the formulation is simple, the expressions are explicit, and all kinds of actions are duly considered in the stiffness matrices. The robustness of the procedure is demonstrated in the solution of several benchmark problems involving the postbuckling response.

This chapter represents the application of the rigid body rule to the derivation of the geometric stiffness matrix, which is mainly based on the paper by Yang *et al.* (2007), but with modifications to make the expressions consistent with the style of presentation of the book. The geometric stiffness derived for the rigid beam element can perform virtually the same as the geometric stiffness derived for the elastic space frame element in Chapter 6. For pedagogical reasons, the materials of this chapter will be presented in a way that is easy for the readers to follow, even though there may be some replications with previous chapters.

8.2 Brief on Fundamentals for Three-Dimensional Beam

Before we proceed to derive the rigid beam for the 3D beam or space frame element, a summary of the fundamentals, i.e., strains, displacements, and stress resultants, for the three-dimensional beam of bisymmetric solid cross-sections is first given. With the reproduction of all these materials, we believe that the chapter can be comprehended in an easier, independent way.

8.2.1 *Virtual work equation, strains and displacements*

In an incremental nonlinear analysis, three configurations are used to denote the initial deformed configuration C_0, last deformed (but known) configuration C_1, and current deformed (unknown) configuration C_2 of a beam element or solid. It is true that the beam element will be generally *deformed* or *curved* at the C_1 configuration. Mathematically, it would be very complicated and cumbersome to start the nonlinear analysis from a deformed and curved beam element at C_1, because this means we need to abandon the straight beam theory and adopt instead the curved beam theory right from the starting point of C_1 This would be infeasible for engineering applications, as it would be extremely complicated from the point of mathematics.

Fortunately, the deformations of the beam element within each incremental step are generally *small*, and so too are the strains and displacement increments generated during the incremental step. Under such a condition, it is possible to treat the beam element as a *straight line* at C_1 and to attach to this straight line an orthogonal coordinate axis system for describing the behavior of the beam. Such a treatment has been referred to as the concept of *convected coordinates* (Belytschko and Hsieh 1974), which is good for problems with small displacements and rotations at each incremental step. Consequently, the beam the beam element at C_1 can be shown as a straight element in Figure 8.1, with x denoting the centroidal axis and (y, z) the two principal axes of the cross-section.

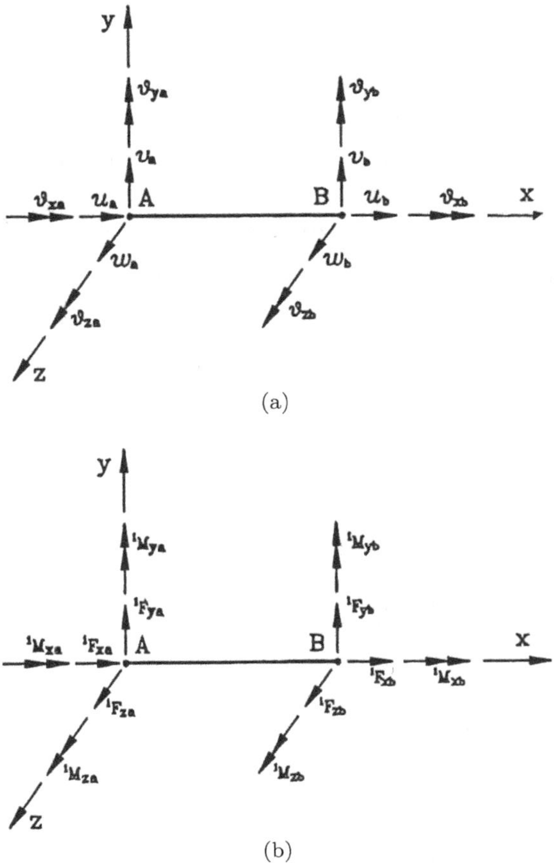

Fig. 8.1. Space frame element: (a) nodal degrees of freedom; (b) initial nodal forces.

Based on the UL formulation, the incremental virtual work equation for a 3D beam at C_2, but with reference to C_1, can be expressed in a linearized form as in equations (5.78) or (6.3), or as reproduced below:

$$\int_V (Ee_{xx}\delta e_{xx} + 4Ge_{xy}\delta e_{xy} + 4Ge_{xz}\delta e_{xz})dV + \int_V ({}^1\tau_{xx}\delta\eta_{xx} + 2{}^1\tau_{xy}\delta\eta_{xy} + 2{}^1\tau_{xz}\delta\eta_{xz})dV = {}^2R - {}^1R \quad (8.1)$$

in which the subscript "1" has been omitted for clarity; E and G denote the moduli of elasticity and rigidity, respectively; ${}^1\tau_{ij}$ is the Cauchy stress existing at C_1; e_{ij} and η_{ij} the linear and nonlinear components of the strain increments, the factors of 4 and 2 are added to account for the symmetry of shear strains, i.e., $e_{xy} = e_{yx}, \ldots, \eta_{xy} = \eta_{yx}, \ldots$, and δ denotes the variation of the quantity following.

The linear and nonlinear components of the strain increments can be expressed with reference to C_1 as

$$e_{xx} = u_{x,x} \quad (8.2)$$

$$e_{xy} = \frac{1}{2}(u_{x,y} + u_{y,x}) \quad (8.3)$$

$$e_{xz} = \frac{1}{2}(u_{x,z} + u_{z,x}) \quad (8.4)$$

and

$$\eta_{xx} = \frac{1}{2}(u_{x,x}^2 + u_{y,x}^2 + u_{z,x}^2) \quad (8.5)$$

$$\eta_{xy} = \frac{1}{2}(u_{x,y}u_{x,x} + u_{y,y}u_{y,x} + u_{z,y}u_{z,x}) \quad (8.6)$$

$$\eta_{xz} = \frac{1}{2}(u_{x,z}u_{x,x} + u_{y,z}u_{y,x} + u_{z,z}u_{z,x}) \quad (8.7)$$

where (u_x, u_y, u_z) are the displacements of a generic point $N(x, y)$ at cross-section x of the beam (see Figure 8.2). By allowing the surface

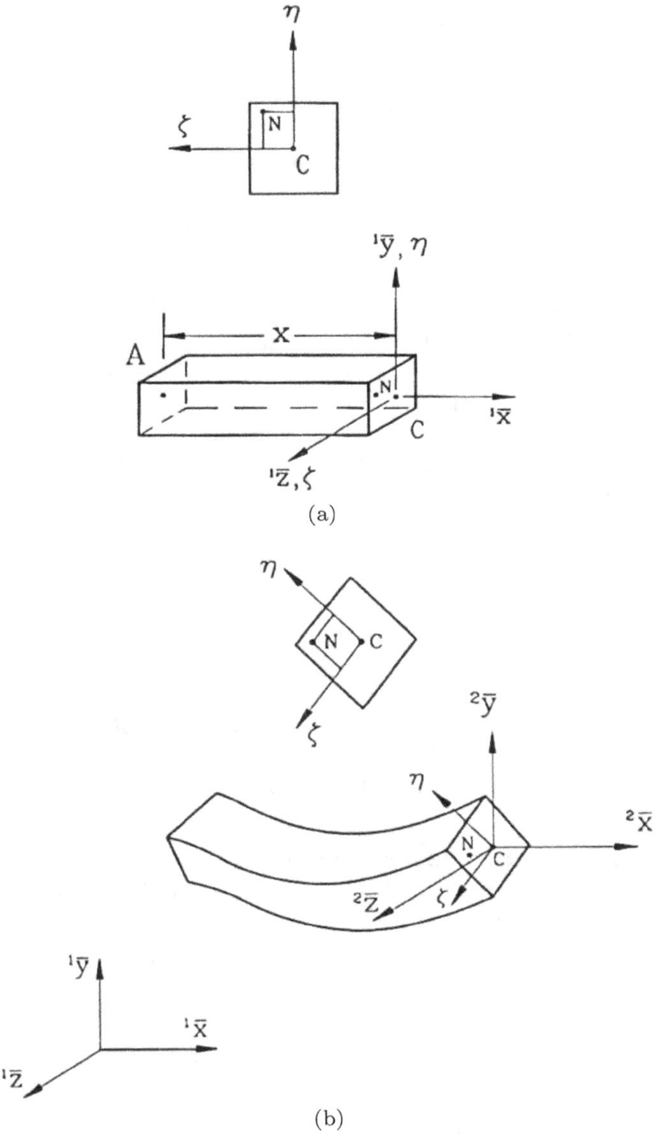

Fig. 8.2. Coordinates of a generic point N at section x at: (a) C_1, (b) C_2.

tractions ${}^1t_k$ and ${}^2t_k$ to act only at the two end sections, the external virtual works ${}^1R$ and ${}^2R$ at C_1 and C_2, respectively, can be written as

$$ {}^1R = \int_S ({}^1t_x \delta u_x + {}^1t_y \delta u_y + {}^1t_z \delta u_z) dS \tag{8.8a} $$

$$ {}^2R = \int_S ({}^2t_x \delta u_x + {}^2t_y \delta u_y + {}^2t_z \delta u_z) dS \tag{8.8b} $$

where S is the surface area of the element at C_1.

Based on the Bernoulli-Euler hypothesis of plane cross sections remaining plane and normal to the centroidal axis of the beam after deformation, the (incremental) displacements (u_x, u_y, u_z) of the generic point N from C_1 to C_2 involved in the strain expressions of equations (8.2)–(8.7) can be related to the displacements (u, v, w) of the centroid of the same cross-section as

$$ u_x = u - yv' - zw' \tag{8.9} $$

$$ u_y = v - z\theta_x \tag{8.10} $$

$$ u_z = w + y\theta_x \tag{8.11} $$

where θ_x is the angle of twist and a prime denotes differentiation with respect the x-axis. In an incremental nonlinear analysis, it is crucial that the coordinates of the generic point N in the deformed C_2 configuration be correctly specified based on the above hypothesis of in-plane rigid cross sections, as will be given in the following section.

For the beam element of length L shown in Figure 8.1, there is a total of 12 degrees of freedom (DOFs), i.e., three displacements and three rotations at each of the two nodes. The element displacement vector $\{u\}$ is

$$ \{u\}^T = \langle u_a \quad v_a \quad w_a \quad \theta_{xa} \quad \theta_{ya} \quad \theta_{za} \quad u_b \quad v_b \quad w_b \quad \theta_{xb} \quad \theta_{yb} \quad \theta_{zb} \rangle \tag{8.12} $$

Correspondingly, the nodal forces $\{{}^1f\}$ and $\{{}^2f\}$ acting on the element at C_1 and C_2 are

$$ \begin{aligned} \{{}^1f\}^T = \langle {}^1F_{xa} \quad {}^1F_{ya} \quad {}^1F_{za} \quad {}^1M_{xa} \quad {}^1M_{ya} \quad {}^1M_{za} \quad {}^1F_{xb} \quad {}^1F_{yb} \\ {}^1F_{zb} \quad {}^1M_{xb} \quad {}^1M_{yb} \quad {}^1M_{zb} \rangle \end{aligned} \tag{8.13} $$

$$\{^2f\}^T = \langle {}^2F_{xa} \quad {}^2F_{ya} \quad {}^2F_{za} \quad {}^2M_{xa} \quad {}^2M_{ya} \quad {}^2M_{za} \quad {}^2F_{xb} \quad {}^2F_{yb} \\ {}^2F_{zb} \quad {}^2M_{xb} \quad {}^2M_{yb} \quad {}^2M_{zb}\rangle \tag{8.14}$$

where all the nodal forces and moments should be interpreted as the stress resultants of the cross-section, as will be elaborated in the following section.

8.2.2 *Stress resultant definitions*

Let us cut a beam at section x and consider its left portion as in Figure 8.2(a). Also, let us attach two sets of coordinates to the centroid C of the cross-section. The first set is the $\eta - \zeta$ coordinates embedded at the cross-section, with η and ζ denoting the two principal axes. The second set is assumed to have an origin fixed at the centroid C, referred to as the ${}^1\bar{x} - {}^1\bar{y} - {}^1\bar{z}$ axes at C_1, with $({}^1\bar{y}, {}^1\bar{z})$ coincident with (η, ζ) (see Figure 8.2(a)), and as the ${}^2\bar{x} - {}^2\bar{y} - {}^2\bar{z}$ axes at C_2, with $({}^2\bar{y}, {}^2\bar{z})$ parallel to $({}^1\bar{y}, {}^1\bar{z})$ (see Figure 8.2(b)).

By the conditions of equilibrium, the axial force $({}^1F_x)$, shear forces $({}^1F_y, {}^1F_z)$, torque $({}^1M_x)$, and bending moments $({}^1M_y, {}^1M_z)$ acting on section x at C_1 can be related to the Cauchy stresses acting on the same section at the same configuration as follows:

$${}^1F_x = \int_A {}^1\tau_{xx}\, dA \tag{8.15}$$

$${}^1F_y = \int_A {}^1\tau_{xy}\, dA \tag{8.16}$$

$${}^1F_z = \int_A {}^1\tau_{xz}\, dA \tag{8.17}$$

$${}^1M_x = \int_A ({}^1\tau_{xz}y - {}^1\tau_{xy}z) dA \tag{8.18}$$

$${}^1M_y = \int_A {}^1\tau_{xx}z\, dA \tag{8.19}$$

$${}^1M_z = -\int_A {}^1\tau_{xx}y\, dA \tag{8.20}$$

where A denotes the cross-sectional area of the beam, and ${}^1\tau_{xx}$, ${}^1\tau_{xy}$, and ${}^1\tau_{xz}$ are the Cauchy stresses at C_1. Similarly, the axial force

(^2F_x), shear forces ($^2F_y, ^2F_z$), torque (^2M_x), and bending moments ($^2M_y, ^2M_z$) acting on section x at C_2 can be related to the second Piola–Kirchhoff stresses acting on the same section at C_2 but referred to C_1 as follows:

$$^2F_x = \int_A {}^2S_{xx}\, dA \tag{8.21}$$

$$^2F_y = \int_A {}^2S_{xy}\, dA \tag{8.22}$$

$$^2F_z = \int_A {}^2S_{xz}\, dA \tag{8.23}$$

$$^2M_x = \int_A ({}^2S_{xz} \cdot {}^2\overline{y} - {}^2S_{xy} \cdot {}^2\overline{z})dA \tag{8.24}$$

$$^2M_y = \int_A ({}^2S_{xx} \cdot {}^2\overline{z} - {}^2S_{xz} \cdot {}^2\overline{y})dA \tag{8.25}$$

$$^2M_z = \int_A ({}^2S_{xy} \cdot {}^2\overline{x} - {}^2S_{xx} \cdot {}^2\overline{y})dA \tag{8.26}$$

where $^2S_{xx}$, $^2S_{xy}$, and $^2S_{xz}$ the second Piola–Kirchhoff stresses acting on section x at C_2, with reference to the C_1 configuration.

Conventionally, a *conservative force* is defined such that the work done by the force is path-independent. This can be realized by defining the force to remain unchanged in the direction of loading during the deformation of the solid upon which it is acting. However, the same is not true for a *conservative moment*, even though it is generated by a mechanism that is composed of a couple or couples of conservative forces. As demonstrated in Section 5.3, *induced moments* may be generated for a conservative moment, such as quaasitantial or semitangential moments, when subjected to rotations in the 3D space. For the reason stated, we shall not talk about the forces ($^2F_x, ^2F_y, ^2F_z$) at C_2, as they are self-explanatory. Instead, we shall focus exclusively on the moments ($^2M_x, ^2M_y, ^2M_z$) at C_2 in the following derivation.

For the element to deform from C_1 to C_2, the incremental displacements (u_x, u_y, u_z) of the generic point N were given in equations (8.9)–(8.11). Consequently, the coordinates of point N at C_2

are (see Figure 8.2(b)):

$$ {}^2\bar{x} = -y\theta_z + z\theta_y \tag{8.27a} $$

$$ {}^2\bar{y} = y - z\theta_x \tag{8.27b} $$

$$ {}^2\bar{z} = z + y\theta_x \tag{8.27c} $$

where it should be noted that the angles of rotation are defined as the derivatives of displacements, i.e., $\theta_y = -w'$ and $\theta_z = v'$. Meanwhile, the second Piola-Kirchhoff stress can be expressed in an incremental form as

$$ {}^2S_i = {}^1\tau_i + S_i \tag{8.28} $$

where the subscript i can be xx, xy, or xz, ${}^1\tau_i$, denotes the Cauchy stresses existing at C_1 and S_i the updated Kirchhoff stress increments. By substituting equations (8.27) and (8.28) into equations (8.24)–(8.26), along with the use of the stress resultant definitions in equations (8.18) and (8.20), the following can be derived for the moments at C_2:

$$ {}^2M_x = \int_A ({}^2S_{xz} \cdot y - {}^2S_{xy} \cdot z) dA \tag{8.29} $$

$$ {}^2M_y = \int_A {}^2S_{xx} \cdot z dA - {}^1M_z\theta_x + \frac{1}{2}{}^1M_x\theta_z \tag{8.30} $$

$$ {}^2M_z = -\int_A {}^2S_{xx} \cdot y dA + {}^1M_y\theta_x - \frac{1}{2}{}^1M_x\theta_y \tag{8.31} $$

where for bisymmetric cross sections, the torsional parameter α in equation (5.92) is equal to ½, namely,

$$ \int_A {}^1\tau_{xz} y dA = -\int_A {}^1\tau_{xy} z dA = \frac{1}{2}{}^1M_x \tag{8.32} $$

The expressions in equations (8.29)–(8.32) are derived based merely on the hypothesis of plane cross sections in equations (8.9)–(8.11) and the stress resultant definitions in equations (8.24)–(8.26). Evidently, the torque ${}^1M_x$ should be interpreted as the *semitangential moment*, as indicated by the terms $\frac{1}{2}{}^1M_x\theta_z$ and $\frac{1}{2}{}^1M_x\theta_y$ in equations (8.30) and (8.31), and the bending moments ${}^1M_z$ and ${}^1M_y$ as the quasitangential moments, as indicated by the terms $-{}^1M_z\theta_x$ and ${}^1M_y\theta_x$ in the same equations.

8.3 Energy Components of Incremental Virtual Work

To assist in the derivation of the *rigid beam element*, each individual term in the incremental virtual work equation (8.1) will be derived in this section.

8.3.1 *Strain energy*

With the linear strain components given in equations (8.2)–(8.4) and the displacements in equations (8.9)–(8.11), we can obtain the strain energy δU in variational form from the first integral of equation (8.1) as in equation (5.81) or

$$\begin{aligned} \delta U &= \int_V (E e_{xx} \delta e_{xx} + 4G e_{xy} \delta e_{xy} + 4G e_{xz} \delta e_{xz}) dV \\ &= \int_0^L \left(EAu'\delta u' + EI_y w''\delta w'' + EI_z v''\delta v'' + GJ\theta'_x \delta\theta'_x\right) dx \end{aligned} \tag{8.33}$$

where L is the length of the beam element, A denotes the cross-sectional area, J the torsional constant, and I_y and I_z the moments of inertia about the y- and z-axis, respectively. From the strain energy term δU, a 12×12 elastic matrix $[k_e]$ can be derived:

$$\delta U = \{u\}^T [k_e] \{u\} \tag{8.34}$$

which has been made available in equations (2.89) or (6.18). However, for the case of rigid displacements to be considered in the following part of this chapter, the strain energy term δU simply vanishes, which therefore is not the focus of the present discussion.

8.3.2 *Potential energy due to initial stresses*

By substituting the nonlinear strain components in equations (8.5)–(8.7) into the second integral of equation (8.1), along with the use of the displacements in equations (8.9)–(8.11) and the definitions for the member forces and moments in equations (8.15)–(8.20), the potential energy δV in variational form due to the initial stresses can be derived as equation (5.94) or

$$\begin{aligned}\delta V &= \int_V \left({}^1\tau_{xx}\delta\eta_{xx} + 2\,{}^1\tau_{xy}\delta\eta_{xy} + 2\,{}^1\tau_{xz}\delta\eta_{xz}\right)dV \\ &= \frac{1}{2}\int_0^L \Bigg[{}^1F_x\delta(u'^2 + v'^2 + w'^2) \\ &+ {}^1F_x\left(\frac{I_y}{A}\delta w''^2 + \frac{I_z}{A}\delta v''^2\right) + \bar{K}\delta\theta_x'^2\Bigg]dx \\ &+ \int_0^L \big[-{}^1M_z\delta(w'\theta_x') - {}^1M_y\delta(v'\theta_x') \\ &- {}^1M_y\delta(u'w'') + {}^1M_z\delta(u'v'')\big]dx \\ &+ \int_0^L \big[{}^1F_y\delta(w'\theta_x - u'v') - {}^1F_z\delta(v'\theta_x + u'w')\big]dx \\ &+ \frac{1}{2}\int_0^L \big[{}^1M_x\delta(v''w') - {}^1M_x\delta(v'w'')\big]dx \end{aligned} \tag{8.35}$$

where higher-order terms (those that cannot be physically interpreted) have been neglected and $\bar{K}$ denotes the Wagner coefficient,

$$\bar{K} = {}^1F_x\frac{I_y + I_z}{A} = {}^1F_x\frac{J}{A} \tag{8.36}$$

with $J = I_y + I_z$ for bisymmetrical cross-sections.

8.3.3 *Virtual work increment*

By treating the surface tractions at C_1 as the Cauchy stresses, i.e., ${}^1t_x = {}^1\tau_{xx}$, ${}^1t_y = {}^1\tau_{xy}$ and ${}^1t_z = {}^1\tau_{xz}$, and by using the displacements in equations (8.9)–(8.11) and the definitions for forces and moments in equations (8.15)–(8.20), the external virtual work ${}^1R$ at C_1, as given in equation (8.8a), can be written as

$${}^1R = \{\delta u\}^T\{{}^1f\} \tag{8.37}$$

where the displacements $\{u\}$ and the initial forces $\{{}^1f\}$ were defined in equations (8.12) and (8.13), respectively.

Similarly, by treating the surface tractions at C_2 as the second Piola-Kirchhoff stresses, i.e., ${}^2t_x = {}^2S_{xx}$, ${}^2t_y = {}^2S_{xy}$ and ${}^2t_z = {}^2S_{xz}$, further expressing the stresses in incremental form as in equation (8.28), and using the displacement field in equations (8.9)–(8.11) and the definitions for forces and moments in equations (8.21)–(8.26), we can derive from equation (8.8b) the external virtual work ${}^2R$ at C_2 as in equation (5.99) or

$$ {}^2R = \{\delta u\}^T\{{}^2f\} + \left[\begin{matrix} \left({}^1M_z\theta_x - \frac{1}{2}{}^1M_x\theta_z\right)\cdot\delta\theta_y \\ +\left(-{}^1M_y\theta_x + \frac{1}{2}{}^1M_x\theta_y\right)\cdot\delta\theta_z \end{matrix} \right]\Bigg|_0^L \tag{8.38} $$

where the C_2 force vector $\{{}^2f\}$ was defined in equation (8.14) and the torsional parameter α has been taken as $1/2$ for bisymmetrical cross sections in the derivation. Clearly, the *semitangential* and *quasitangential* properties of the torque ${}^1M_x$ and bending moments $({}^1M_y, {}^1M_z)$, respectively, are maintained in the preceding expression.

Consequently, the *virtual work increment* (V.W.I.) for the beam element is

$$ \begin{aligned} V.W.I. &= {}^2R - {}^1R \\ &= \{\delta u\}^T\left(\{{}^2f\} - \{{}^1f\}\right) + \left[\begin{matrix} \left({}^1M_z\theta_x - \frac{1}{2}{}^1M_x\theta_z\right)\cdot\delta\theta_y \\ +\left(-{}^1M_y\theta_x + \frac{1}{2}{}^1M_x\theta_y\right)\cdot\delta\theta_z \end{matrix} \right]\Bigg|_0^L \end{aligned} \tag{8.39} $$

where the boundary terms have been the source of the *induced moment* effect in Section 6.3.3. In this chapter, the boundary terms (i.e., induced moment effect) will be directly included as part of the geometric stiffness matrix for the rigid beam element.

8.4 Geometric Stiffness for the Rigid Beam Element

As was demonstrated above, the elastic stiffness matrix $[k_e]$ has been derived from the strain energy δU without any difficulty. The matrix $[k_e]$ is rigid-body-qualified since it won't produce any forces upon rigid rotations. Consequently, it will just be adopted without any further discussion in later analysis.

In the formulation for the beam elements, the axial displacement u and the angle of twist θ_x are often interpolated by linear functions, and the transverse displacements v and w by cubic functions, as in equations (6.9)–(6.12), i.e.,

$$u = \{n_1\}^T\{\bar{u}\} \tag{8.40}$$

$$v = \{n_3\}^T\{\bar{v}\} \tag{8.41}$$

$$w = \{n_3\}^T\{\bar{w}\} \tag{8.42}$$

$$\theta_x = \{n_1\}^T\{\bar{\theta}_x\} \tag{8.43}$$

where the linear and cubic interpolation functions $\{n_1\}$ and $\{n_3\}$ are

$$\langle n_1\rangle = \langle (1-\imath) \quad \imath\rangle \tag{8.44a}$$

$$\langle n_3\rangle = \langle (1-3\imath^2+2\imath^3) \quad (\imath-2\imath^2+\imath^3) \quad (3\imath^2-2\imath^3) \quad (\imath^3-\imath^2)\rangle \tag{8.44b}$$

with $\imath = x/L$, and the displacement vectors are

$$\langle\bar{u}\rangle = \langle u_a \quad u_b\rangle \tag{8.45a}$$

$$\langle\bar{v}\rangle = \langle v_a \quad L\theta_{za} \quad v_b \quad L\theta_{zb}\rangle \tag{8.45b}$$

$$\langle\bar{w}\rangle = \langle w_a \quad -L\theta_{ya} \quad w_b \quad -L\theta_{yb}\rangle \tag{8.45c}$$

$$\langle\bar{\theta_x}\rangle = \langle\bar{\theta}_{xa} \quad \bar{\theta}_{xb}\rangle \tag{8.45d}$$

where each nodal degree of freedom has been defined in Figure 8.1(a).

Using the *linear interpolation* function for the axial and twisting displacements and the *cubic interpolation* function for the transverse displacements, the geometric stiffness $[k_g]$ has to be qualified by the rigid body test, as was demonstrated in Section 6.4. In addition, it was demonstrated in Chapter 7 that a rigid-body-qualified geometric stiffness is all that we need for incremental-iterative nonlinear analysis. In the following, we shall demonstrate that a rigid-body-qualified geometric stiffness can be derived simply by using the rigid body mode shape, i.e., *linear interpolation* function, for all the displacements (i.e., axial, twisting, and transverse displacements) of the beam element in the stiffness derivation.

For an element of length L subjected only to nodal forces, the bending moments at section x can be related to those at the two ends a and b as

$$ {}^1M_y = -{}^1M_{ya}\left(1-\frac{x}{L}\right) + {}^1M_{yb}\left(\frac{x}{L}\right) \tag{8.46a} $$

$$ {}^1M_z = -{}^1M_{za}\left(1-\frac{x}{L}\right) + {}^1M_{zb}\left(\frac{x}{L}\right) \tag{8.46b} $$

where the sign convention as the one in Figure 2.3 has been used for the nodal quantities. Further, based on the conditions of equilibrium, the following can be written

$$ {}^1F_x = -{}^1F_{xa} = {}^1F_{xb} \tag{8.47a} $$

$$ {}^1F_y = -{}^1F_{ya} = {}^1F_{yb} = -\frac{{}^1M_{za} + {}^1M_{zb}}{L} \tag{8.47b} $$

$$ {}^1F_z = -{}^1F_{za} = {}^1F_{zb} = \frac{{}^1M_{ya} + {}^1M_{yb}}{L} \tag{8.47c} $$

$$ {}^1M_x = -{}^1M_{xa} = {}^1M_{xb} \tag{8.47d} $$

For a *rigid rotation*, the axial and angular displacements can be expressed as

$$ u = u_a\left(1-\frac{x}{L}\right) + u_b\left(\frac{x}{L}\right) \tag{8.48a} $$

$$ \theta_x = \theta_{xa}\left(1-\frac{x}{L}\right) + \theta_{xb}\left(\frac{x}{L}\right) \tag{8.48b} $$

subject to the constraints:

$$ u_a = u_b \tag{8.49a} $$

$$ \theta_{xa} = \theta_{xb} \tag{8.49b} $$

The lateral and rotational displacements can be expressed as

$$ v = v_a\left(1-\frac{x}{L}\right) + v_b\left(\frac{x}{L}\right) \tag{8.50a} $$

$$ w = w_a\left(1-\frac{x}{L}\right) + w_b\left(\frac{x}{L}\right) \tag{8.50b} $$

$$ \theta_y = \theta_{ya}\left(1-\frac{x}{L}\right) + \theta_{yb}\left(\frac{x}{L}\right) \tag{8.50c} $$

$$ \theta_z = \theta_{za}\left(1-\frac{x}{L}\right) + \theta_{zb}\left(\frac{x}{L}\right) \tag{8.50d} $$

subject to the constraints:

$$\theta_{ya} = \theta_{yb} \tag{8.51a}$$

$$\theta_{za} = \theta_{zb} \tag{8.51b}$$

For the *rigid beam*, the strain energy vanishes, i.e., $\delta U = 0$. Consequently, the incremental virtual work equation (8.1) for the 3D beam reduces to

$$\delta V = V.W.I \tag{8.52}$$

By the use of equations (8.33), (8.35), (8.37), and (8.38), we may write

$$\begin{aligned}
&\frac{1}{2}\int_0^L \left[{}^1F_x\delta(u'^2 + v'^2 + w'^2) + {}^1F_x\left(\frac{I_y}{A}\delta w''^2 + \frac{I_z}{A}\delta v''^2\right) + \bar{K}\delta\theta_x'^2\right] dx \\
&\quad + \int_0^L \left[-{}^1M_z\delta(w'\theta_x') - {}^1M_y\delta(v'\theta_x') - {}^1M_y\delta(u'w'')\right. \\
&\quad \left. + {}^1M_z\delta(u'v'')\right] dx \\
&\quad + \int_0^L \left[{}^1F_y\delta(w'\theta_x - u'v') - {}^1F_z\delta(v'\theta_x + u'w')\right] dx \\
&\quad + \frac{1}{2}\int_0^L \left[{}^1M_x\delta(v''w') - {}^1M_x\delta(v'w'')\right] dx \\
&= \{\delta u\}^T\left(\{{}^2f\} - \{{}^1f\}\right) + \left[\begin{array}{c} \left({}^1M_z\theta_x - \frac{1}{2}{}^1M_x\theta_z\right)\cdot\delta\theta_y \\ + \left(-{}^1M_y\theta_x + \frac{1}{2}{}^1M_x\theta_y\right)\cdot\delta\theta_z \end{array}\right]\Bigg|_0^L
\end{aligned} \tag{8.53}$$

Since this is a variational problem with constraints, the following points must be considered in deriving each of the terms involved in equation (8.53) using the displacement field in equations (8.48)–(8.51):

(1) The following variational terms are equal to zero: $\delta u'^2 = 0$, $\delta\theta_x'^2 = 0$, $\delta v''^2 = 0$, $\delta w''^2 = 0$, $\delta(u'w'') = 0$, $\delta(u'v'') = 0$, as a direct consequence of the rigid displacement field.

(2) The variation of a quantity that is zero in the rigid displacement field need *not* be zero. For instance, $u' = (u_b - u_a)/L = 0$ according to equation (8.49a), but $\delta u' = (\delta u_b - \delta u_a)/L$ is not equal to zero, since $\delta u_b \neq \delta u_a$.

(3) Even though some of the variational terms in equation (8.53) are known to be zero, such as $u'\delta v'$, $u'\delta w'$, $\theta'_x \delta v'$, $\theta'_x \delta w'$, they are kept in the derivation for the sake of symmetry of the related entries in the stiffness matrix.

(4) The second derivatives v'' and w'' should be treated as θ'_z and $-\theta'_y$, respectively, and interpreted by equations (8.50c) and (8.50d) in order to capture the property of rotations.

With the above considerations in mind, we may substitute the force expressions in equations (8.46)–(8.47) and the displacement field in equations (8.48) and (8.50) into the virtual work equation for the rigid beam in equation (8.53). By taking into the arbitrary nature of the variations, we can derive the stiffness equation for the *rigid beam* element as

$$[k_g]_{r.b.}\{u\} = \{^2f\} - \{^1f\} \tag{8.54}$$

where the subscript "r.b." is used to denote the geometric stiffness for the rigid beam, to distinguish it from that for the TPE to be derived latter. The geometric stiffness in equation (8.54) is

$$[k_g]_{r.b.} = \begin{bmatrix} [g] & [h_a] & -[g] & [h_b] \\ [h_a]^T & [i_a] & -[h_a]^T & [0] \\ -[g] & -[h_a] & [g] & -[h_b] \\ [h_b]^T & [0] & -[h_b]^T & [i_b] \end{bmatrix} \tag{8.55}$$

where $[0]$ is a 3×3 null matrix containing all zero entries and

$$[g] = \begin{bmatrix} 0 & -\frac{^1F_{yb}}{L} & -\frac{^1F_{zb}}{L} \\ -\frac{^1F_{yb}}{L} & \frac{^1F_{xb}}{L} & 0 \\ -\frac{^1F_{zb}}{L} & 0 & \frac{^1F_{xb}}{L} \end{bmatrix} \tag{8.56}$$

$$[h_b] = \begin{bmatrix} 0 & 0 & 0 \\ \frac{^1M_{yb}}{L} & -\frac{^1M_{xb}}{2L} & 0 \\ \frac{^1M_{zb}}{L} & 0 & -\frac{^1M_{xb}}{2L} \end{bmatrix} \tag{8.57}$$

$$[i_b] = \begin{bmatrix} 0 & 0 & 0 \\ -^1M_{zb} & 0 & \frac{^1M_{xb}}{2} \\ ^1M_{yb} & -\frac{^1M_{xb}}{2} & 0 \end{bmatrix} \tag{8.58}$$

The submatrices $[h_a]$ and $[i_a]$ associated with end a can be obtained by replacing the terms $(^1M_{xb}, ^1M_{yb}, ^1M_{zb})$ in the submatrices $[h_b]$ and $[i_b]$ by the terms $(^1M_{xa}, ^1M_{ya}, ^1M_{za})$, respectively. The matrix as presented in equation (8.55) is asymmetric, due to the asymmetry of the submatrices $[i_a]$ and $[i_b]$ originating from the boundary terms in equation (8.53).

One advantage with the rigid beam approach is that the geometric stiffness derived is simple and explicit. No numerical integration is needed in the derivation. Particularly, the submatrices $[i_a]$ and $[i_b]$ in equation (8.55) represents the same *induced moment effect* as that for the elastic beam in equations (6.34) and (6.35). But in this chapter, they are included as part of the geometric stiffness matrix, but not independently identified as the induced moment matrix. In other words, the rigid geometric stiffness $[k_g]_{r.b.}$ herein is equivalent to the $[k_g] + [k_i]$ matrix in Chapter 6.

Of interest to note is that the effects of all kinds of actions, i.e., the axial forces, shear forces, bending moments, and torques undergoing the rigid rotations are considered in the $[k_g]_{r.b.}$ matrix. Moreover, this matrix reduces to the one for the truss element acted upon be a pair of axial forces undergoing the rigid rotations.

8.5 Incremental-Iterative Nonlinear Analysis Using the Rigid Beam Element

The geometric stiffness matrix $[k_g]_{r.b.}$ derived in equation (8.55) is born to be qualified by the rigid body test, as the *linear (rigid) displacement field* (with constraints) has been adopted in the interpolation of all the displacements of the rigid element. This is different from the use of the *cubic polynomial displacement field* for interpolating the transverse displacements of the elastic beam element that leads to the geometric stiffness matrix $[k_g]$ and induced moment matrix $[k_i]$, in equations (6.30) and (6.33), respectively.

Because of its rigid-body qualified property, the geometric stiffness matrix $[k_g]_{r.b.}$ (including already the induced moment matrix $[k_i]$) will be used to replace the geometric stiffness matrix $[k_g]$ and induced moment matrix $[k_i]$ in the incremental stiffness equation for the space frame element in equation (6.37), in recognition that the matrix $[k_g]_{r.b.}$ can account for most of the functions of the geometric stiffness matrix $[k_g]$ in the incremental-iterative analysis. In other words, we will use the following incremental stiffness equation for the space frame element in this chapter:

$$[k_e]\{u\} + [k_g]_{r.b.}\{u\} = \{{}^2f\} - \{{}^1f\} \tag{8.59}$$

where the elastic stiffness matrix $[k_e]$ is exactly the one given in equation (2.89) or (6.18). In the numerical examples to be presented later, it will be demonstrated that the rigid geometric stiffness $[k_g]_{r.b.}$, as a substitute of the elastic geometric stiffness $[k_g]$ derived in Chapter 6, can perform equally well in the solution of nonlinear space frame problems. It is based on such a fact that the rigid beam element derived in Section 8.4 will be used as the basis for deriving the rigid TPE in what follows.

8.6 Geometric Stiffness Matrix for the Triangular Plate Element

Let us consider the TPE shown in Figure 8.3, which has three translational and three rotational DOFs at each of the three tip nodes (Yang *et al.*, 1999). We choose this element simply because it is compatible with the 12-DOF beam element derived in Chapter 6, and can be easily assembled with the space frame elements for general applications.

In Section 8.4, the geometric stiffness matrix for the rigid beam has been explicitly given in terms of the nodal forces and element length. Thus, as far as the *rigid body behavior* of an element is concerned, only the initial forces acting on the element and the external shape of the element need to be considered. The elastic properties that are essential to the deformation of the element, such as Young's

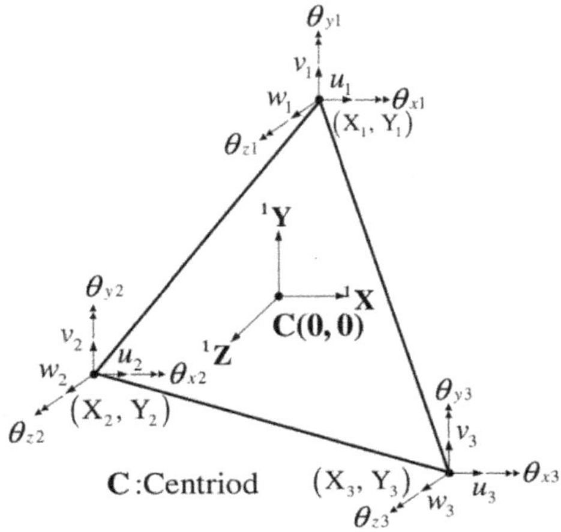

Fig. 8.3. Three-node TPE.

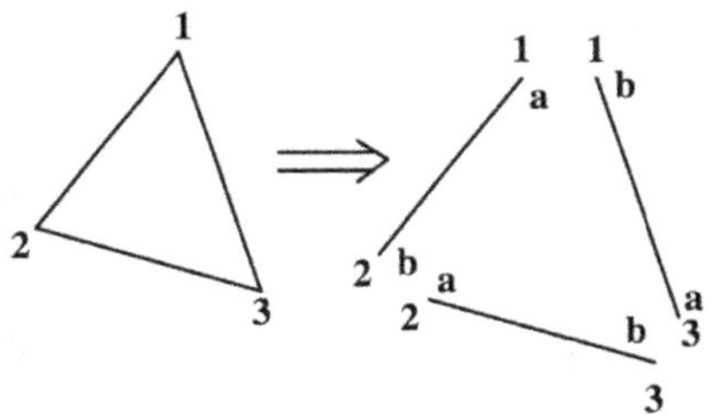

Fig. 8.4. TPE treated as the composite of three rigid beams.

modulus, cross-sectional area, and moments of inertia, can be totally ignored. Based on such an idea, the rigid behavior of the TPE can be simulated as if it were composed of three rigid beams lying along the three sides of the element, as shown in Figure 8.4. It is in this sense that the geometric stiffness matrix for the rigid TPE will be derived.

Figure 8.5 shows the nodal forces acting on each of the three beam elements, named as beam 12, beam 23, and beam 31. In this figure, ${}^1F_k^{ij}$ and ${}^1M_k^{ij}$ denote the nodal force and nodal moment, respectively, acting on beam ij at the C_1 configuration, with the right subscript k denoting the direction and the nodal point on which the force or moment is acting. In the following, we shall show how to determine the nodal forces acting on each beam, based on the equilibrium equations for each node and each beam, and for the whole TPE.

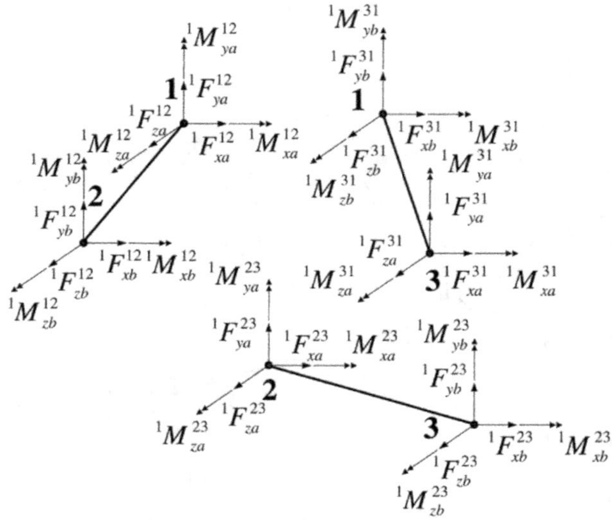

Fig. 8.5. The forces and moments acting on each of the three rigid beam elements.

8.6.1 *Equations of equilibrium for each node of the TPE*

There are three forces and three moments acting at each node of the TPE. The following is the vector of forces $\{^1f\}_{\mathrm{TPE}}$ for the element at the C_1 configuration:

$$
\begin{aligned}
\{^1f\}^T_{\mathrm{TPE}} = \langle {}^1F_{x1} \quad {}^1F_{y1} \quad {}^1F_{z1} \quad {}^1M_{x1} \quad {}^1M_{y1} \quad {}^1M_{z1} \\
{}^1F_{x2} \quad {}^1F_{y2} \quad {}^1F_{z2} \quad {}^1M_{x2} \quad {}^1M_{y2} \quad {}^1M_{z2} \\
{}^1F_{x3} \quad {}^1F_{y3} \quad {}^1F_{z3} \quad {}^1M_{x3} \quad {}^1M_{y3} \quad {}^1M_{z3} \rangle
\end{aligned} \tag{8.60}
$$

There are six equations of equilibrium at each node of the TPE, three for forces and another three for moments. With reference to Figure 8.5, they can be written as follows:

(a) Node 1

For the forces:

$$
{}^1F^{12}_{xa} + {}^1F^{31}_{xb} = {}^1F_{x1} \tag{8.61a}
$$

$$ {}^{1}F_{ya}^{12} + {}^{1}F_{yb}^{31} = {}^{1}F_{y1} \tag{8.61b} $$

$$ {}^{1}F_{za}^{12} + {}^{1}F_{zb}^{31} = {}^{1}F_{z1} \tag{8.61c} $$

For the moments:

$$ {}^{1}M_{xa}^{12} + {}^{1}M_{xb}^{31} = {}^{1}M_{x1} \tag{8.61d} $$

$$ {}^{1}M_{ya}^{12} + {}^{1}M_{yb}^{31} = {}^{1}M_{y1} \tag{8.61e} $$

$$ {}^{1}M_{za}^{12} + {}^{1}M_{zb}^{31} = {}^{1}M_{z1} \tag{8.61f} $$

(b) Node 2

For the forces:

$$ {}^{1}F_{xa}^{23} + {}^{1}F_{xb}^{12} = {}^{1}F_{x2} \tag{8.62a} $$

$$ {}^{1}F_{ya}^{23} + {}^{1}F_{yb}^{12} = {}^{1}F_{y2} \tag{8.62b} $$

$$ {}^{1}F_{za}^{23} + {}^{1}F_{zb}^{12} = {}^{1}F_{z2} \tag{8.62c} $$

For the moments:

$$ {}^{1}M_{xa}^{23} + {}^{1}M_{xb}^{12} = {}^{1}M_{x2} \tag{8.62d} $$

$$ {}^{1}M_{ya}^{23} + {}^{1}M_{yb}^{12} = {}^{1}M_{y2} \tag{8.62e} $$

$$ {}^{1}M_{za}^{23} + {}^{1}M_{zb}^{12} = {}^{1}M_{z2} \tag{8.62f} $$

(c) Node 3

For the forces:

$$ {}^{1}F_{xa}^{31} + {}^{1}F_{xb}^{23} = {}^{1}F_{x3} \tag{8.63a} $$

$$ {}^{1}F_{ya}^{31} + {}^{1}F_{yb}^{23} = {}^{1}F_{y3} \tag{8.63b} $$

$$ {}^{1}F_{za}^{31} + {}^{1}F_{zb}^{23} = {}^{1}F_{z3} \tag{8.63c} $$

For the moments:

$$ {}^{1}M_{xa}^{31} + {}^{1}M_{xb}^{23} = {}^{1}M_{x3} \tag{8.63d} $$

$$ {}^{1}M_{ya}^{31} + {}^{1}M_{yb}^{23} = {}^{1}M_{y3} \tag{8.63e} $$

$$ {}^{1}M_{za}^{31} + {}^{1}M_{zb}^{23} = {}^{1}M_{z3} \tag{8.63f} $$

8.6.2 *Equations of equilibrium of beam ij*

By letting (X_i, Y_i) denote the coordinates of node I on the (X, Y) plane, all the forces and moments acting on beam ij (with $ij = 12, 23,$ or 31) must satisfy the following conditions of equilibrium for the beam:

(a) Equilibrium of forces:

$$^1F_{xa}^{ij} + {^1F_{xb}^{ij}} = 0 \tag{8.64a}$$

$$^1F_{ya}^{ij} + {^1F_{yb}^{ij}} = 0 \tag{8.64b}$$

$$^1F_{za}^{ij} + {^1F_{zb}^{ij}} = 0 \tag{8.64c}$$

(b) Equilibrium of moments:

$$(Y_i{^1F_{za}^{ij}} + Y_j{^1F_{zb}^{ij}}) + {^1M_{xa}^{ij}} + {^1M_{xb}^{ij}} = 0 \tag{8.64d}$$

$$-(X_i{^1F_{za}^{ij}} + X_j{^1F_{zb}^{ij}}) + {^1M_{ya}^{ij}} + {^1M_{yb}^{ij}} = 0 \tag{8.64e}$$

$$-(Y_i{^1F_{xa}^{ij}} + Y_j{^1F_{xb}^{ij}}) + (X_i{^1F_{ya}^{ij}} + X_j{^1F_{yb}^{ij}}) + {^1M_{za}^{ij}} + {^1M_{zb}^{ij}} = 0 \tag{8.64f}$$

8.6.3 *Equations of equilibrium for the TPE*

The conditions of equilibrium must be obeyed by the TPE as a whole, as given below:

(a) Equilibrium of forces:

$$\sum_{k=1}^{3} {^1F_{xk}} = 0 \tag{8.65a}$$

$$\sum_{k=1}^{3} {^1F_{yk}} = 0 \tag{8.65b}$$

$$\sum_{k=1}^{3} {^1F_{zk}} = 0 \tag{8.65c}$$

(b) Equilibrium of moments:

$$\sum_{k=1}^{3} (Y_k{^1F_{zk}} + {^1M_{xk}}) = 0 \tag{8.65d}$$

$$\sum_{k=1}^{3}(-X_k{}^1F_{zk} + {}^1M_{yk}) = 0 \tag{8.65e}$$

$$\sum_{k=1}^{3}(-Y_k{}^1F_{xk} + X_k{}^1F_{yk} + {}^1M_{zk}) = 0 \tag{8.65f}$$

Now we can solve equations (8.61)–(8.65) to obtain the nodal forces for beam ij as

$$^1F_{xa}^{ij} = \frac{1}{3}{}^1F_{xij} + f_x \tag{8.66a}$$

$$^1F_{ya}^{ij} = \frac{1}{3}{}^1F_{yij} + f_y \tag{8.66b}$$

$$^1F_{za}^{ij} = \frac{1}{3}{}^1F_{zij} + f_z \tag{8.66c}$$

$$^1F_{xb}^{ij} = -{}^1F_{xa}^{ij} \tag{8.66d}$$

$$^1F_{yb}^{ij} = -{}^1F_{ya}^{ij} \tag{8.66e}$$

$$^1F_{zb}^{ij} = -{}^1F_{za}^{ij} \tag{8.66f}$$

and the nodal moments as

$$^1M_{xa}^{ij} = \frac{1}{3}{}^1M_{xij} + \frac{1}{3}Y_{ij}{}^1F_{zj} + (m_x - Y_i f_z) \tag{8.67a}$$

$$^1M_{ya}^{ij} = \frac{1}{3}{}^1M_{yij} - \frac{1}{3}X_{ij}{}^1F_{zj} + (m_y + X_i f_z) \tag{8.67b}$$

$$^1M_{za}^{ij} = \frac{1}{3}{}^1M_{zij} - \frac{1}{3}(Y_{ij}{}^1F_{xj} - X_{ij}{}^1F_{yj}) + [m_z + (Y_i f_x - X_i f_y)] \tag{8.67c}$$

$$^1M_{xb}^{ij} = -{}^1M_{xa}^{ij} - Y_{ij}{}^1F_{za}^{ij} \tag{8.67d}$$

$$^1M_{yb}^{ij} = -{}^1M_{ya}^{ij} + X_{ij}{}^1F_{za}^{ij} \tag{8.67e}$$

$$^1M_{zb}^{ij} = -{}^1M_{za}^{ij} + Y_{ij}{}^1F_{xa}^{ij} - X_{ij}{}^1F_{ya}^{ij} \tag{8.67f}$$

where a variable with right subscripts ij denotes the difference of the two associated quantities, e.g., ${}^1M_{xij} = {}^1M_{xi} - {}^1M_{xj}, X_{ij} = X_i - X_j$, and $f_x, f_y, f_z, m_x, m_y, m_z$ are the constants to be determined.

One feature with the rigid element is that there are more unknowns than the equations that can be utilized. For instance, the total number of equations given in equations (8.61)–(8.67) is 42. By deducting the 12 dependent equations, the total number of equations that can be used is 30, but the total number of unknown forces in equations (8.66)–(8.67) is 36, with 12 for each of the three beam elements. This means that the force distribution of a *rigid element* is not unique, unlike that of an *elastic element*, and that six more conditions should be made in order to obtain a solution. Following Yang *et al.* (2007), we simply let the six variables f_x, f_y, f_z, m_x, m_y, m_z equal to zero in the present formulation.

8.6.4 *Assembly of the geometric stiffness matrix for the TPE*

All the nodal forces and moments solved and given in equations (8.66) and (8.67) have been expressed with reference to the coordinates of the TPE, as indicated in Figure 8.5. They have to be transformed into the local coordinates of each element in order to compute the geometric stiffness matrix $[k_g]_{r.b.}$ of the element. Let $\{{}^1f^{ij}\}$ denote the nodal forces of element ij in the global coordinates of the TPE and $\{{}^1\bar{f}^{ij}\}$ in the local coordinates of element ij,

$$\begin{aligned}\{{}^1f^{ij}\}^T = \langle {}^1F^{ij}_{xa} \quad {}^1F^{ij}_{ya} \quad {}^1F^{ij}_{za} \quad {}^1M^{ij}_{xa} \quad {}^1M^{ij}_{ya} \quad {}^1M^{ij}_{za} \\ {}^1F^{ij}_{xb} \quad {}^1F^{ij}_{yb} \quad {}^1F^{ij}_{zb} \quad {}^1M^{ij}_{xb} \quad {}^1M^{ij}_{yb} \quad {}^1M^{ij}_{zb}\rangle\end{aligned} \tag{8.68}$$

$$\begin{aligned}\{{}^1\bar{f}^{ij}\}^T = \langle {}^1\bar{F}^{ij}_{xa} \quad {}^1\bar{F}^{ij}_{ya} \quad {}^1\bar{F}^{ij}_{za} \quad {}^1\bar{M}^{ij}_{xa} \quad {}^1\bar{M}^{ij}_{ya} \quad {}^1\bar{M}^{ij}_{za} \\ {}^1\bar{F}^{ij}_{xb} \quad {}^1\bar{F}^{ij}_{yb} \quad {}^1\bar{F}^{ij}_{zb} \quad {}^1\bar{M}^{ij}_{xb} \quad {}^1\bar{M}^{ij}_{yb} \quad {}^1\bar{M}^{ij}_{zb}\rangle\end{aligned} \tag{8.69}$$

The following is the transformation between the two sets of forces for element ij:

$$\{{}^1\bar{f}^{ij}\} = [T]\{{}^1f^{ij}\} \tag{8.70}$$

where the transformation matrix $[T]$ is

$$[T] = \begin{bmatrix} [R] & [0] & [0] & [0] \\ [0] & [R] & [0] & [0] \\ [0] & [0] & [R] & [0] \\ [0] & [0] & [0] & [R] \end{bmatrix} \tag{8.71}$$

with

$$[R] = \begin{bmatrix} -\frac{X_{ij}}{L_{ij}} & -\frac{Y_{ij}}{L_{ij}} & 0 \\ \frac{Y_{ij}}{L_{ij}} & -\frac{X_{ij}}{L_{ij}} & 0 \\ 0 & 0 & 1 \end{bmatrix} \tag{8.72}$$

Here, L_{ij} denotes the length of element ij. By substituting the nodal forces $\{^1\bar{f}^{ij}\}$ into equation (8.55), the geometric stiffness matrix $[k_g]_{beam\_ij}$ for element ij can be obtained. Finally, by assembling the geometric stiffness matrices for the three elements following the standard finite element procedure, the geometric stiffness matrix $[k_g]_{TPE}$ for the rigid TPE can be obtained:

$$[k_g]_{TPE} = \sum_{ij=12,23,31} [T]^T [k_g]_{beam\_ij}[T] \tag{8.73}$$

which is of dimension 18×18 with 6 degrees of freedom for each node, i.e., three translations and three rotations. Details of the $[k_g]_{TPE}$ has been explicitly given in Appendix C. Although the details of the $[k_g]_{TPE}$ matrix have been given in Appendix C, they are so lengthy and cumbersome. From the point of computer programming, the coding of the $[k_g]_{TPE}$ matrix may be error-pone. One suggestion herein is to construct the $[k_g]_{TPE}$ matrix instead by directly assembling the stiffness matrices $[k_g]_{beam\_ij}$ for the three beam elements ij, following the standard element assembly procedure. By so doing, the human effort is reduced to the minimum.

Consequently, the incremental stiffness equation for the rigid TPE can be augmented from equation (8.54) for the beam element, i.e., by the assembly of the stiffness matrix equations for the three rigid elements, as follows:

$$[k_g]_{TPE}\{u\}_{TPE} = \{^2f\}_{TPE} - \{^1f\}_{TPE} \tag{8.74}$$

where both the displacement vector $\{u\}_{TPE}$ and force vectors $\{^2f\}_{TPE}$ and $\{^1f\}_{TPE}$ for the TPE are of dimension 18×1.

It is easy to verify that the geometric stiffness matrix $[k_g]_{TPE}$ derived above for the TPE passes the rigid body test described in Chapter 3, in that the resulting forces $\{^2f\}_{TPE}$ computed from equation (8.74) for the TPE undergoing a rigid rotation to maintain a magnitude equal to that of the initial forces $\{^1 f\}_{TPE}$, but with the directions of the nodal actions rotated by an angle equal to the rigid rotation.

8.7 Joint Equilibrium Conditions in the Rotated Position

Both the geometric stiffness matrix $[k_g]_{r.b.}$ for the rigid beam, as given in equation (8.55), and the matrix $[k_g]_{TPE}$ derived for the TPE in equation (8.73), as detailed in Appendix C, are asymmetric. The asymmetry originates from the submatrices $\lceil i_a \rceil$ and $\lceil i_a \rceil$ for $[k_g]_{r.b.}$ and $[I_1], [I_2]$ and $[I_3]$ for $[k_g]_{TPE}$, which relates to the behavior of nodal moments undergoing 3D rotations. They play a role similar to the *induced moment matrix* for the 3D beam element in equation (6.33). Such a property of asymmetry is restricted to the element level, but not on the structure level, which adds little extra effort to the nonlinear analysis. To explain such a fact, let us decompose the geometric stiffness matrices for the rigid beam in equation (8.55) and for the TPE in equation (C.1) into the symmetric and anti-symmetric parts:

$$[k_g]_{r.b.} = [k_g]_{r.b.}^{sym} + [k_g]_{r.b.}^{anti-sym} \tag{8.75}$$

$$[k_g]_{TPE} = [k_g]_{TPE}^{sym} + [k_g]_{TPE}^{anti-sym} \tag{8.76}$$

where the anti-symmetric parts are similar to the one given in equation (6.61) for the 3D beam and can be given as follows:

$$[k_g]_{r.b.}^{anti-sym} = \begin{bmatrix} [0] & [0] & [0] & [0] \\ [0] & [A_a] & [0] & [0] \\ [0] & [0] & [0] & [0] \\ [0] & [0] & [0] & [A_b] \end{bmatrix} \tag{8.77}$$

$$[k_g]_{TPE}^{anti-sym} = \begin{bmatrix} [0] & [0] & [0] & [0] & [0] & [0] \\ [0] & [A_1] & [0] & [0] & [0] & [0] \\ [0] & [0] & [0] & [0] & [0] & [0] \\ [0] & [0] & [0] & [A_2] & [0] & [0] \\ [0] & [0] & [0] & [0] & [0] & [0] \\ [0] & [0] & [0] & [0] & [0] & [A_3] \end{bmatrix} \tag{8.78}$$

The asymmetry in the preceding two matrices is caused by the submatrix $[A_k]$, which is a function of the nodal moments,

$$[A_k] = \frac{1}{2}\begin{bmatrix} 0 & {}^1M_{zk} & -{}^1M_{yk} \\ -{}^1M_{zk} & 0 & {}^1M_{xk} \\ {}^1M_{yk} & -{}^1M_{xk} & 0 \end{bmatrix} \tag{8.79}$$

with the right subscript $k = a, b$ for the rigid beam element, and $k = 1, 2, 3$ for the TPE.

Following Yang and Kuo (1992), it has been demonstrated in Section 6.5 for the three-dimensional beam element and that the anti-symmetric parts of the geometric stiffness matrices of all beam elements meet at the same joint will cancel out, once the conditions of equilibrium for the joint in the rotated configuration C_2 are enforced. As a result, the stiffness matrix assembled for the structure turns out to be symmetric. The same is also true for the plates and shells represented by the TPE, as will be proved below.

By making use of the permutation symbol e_{ijk} (Malvern, 1969), the antisymmetric matrix $[A]$ in equations (8.77) and (8.78) can be expressed as follows:

$$A_{ij} = \frac{1}{2} e_{ijk} {}^1M_k \quad (i, j, k = 1, 2, 3) \tag{8.80}$$

where it is implied that ${}^1M_1 = {}^1M_x, {}^1M_2 = {}^1M_y$, etc.

Let $[\Phi]$ denote the transformation matrix for the element from the local coordinates to the global coordinates. Then,

$$[\Phi]^T[\Phi] = [I] \tag{8.81}$$

For the identity matrix $[I]$,

$$\Phi_{pi}\Phi_{pj} = \delta_{ij} \quad (i, j, p = 1, 2, 3) \tag{8.82}$$

where δ_{ij} is the Kronecker delta, which equals one for $i = j$, and zero otherwise. Based on the tensorial operations, the following relation can be shown to be valid:

$$e_{ijk}\Phi_{pi}\Phi_{qj}\Phi_{rk} = e_{pqr} \quad (i, j, k, p, q, r = 1, 2, 3) \tag{8.83}$$

The nodal moments $\{M\}$ in the element coordinates can be transformed to the structure coordinates as

$$\{\bar{M}\} = [\Phi]\{M\} \tag{8.84}$$

where $\{\bar{M}\}$ denotes the nodal moments in the structural coordinates, or

$$^1\bar{M}_p = \Phi_{pi}{}^1M_i \quad (i, p = 1, 2, 3) \tag{8.85}$$

Similarly, the antisymmetric matrix $[A]$ can be transformed to the global coordinates as

$$[\bar{A}] = [\Phi][A][\Phi]^T \tag{8.86}$$

or

$$\bar{A}_{pq} = \Phi_{pi}A_{ij}\Phi_{qj} \quad (i, j, p, q = 1, 2, 3) \tag{8.87}$$

Substituting equation (8.80) into (8.87) yields

$$\bar{A}_{pq} = \Phi_{pi}\Phi_{qj}A_{ij} = \frac{1}{2}e_{ijk}\Phi_{pi}\Phi_{qj}{}^1M_k \quad (i, j, k, p, q, r = 1, 2, 3) \tag{8.88}$$

By the following relations:

$$e_{ijk}\Phi_{pi}\Phi_{qj} = e_{ijt}\Phi_{pi}\Phi_{qj}\delta_{tk} = e_{ijt}\Phi_{pi}\Phi_{qj}\Phi_{rt}\Phi_{rk} = e_{pqr}\Phi_{rk} \tag{8.89}$$

the preceding equation reduces to

$$\bar{A}_{pq} = \frac{1}{2}e_{pqr}\Phi_{rk}{}^1M_k \quad (k, p, q, r = 1, 2, 3) \tag{8.90}$$

In view of equation (8.85),

$$\bar{A}_{pq} = \frac{1}{2}e_{pqr}{}^1\bar{M}_r \quad (p, q, r = 1, 2, 3) \tag{8.91}$$

Next, consider a *nodal point* of the plate/shell structure where n TPE elements lying along different directions are connected. For the nodal point to be in equilibrium at the deformed configuration C_2, the sum of moments exerted by all the elements meeting at the node along the three global axes must be equal to zero,

$$\sum_{s=1}^{n} {}^1\bar{M}_r^{(s)} = 0 \quad (r = 1, 2, 3) \tag{8.92}$$

Consequently, we can derive the following:

$$\sum_{s=1}^{n} {}^1\bar{A}_{pq}^{(s)} = \sum_{s=1}^{n} \frac{1}{2} e_{pqr} {}^1\bar{M}_r^{(s)} = \frac{1}{2} e_{pqr} \sum_{s=1}^{n} {}^1\bar{M}_r^{(s)} = 0 \tag{8.93}$$

where $p, q, r = 1, 2, 3$. The implication here is that when the antisymmetric matrices $[\bar{A}]$, which are referred to as the global coordinates, are summed over all the TPE elements that share a common node, the resulting anti-symmetric stiffness associated with that joint is nothing but zero. Thus, by enforcing the conditions of equilibrium to be satisfied by all the TPE elements connected to the same node, only the symmetric portion of the induced moment matrix, i.e., the $[k_g]_{TPE}^{sym}$ matrix in equation (8.76), associated with that node will be retained after the element assembly. Such a symmetric portion of the matrix in equation (8.76) has been referred to as the *joint moment matrix* $\lceil k_j \rceil$ in Section 6.5 following Yang and Kuo (1992). Hence, the total stiffness matrix for the structure (i.e., plates/shells) in the rotated configuration C_2 is symmetric, although the element stiffness matrices are asymmetric.

8.8 Incremental-Iterative Nonlinear Analysis Using the Rigid TPE

One important feature of the rigid-plate geometric stiffness $[k_g]_{TPE}$ is that it is rigid body qualified and is therefore suitable for use in the incremental-iterative nonlinear analysis. Parallel to the incremental equation (8.59) that uses the rigid geometric stiffness $[k_g]_{r.b.}$ to replace the elastic geometric stiffness $[k_g]$ for the beam element, we shall augment the incremental equation for the TPE in equation (8.74) to include the elastic stiffness matrix $[k_e]_{TPE}$ as follows:

$$[k_e]_{TPE}\{u\}_{TPE} + [k_g]_{TPE}\{u\}_{TPE} = \{{}^2f\}_{TPE} - \{{}^1f\}_{TPE} \tag{8.94}$$

Obviously, in the above context, we have not derived any elastic stiffness matrix $[k_e]_{TPE}$ for the TPE.

Judging from the fact that derivation of the elastic stiffness matrix $[k_e]_{TPE}$ for the 3-node TPE is by itself a complicated and challenging task. We shall not devote our effort to derive such a matrix for the

TPE in this book, but choose to adopt the best available elastic stiffness matrix in the literature. In this regard, we shall construct the elastic stiffness matrix $[k_e]_{TPE}$ as the composition of Cook's (1987) *plane hybrid element* to simulate the membrane actions and the *HSM element* of Batoz *et al.* (1980) to simulate the bending actions. In the numerical examples, it will be demonstrated that using the rigid-plate geometric stiffness $[k_g]_{TPE}$ and the composed elastic stiffness $[k_e]_{TPE}$ allows us to well solve the nonlinear and postbuckling responses of various plate and shell problems.

8.9 Predictor and Corrector for Incremental-Iterative Nonlinear Analysis

Only incremental-iterative nonlinear analysis of the updated Lagrangian type is of concern in this book. The idea is that if the rigid rotation effects are fully taken into account at each stage of analysis, then the remaining effects of natural deformations can be treated using the small-deformation linearized theory (Yang *et al.*, 2002). In this regard, two major stages can be identified, as described in Chapter 7 (Yang and Leu, 1991). The first is the *predictor* or trial stage, which relates to solution of the structural displacements $\{U\}$, given the load increments $\{^2P\} - \{^1P\}$, as indicated by the following equation:

$$[K]\{U\} = \{^2P\} - \{^1P\} \tag{8.95}$$

where $[K]$ is the total stiffness of the structure, and $\{^1P\}$ and $\{^2P\}$ denote the applied loads acting on the structure at C_1 and C_2, respectively. Once the structural displacement increments $\{U\}$ are solved and made available, the displacement increments $\{u\}$ for each element can be computed accordingly. It is known that the predictor affects only the number of iterations or the speed of convergence, as described in Chapter 7 (Yang and Leu, 1991). Therefore, the structural stiffness matrix $[K]$ is allowed to be approximate, but to the limit that the direction of iterations is not misguided. In addition to the elastic stiffness matrix, this will require the use of rigid-qualified geometric stiffness matrices, such as $[k_g]_{r.b.}$ and $[k_g]_{TPE}$ for the three-dimensional rigid beam and TPE elements, that are capable of simulating the rigid rotational behavior of the initially stressed elements.

The *corrector stage* is concerned with the recovery of the element forces $\{^2_2f\}$ at C_2 and with reference to C_2. For incremental analysis of the updated Lagrangian type, two contributions need to be considered. The first is the initial nodal forces $\{^1_1f\}$ existing at C_1 and expressed with reference to C_1. According to the rigid body rule, initial nodal forces $\{^1_1f\}$ will be rotated by an angle equal to the rigid rotation (with no limit on the magnitude of rotation) during the incremental step from C_1 to C_2, while their magnitude remains unchanged. Thus, we can directly treat the initial forces $\{^1_1f\}$ as the forces acting at C_2 and expressed with respect to C_2. As the rigid rotation effect is already taken into account, the force increments $\{\Delta f\}$ can be computed from the displacement increments $\{u\}$ using the elastic stiffness matrix $[k_e]$ alone based on the small-deformation linearized theory, i.e.,

$$\{\Delta f\} = [k_e]\{u\} \tag{8.96}$$

Summing up the above two effects yields the total element forces $\{^2_2f\}$ at C_2 as

$$\{^2_2f\} = \{^1_1f\} + \{\Delta f\} \tag{8.97}$$

where the reference configuration is known to be C_2.

As the structural and element displacements, i.e., $\{U\}$ and $\{u\}$, are made available from the predictor stage in equation (8.95), we can compute the nodal coordinates and element orientations for the structure and each of its elements in the displaced configuration C_2, following basically the procedure described in Chapter 7. Here, the nodal rotations generated at each incremental (or iterative) step are assumed to be so small that the *law of commutativity* applies. Then, by summing the element forces $\{^2_2f\}$ computed from equation (8.97) for each node of the structure at C_2 and by comparing them with the total applied loads $\{^2P\}$, the unbalanced forces $\{R\}$ for the structure at C_2 can be computed. Another iteration involving the predictor and corrector should be conducted to eliminate the unbalanced forces, should they be greater than present tolerances.

In a UL-type incremental-iterative nonlinear analysis, the accuracy of the solution is governed primarily by the corrector; while the predictor can only affect the speed of convergence or the number of iterations, as described in Chapter 7. For the purpose of investigating the capacity of the rigid geometric stiffness matrices for the rigid beam and TPE derived in this chapter, the following combinations of predictor and corrector will be numerically tested in the analysis.

(a) For *space frames*:

- *Predictor*: The structural stiffness matrix $[K]$ in equation (8.95) is assembled using either of the following:

  (P1) $[k_e] + [k_g]_{r.b.}$, where the elastic stiffness matrix $[k_e]$ given in equation (2.89) or (6.18) and $[k_g]_{r.b.}$ is the geometric stiffness for the rigid beam, given in equation (8.55).

  (P2) $[k_e] + [k_g] + [k_i]$, where the geometric stiffness matrix $[k_g]$ and the induced moment matrix $[k_i]$ were given in equations (6.30) and (6.33), respectively.

  (P3) $[k_e]$ matrix only.

- *Corrector*: The part of the initial nodal forces $\{^1_1f\}$ is updated by the rigid body rule, by treating the initial nodal forces $\{^1_1f\}$ existing at C_1 and referred to C_1 as those acting at C_2 and referred to C_2, i.e., $\{^2_2f\} \equiv \{^1_1f\}$.

  (C1) The force increment is calculated as $\{\Delta f\} = [k_e]\{u\}$ as in equation (8.96).

  (C2) $\{\Delta f\} = ([k_e] + [k_g] + [k_i])\,\{u_n\}$, where the geometric stiffness matrix $[k_g]$ and the induced moment matrix $[k_i]$ are those given in equations (6.30) and (6.33), respectively, and $\{u_n\}$ the natural deformations of the element, computed using the procedure in Yang and Chiou (1987).

(b) For *plates and shells*: The TRIC element developed by Argyris *et al.* (1979) will be used to yield solutions for comparison. The following are the combinations to be used:

- *Predictor*: The structural stiffness matrix $[K]$ in equation (8.95) is assembled using either of the following:

  (P1) $[k_e]_{TPE} + [k_g]_{TPE}$, where the elastic stiffness $[k_e]_{TPE}$ is composed of Cook's (1987) *plane hybrid element* and the *HSM element* of Batoz *et al.* (1980), and $[k_g]_{TPE}$

is the one derived in this chapter in equation (8.73) or in Appendix C.

(P2) $[k_e]_{TRIC} + [k_g]_{TRIC}$, where both the stiffness matrices are given in Argyris *et al.* (1979).

(P3) $[k_e]_{TPE}$ matrix only.

- *Corrector*: The part of the initial nodal forces $\{^1_1 f\}$ is updated by the rigid body rule, by treating the initial nodal forces $\{^1_1 f\}$ existing at C_1 and referred to C_1 as those acting at C_2 and referred to C_2, i.e., $\{^2_2 f\} \equiv \{^1_1 f\}$.

(C1) $\{\Delta f\} = [k_e]_{TPE}\{u\}$, as proposed in equation (8.96) of this chapter.

(C2) $\{\Delta f\} = [k_e]_{TRIC}\{u\}$.

8.10 Numerical Examples

The *generalized displacement control* (GDC) method devised by Yang and Shieh (1990) and detailed in Chapter 7 is adopted for tracing the nonlinear load–deflection response of the problems studied. With the aid of the *general stiffness parameter* (GSP), this method can automatically adjust the load increment sizes to reflect the variation in stiffness of the structure, while reversing the direction of loading when passing a limit point. This method can easily deal with the various critical points encountered in the postbuckling response of a structure. Since all the analyses are allowed to self-adjust their load increments by the GDC method, the total computation time consumed by each combination scheme of predictor and corrector listed in Section 8.9 may not truly reflect its efficiency of computation.

8.10.1 *Single beam subjected to an end moment*

Figure 8.6 shows a single beam restrained against the rotations about the x and y axes on the left end and against the rotations about the y and z axes on the right end. The following data are adopted: length $L = 100$ mm, cross-sectional area $A = 0.18\,\text{cm}^2$, torsional constant $J = 2.16\,\text{mm}^4$, moments of inertia $I_y = 0.54\,\text{mm}^4$, $I_z = 1{,}350\,\text{mm}^4$, Young's modulus $E = 71{,}240\,\text{N/mm}^2$, and shear modulus $G = 27{,}190\,\text{N/mm}^2$. The beam is subjected to a moment of M_{za}

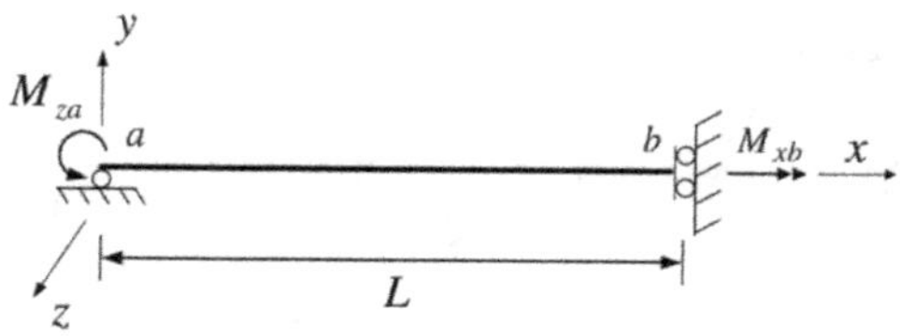

Fig. 8.6. Single beam subjected to an end moment.

at the left end, which implies uniform bending on the beam. To trigger the lateral buckling, a perturbation of $M_{xb} = 0.01\,M_{za}$ is applied at the right end. The beam is discretized into 10 elements. The theoretical critical moment for the beam under uniform moment is $M_{cr} = \pm\frac{\pi\sqrt{EI_yGJ}}{L} = \pm 1{,}493.3\,\text{N}\cdot\text{mm}$ (Yang and Kuo, 1992, 1994).

The numerical results obtained by the various schemes of predictor and corrector have been plotted in Figure 8.7, along with the computation time (using the notebook typically available in 2006) listed in Table 8.1. The peak load in Figure 8.7 is 1,463.0 N mm. As can be seen, the proposed scheme P1C1 (using the rigid geometric stiffness) can be used to yield solutions that are as accurate as the one using the elastic geometric stiffness derived in Chapter 6, though at the cost of longer computation time. As expected, the computation time for P2C2 (which uses $[k_e] + [k_g] + [k_i]$ both in the predictor and in calculating the force increments $\{\Delta f\}$) is the smallest. It is interesting that the same problem can be solved using the elastic stiffness matrix $[k_e]$ alone in all stages of analysis, as indicated by P3C1 in Table 8.1 (not shown in Figure 8.7). Of course, this is achieved at the cost of much longer computation time.

8.10.2 *Hinged angled frame in pure bending*

Figure 8.8 shows an angled frame under uniform bending, i.e., with $M_{za} = M_{zc}$ (N.mm), and a shear load F_{zb} (N) of magnitude $5 \times 10^{-5} M_{za}5$ (Argyris *et al.*, 1979, Yang and Kuo, 1992). The length L of each member is 240 mm and the other properties are the same as those of the example in Section 8.10.1. Due to the symmetry of the frame, only the left half modeled by 10 elements is analyzed. The member ab is restrained against rotations about the x, y-axes and translations along the y, z-axes. The translation along the x-axis at

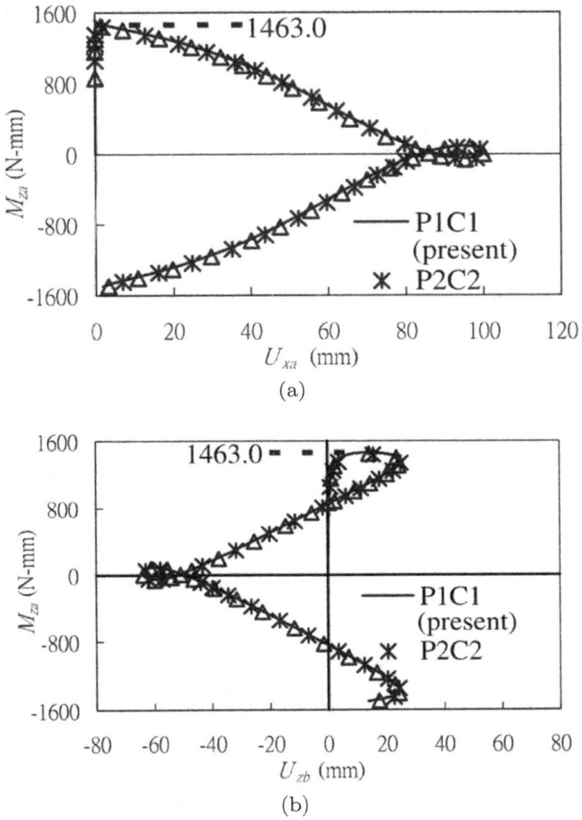

Fig. 8.7. Moment–displacement curves for a single beam: (a) U_{xa}, (b) U_{zb}.

Table 8.1. Running time for analysis of the example in Section 8.10.1.

Combination	P1C1	P2C2	P3C1
CPU time(s)	55.99	38.02	474.33

node b is also restrained. The theoretical critical load for the frame is $M_{cr} = \pm\frac{\pi\sqrt{EI_yGJ}}{L} = \pm 622.2\,\text{N}\cdot\text{mm}$.

From the results plotted in Figure 8.9, we observe that the proposed scheme P1C1 can be reliably used to trace the postbuckling

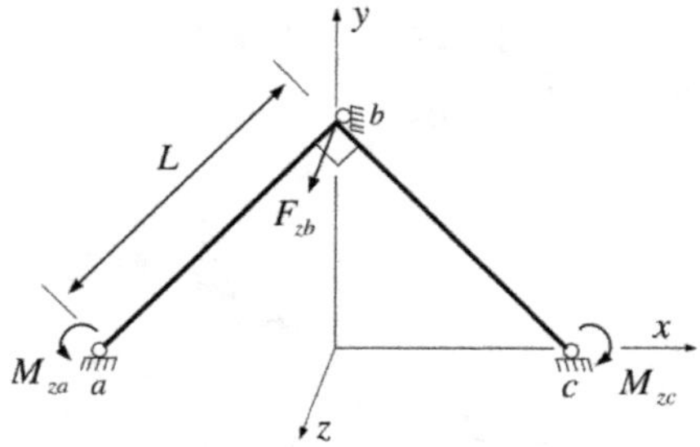

Fig. 8.8. Hinged angled frame in pure bending.

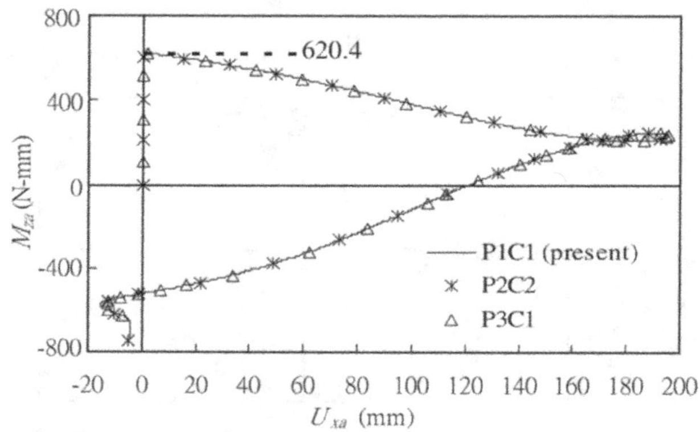

(a)

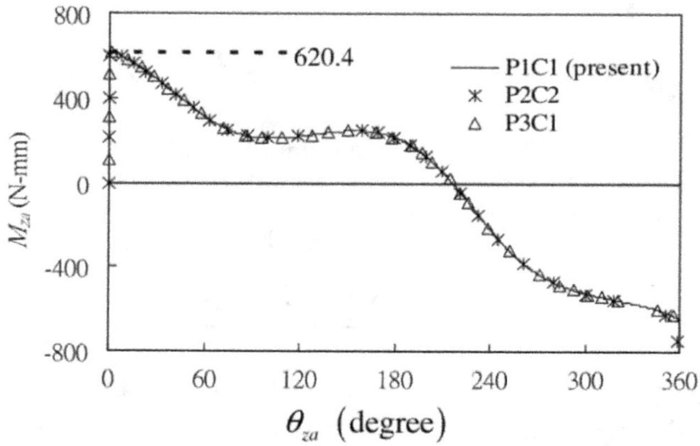

(b)

Fig. 8.9. Moment–displacement curves for hinged angled frame: (a) U_{xa}, (b) θ_{za}.

Table 8.2. Running time for analysis of the example in Section 8.10.2.

Combination	P1C1	P2C2	P3C1
CPU time(s)	116.91	109.25	1,081.13

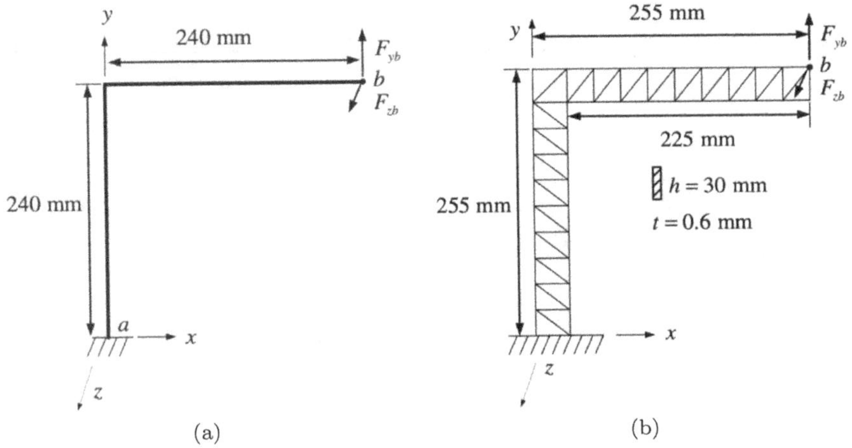

Fig. 8.10. Right-angled frame with fixed support: (a) beam element; (b) plate element.

behavior of the frame, though at the expense of slightly longer computation time, compared with the one using the complete geometric stiffness (P2C2). Of interest is that using the $[k_e]$ matrix alone in both the predictor and corrector, as indicated by P3C1, can arrive exactly at the same solution, but at the expense of much longer computation time (see Table 8.2).

8.10.3 *Right-angled frame subjected to end forces*

The right-angled frame subjected to an in-plane load F_{yb} at the free end in Figure 8.10(a) was studied (Argyris *et al.*, 1979), for which the critical load is $F_{yb,cr} = 1.088\,\text{N}$. The same material and geometry data as those used in the example of Section 8.10.2 are adopted. Each member of the frame is modeled by 10 beam elements. The entire frame was also modeled by 38 triangular elements

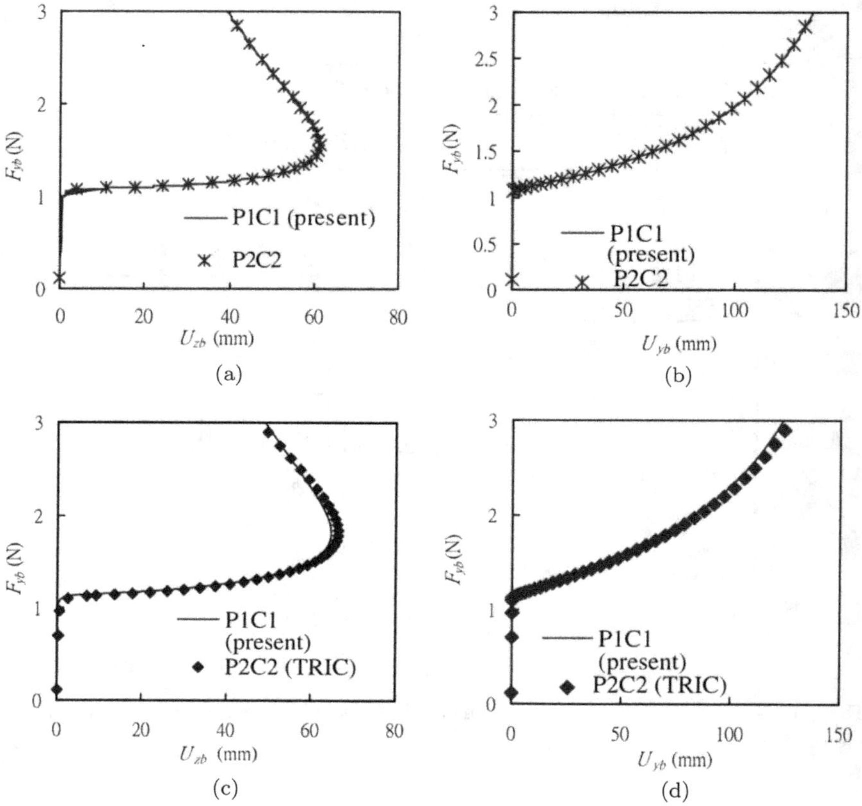

Fig. 8.11. Load–displacement curves for angled frame: (a) and (b) beam approach, (c) and (d) plate approach.

as in Figure 8.10(b). An imperfection load F_{zb} equal to one-1,000th of F_{yb} is applied at the free end.

From the results plotted in Figures 8.11(a)–8.11(d), where parts (a) and (b) are based on the beam approach, and parts (c) and (d) on plate approach, it can be seen that the proposed scheme P1C1 using either the beam or TPE approach is capable of tracing the postbuckling response of the angled frame. Table 8.3 indicates that the computation time consumed by the present rigid beam approach is slightly longer than the elastic beam approach. The same is also true for the rigid TPE approach, compared with the TRIC approach. It should be noted that the number of iterations at each incremental

Table 8.3. Running time for analysis of the example in Section 8.10.3.

	Combination		
Approach	P1C1	P2C2	P3C1
Beam(s)	37.59	36.16	Divergent
Plate(s)	1,492.22	1,211.25	Divergent

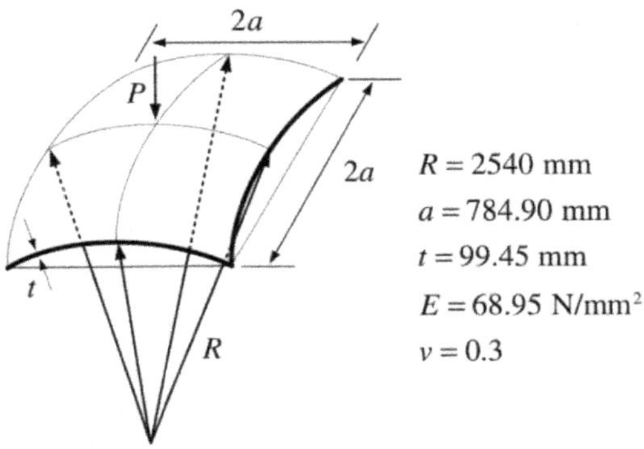

Fig. 8.12. Hinged spherical shell subjected to a central load.

step is about the same for both the TPE and TRIC approaches. However, less time is consumed by the TRIC approach in computing the stiffness matrices, since it considers only the in-plane actions, while the TPE considers all kinds of in-plane and out-of-plane actions.

No convergent solution was obtained by the P3C1 scheme using only the $[k_e]$ matrix in the predictor and corrector, due to the fact that the trial directions produced by the predictor are based on the $[k_e]$ matrix alone deviates too much from the equilibrium path in the region near the bifurcation point.

8.10.4 *Spherical shell subjected to a central load*

Figure 8.12 shows a spherical shell subjected to the central point load P with all edges hinged and immovable (Horrigmoe and Bergan, 1978). The data adopted herein are: half of side length $a = 784.9$ mm,

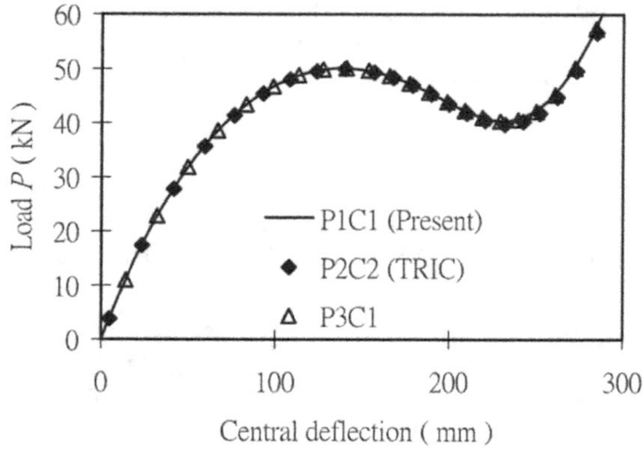

Fig. 8.13. Load–deflection curve for hinged spherical shell.

Table 8.4. Running time for analysis of the example in Section 8.10.4.

Mesh	Combination		
	P1C1	P2C2	P3C1
8×8 (s)	114.39	100.01	184.92

thickness $t = 99.45\,\text{mm}$, Young's modulus $E = 68.95\,\text{N/mm}^2$, Poisson's ratio $\nu = 0.3$, and radius of curvature $R = 2{,}540\,\text{mm}$. Because of symmetry, only one-quarter of the shell is modeled as an 8×8 mesh and analyzed.

The results obtained for the load–deflection response of the central point using different combinations were plotted in Figure 8.13, along with the running time listed in Table 8.4. Basically no distinction can be made between the results obtained by different approaches. Compared with the TRIC approach, slightly higher computation time is required by the rigid TPE approach due to the inclusion of all kinds of in-plane and out-of-plane actions in the stiffness matrices by the TPE approach, though the number of iterations at each incremental step is about the same. Of interest is that using the $[k_e]_{TPE}$ stiffness matrix alone, as indicated by P3C1, is good enough for obtaining the solution, though at the cost of extra computation time.

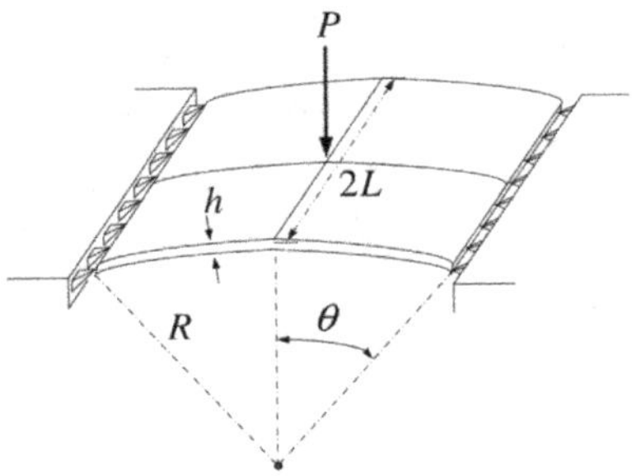

Fig. 8.14. Hinged cylindrical shell.

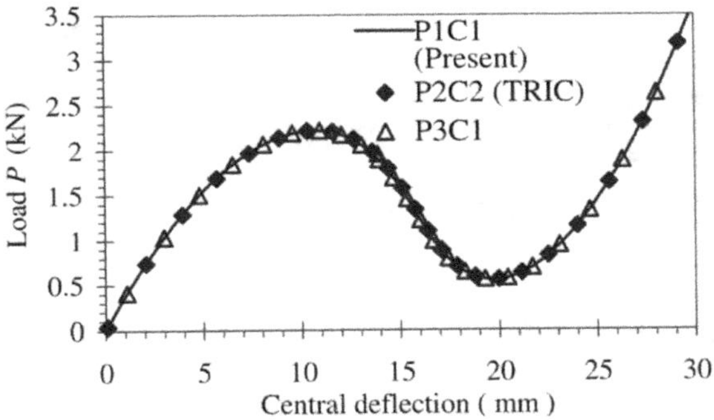

Fig. 8.15. Central deflection of hinged cylindrical shell (thick) with $h = 12.7\,\text{mm}$.

8.10.5 *Hinged cylindrical shell (thick) subjected to central load*

A cylindrical shell subjected to a central load P on the top surface is shown in Figure 8.14 (Surana, 1982). The longitudinal boundaries of the shell are hinged and immovable, while the curved edges are completely free. The following data are adopted: $E = 3.10275\,\text{kN/mm}^2$, $R = 2{,}540\,\text{mm}$, $L = 254\,\text{mm}$, $h = 12.7\,\text{mm}$, $\theta = 0.1\,\text{rad}$, and $\nu = 0.3$. Because of symmetry, only one-quarter of the shell is considered and modeled by an 8×8 mesh.

The results obtained for the central deflection of the shell were plotted in Figure 8.15, along with the computation time of each analysis listed in Table 8.5. Again, no distinction can be made between the results obtained by different approaches. The computation time

Table 8.5. Running time for analysis of the example in Section 8.10.5.

Mesh	Combination		
	P1C1	P2C2	P3C1
8 × 8 (s)	397.42	388.57	668.48

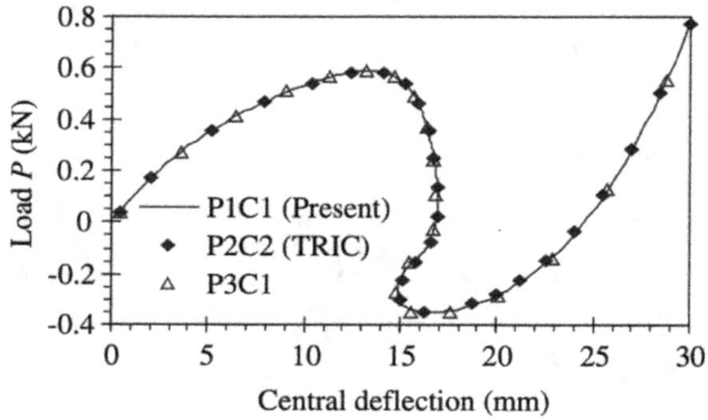

Fig. 8.16. Central deflection of hinged cylindrical shell (thin) with h = 6.35 mm.

consumed by the rigid TPE approach is slightly longer than the TRIC approach for the reasons stated in the example of Section 8.10.3. The P3C1 approach with the $[k_e]_{TPE}$ matrix alone is able to trace the entire load–deflection response, though at the cost of extra computation time.

8.10.6 *Hinged cylindrical shell (thin) subjected to central load*

This example is identical to the one in Section 8.10.5 except that the thickness of the shell is halved to $h = 6.35$ mm (Surana, 1982). The results obtained by using an 8 × 8 mesh have been plotted in Figure 8.16, together with the computation time listed in Table 8.6. A comparison of Figure 8.16 with Figure 8.15 indicates that reducing the thickness of the shell by half has resulted not only in a drastic decline of the limit load capacity but also in a shift of the response curve pattern to embrace a snap-back region. As can be seen

Table 8.6. Running time for analysis of the example in Section 8.10.6.

	Combination		
Mesh	P1C1	P2C2	P3C1
8×8 (s)	197.08	177.78	359.73

from Table 8.6, slightly longer computation time is consumed by the present rigid TPE approach, compared with the TRIC approach, for which the reason was given in the example of Section 8.10.3. Again, this problem can be solved using exclusively $[k_e]_{TPE}$ matrix in each stage of analysis, as indicated by P3C1, though a much longer computation time is required.

8.11 Concluding Remarks

The procedure presented in this chapter is based on the idea that if the rigid rotation effects are fully taken into account at each stage of the incremental-iterative nonlinear analysis, then the remaining effects of natural deformations can be treated using the small-deformation linearized theory. Based on the updated Lagrangian formulation, two stages are considered essential, i.e., the predictor stage for computing the structural displacements given the load increments, and the corrector stage for recovering the element forces. The former affects the number of iterations, but the latter determines the accuracy of the solution.

As for use in the predictor, the geometric stiffness matrix $[k_g]_{r.b.}$ is derived for a three-dimensional rigid beam element from the virtual work equation by using a rigid displacement field, and the geometric stiffness matrix $[k_g]_{TPE}$ for the rigid TPE is assembled from those for the three rigid beams lying along the three sides of the element. One advantage with the present approach is that the geometric stiffness matrices for both the three-dimensional rigid beam and rigid TPE are explicitly given and all kinds of in-plane and out-of-plane actions are taken into account. Using the rigid element concept, the work of derivation is greatly reduced for both elements, as only rigid displacement fields are required.

As for the corrector, the rigid body rule is used to update the initial nodal forces $\{^1 f\}$ existing at C_1 with no limit on the magnitude of rigid rotations. With this, the force increments $\{\Delta f\}$ generated at each incremental step are computed using only the elastic stiffness matrix $[k_e]$ based on the small-deformation linearized theory.

The robustness of the proposed combination of predictor and corrector has been demonstrated in the solution of a number of frame and shell problems involving post-buckling responses. A slightly longer computation time is required for the TPE approach due to the consideration of all kinds of actions, compared with the TRIC approach that considers only in-plane actions, based on the framework of the generalized displacement control method for self-adjustment of load increments. For problems with no abrupt change in the slope of the load–deflection curves, the entire post-buckling response can be traced by using only the elastic stiffness matrix $[k_e]$ in each stage of analysis, though at the cost of extra computation time.

References

Argyris, J.H., Hilpert, O., Malejannakis, G.A., and Scharpf, D.W. (1979). On the geometrical stiffness matrix of a beam in space – a consistent v.w. approach, *Comp. Meth. in Appl. Mech. & Eng.*, 20, 105–131.

Batoz, J.L., Bathe, K.J., and Ho, L.W. (1980). A study of three-node triangular plate bending elements, *Int. J. Numer. Meth. Eng.*, 15, 1711–1812.

Belytschko, T. and Hsieh, B.J. (1974). Nonlinear transient analysis of shells and solids of revolution by convected elements, *AIAA J.*, 12(8), 1031–1035.

Cook, R.D. (1987). A plate hybrid element with rotational d.o.f. and adjustable stiffness, *Int. J. Numer. Meth. Eng.*, 24, 1499–1508.

Horrigmoe, G. and Bergan, P.G. (1978). Numerical analysis of free-form shells by flat finite elements, *Comp. Methods Appl. Mech. Engrg.* 16, 11–35.

Malvern, L.E. (1969). *Introduction to the Mechanics of a Continuous Medium*, Prentice-Hall, Englewood Cliffs, N.J.

Surana, K.S. (1982). Geometrically non-linear formulation for the three dimensional solid-shell transition finite elements, *Comput. Struct.*, 15(2), 549–566.

Yang, Y.B., Chang, J.T., and Yau, J.D. (1999). A simple nonlinear triangular plate element and strategies of computation for nonlinear analysis, *Comp. Meth. in Appl. Mech. & Eng.*, 178, 307–321.

Yang, Y.B. and Kuo, S.R. (1992). Frame buckling analysis with full consideration of joint compatibilities, *J. Eng. Mech.*, ASCE, 118(5), 871–889.

Yang, Y.B. and Kuo, S.R. (1994). *Theory and Analysis of Nonlinear Framed Structures*, Prentice-Hall, Englewood Cliffs, New Jersey.

Yang, Y.B., Kuo, S.R., and Wu, Y.S. (2002). Incrementally small-deformation theory for nonlinear analysis of structural frames, *Eng. Struct.*, 24(6), 783–798.

Yang, Y.B. and Leu, L.J. (1991). Force recovery procedures in nonlinear analysis, *Comput. Struct.*, 41(6), 1255–1261.

Yang, Y.B., Lin, S.P., and Chen, C.S. (2007). Rigid body concept for geometric nonlinear analysis of 3D frames, plates and shells based on the updated Lagrangian formulation, *Comp. Meth. in Appl. Mech. & Eng.*, 196(7), 1178–1192.

Yang, Y.B. and Shieh, M.S. (1990). Solution method for nonlinear problems with multiple critical points, *AIAA J.* 28(12), 2110–2116.

Chapter 9

Analytical Solutions for Buckling of Some Simple Plane Frames

9.1 Introduction

In the previous eight chapters, efforts have been directed toward the derivation of the finite elements for the truss elements, plane and space frame elements, incremental-iterative nonlinear analysis, and rigid triangular plate elements. All of these were focused on the rationality of the formulation for the buckling and incremental-iterative analysis of structures with each parameter strictly defined in the deformed configuration. From the point of physics, the rigid body rule has been employed as an effective tool to examine the natural boundaries and finite elements derived, as well as for the derivation of rigid-body qualified geometric stiffness matrices. Because of this, the above presentations have been referred to as the *physics-based computational method* for the analysis of nonlinear framed structures and plates/shells.

On the other hand, previous research on the stability of structural frames using analytical approaches has been restricted to the *flexural-torsional buckling* of *single members* or the *in-plane buckling* of plane frames (Timoshenko and Gere, 1961; Vlasov, 1961; Chajes, 1974; Simitses, 1976; Chen and Lui, 1987). In general, there is an apparent lack of analytical studies on the *lateral* or *out-of-plane buckling* of plane frames subjected to in-plane loads, though numerical results are available elsewhere (Argyris *et al.*, 1979; Elias, 1986).

While the finite element method has been adopted as a substitute for analytical approaches, its application to the buckling analysis of frames in the three-dimensional sense requires special treatment of the rotational degrees of freedom, as was described in Chapters 5 and 6. For problems of this sort, there is a lack of *benchmark problems* with which the finite element results can be calibrated or compared with. To fill such a gap, Yang and co-workers have presented in 1980–1990s a series of analytical solutions for the out-of-plane buckling of some simple plane frames under either the in-plane loads or torsional loads (Yang and Kuo, 1986, 1987; Yang and Yau, 1987, 1989; Yang and Kuo, 1991a, 1991b; Yau *et al.*, 1992).

Based on the works by Yang and co-workers, the purpose of this chapter is to offer some analytical solutions for the problems of concern, such that the finite element solutions can have a reference to compare with. In Part I, the out-of-plane buckling of some angled frames in uniform bending will be solved. In Part II, an extension will be made of Part I to cover the buckling under torsional loadings, which are three-dimensional in nature. These are a few of the existing works dealing with the lateral or out-of-buckling of two-member frames.

9.2 Part I: Out-Of-Plane Buckling of Angled Frames in Uniform Bending

That the lateral or out-of-plane buckling of a plane frame is different from its in-plane counterpart can be stated as follows. First, the *critical (buckling) load* of the frame is not determined solely by the flexural resistance of the members of which the frame is composed. The effect of torsional resistance has to be considered as well. Secondly, the flexural and torsional actions of individual members are coupled through the equations of equilibrium for structural joints and the governing differential equations for structural members, when they are both written for the buckling configuration. Finally, the rotational properties of internal nodal moments have to be taken into account in setting up the natural boundary conditions for structural joints directly acted upon by loadings of the moment type.

For the purpose of illustration, three angled frames under uniform bending will be studied in the first part based on the paper by Yang and Kuo (1991a). The governing differential equations and natural boundary conditions adopted here are those derived in Section 5.5 for the three-dimensional beam based on the *engineering theory* that considers three components of strains and stresses. The final solutions for critical loads are presented in closed form, which can be used as benchmarks for verifying the validity of various finite element procedures concerning the rotational properties of nodal moments in the three-dimensional space.

9.2.1 *Symmetric frame 1 – Simply supported out of plane*

The angled frame shown in Figure 9.1 is simply supported both in and out of its own plane and is subjected to uniform bending moments M_0. Cross sections of the members are assumed to be identical with negligible warping rigidity. The governing differential equations of equilibrium for the left-hand member of the frame can be obtained from equations (5.106) and (5.107):

$$EI_z v'''' + M_0 \theta_x'' = 0 \tag{9.1}$$

$$GJ\theta_x'' - M_0 v'' = 0 \tag{9.2}$$

where E is Young's modulus, G the shear modulus, v the translation along the y-axis, θ_x the angle of twist about the x-axis, I_z the moment of inertia about the z-axis, J the torsional constant, and M_0 the bending moments applied at the supports about the y-axis (assumed to be constant). It should be noted that where numerical calculations are made in this section, the following data are adopted: $L = 240\,\text{mm}$, $E = 71{,}240\,\text{N/mm}^2$, $G = 27{,}190\,\text{N/mm}^2$, $J = 2.16\,\text{mm}^4$, and $I_z = 0.54\,\text{mm}^4$.

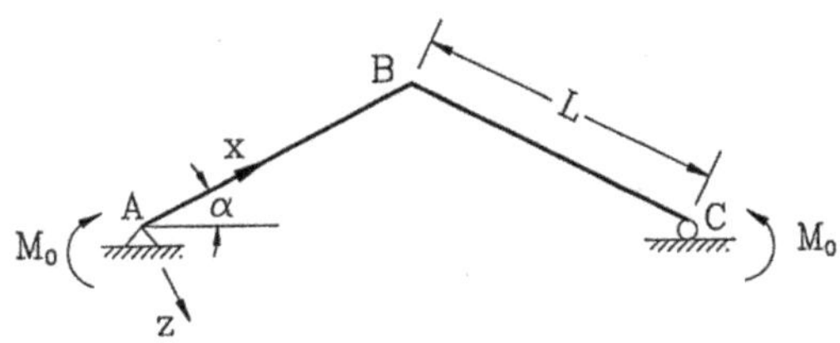

Fig. 9.1. Symmetric frame under uniform bending.

Only buckling of the bifurcation type is considered herein. For problems of this sort, the prebuckling configuration C_1 is identical to the initial (unloaded) configuration C_0. Hence, all the equations derived in Chapter 5 can still be used by noting that $C_1 \equiv C_0$. Consequently, the shear force ${}^2F_y$, torque ${}^2M_x$, and bending moment ${}^2M_Z$ acting on the member in the buckling configuration at C_2 can be obtained from equations (5.109), (5.111), and (5.113) as

$$ {}^2F_y = -EI_z v''' - M_0\theta_x' \tag{9.3} $$

$$ {}^2M_x = GJ\theta_x' - M_0 v' \tag{9.4} $$

$$ {}^2M_Z = EI_z v'' + M_0\theta_x \tag{9.5} $$

where it is realized that the bending moment M_0 is the only nonzero force acting on the member before buckling sets in.

The force-displacement relations in equations (9.3)–(9.5) is featured by the fact that they have been written exactly for the member at the buckling position, rather than the initial position. Consequently, we can see the rotational effect of the bending moment M_0 appearing as the last term in equation (9.5) (see also the right-hand sides of equations (5.111) and (5.113) for clearer expressions), which was referred to as the *quasitangential nature* of the bending moment. This is due to the fact that equations (9.3)–(9.5) have been derived rigorously as the stress resultants at C_2 in Chapter 5. It was pointed out by Yang and Kuo (1991c) that such a rotational behavior of the nodal moments was not included in the conventional analysis, which for the reason stated will be referred to as the "incorrect" solutions in later comparison.

The boundary conditions for the simple support at end A are

$$ (v)_{x=0} = 0, \quad (\theta_x)_{x=0} = 0, \quad (v'')_{x=0} = 0 \tag{9.6} $$

Substituting θ_x'' from equation (9.2) into (9.1) yields

$$ v'''' + \frac{\phi^2}{L^2} v'' = 0 \tag{9.7} $$

in which L denotes the length of each member of the frame and the following definitions are adopted for the parameters:

$$ \phi = \sqrt{\lambda\mu}, \quad \lambda = \frac{M_0 L}{GJ}, \quad \mu = \frac{M_0 L}{EI_z} \tag{9.8} $$

The general solution to equation (9.7) is

$$v = aL\sin\frac{\phi x}{L} + bL\cos\frac{\phi x}{L} + c\cdot\phi x + d\cdot L \tag{9.9}$$

where a, b, and c are integration constants. Through enforcement of the boundary conditions $(v)_{x=0} = 0$ and $(v'')_{x=0} = 0$, the preceding equation reduces to

$$v = aL\sin\frac{\phi x}{L} + c\cdot\phi x \tag{9.10}$$

Further, by substituting the expression for v into equation (9.2), we obtain the following after integration:

$$\theta_x = a\lambda\sin\frac{\phi x}{L} + f\frac{\phi x}{L} + g \tag{9.11}$$

where f and g are integration constants. Similarly, by setting $(\theta_x)_{x=0} = 0$, the preceding equation can be reduced to

$$\theta_x = a\lambda\sin\frac{\phi x}{L} + f\frac{\phi x}{L} \tag{9.12}$$

With the substitution of the expressions for v and θ_x, the forces in equations (9.3)–(9.5) become

$$^2F_y = -M_0\frac{\phi}{L}f \tag{9.13}$$

$$^2M_x = \frac{GJ}{L}(\phi f - \lambda\phi c) \tag{9.14}$$

$$^2M_Z = M_0\frac{\phi x}{L}f \tag{9.15}$$

Both symmetric and antisymmetric buckling are possible for this frame. They can be analyzed by considering only half of the frame, as will be demonstrated below.

- *Symmetric buckling*

To account for symmetric buckling, the continuity conditions at joint B can be written as

$$(\theta_x)_{x=L}\sin\alpha - (v')_{x=L}\cos\alpha = 0 \tag{9.16}$$

$$({}^2F_y)_{x=L} = 0 \tag{9.17}$$

$$({}^2M_x)_{x=L}\cos\alpha + ({}^2M_z)_{x=L}\sin\alpha = 0 \tag{9.18}$$

where the inclination angle α has been defined in Figure 9.1. From the condition of (9.17), we obtain $f = 0$. From the remaining conditions, we can derive the following characteristic equation:

$$\lambda\sin\phi\sin\alpha - \phi\cos\phi\cos\alpha = 0 \tag{9.19}$$

from which the critical loads can be determined:

$$M_{0,cr} = \frac{\sqrt{EI_zGJ}}{L}\tan^{-1}\left(\sqrt{\frac{GJ}{EI_z}}\cot\alpha\right) \quad \text{for } M_0 > 0 \tag{9.20}$$

$$M_{0,cr} = \frac{\sqrt{EI_zGJ}}{L}\left[\tan^{-1}\left(\sqrt{\frac{GJ}{EI_z}}\cot\alpha\right) - \pi\right] \quad \text{for } M_0 < 0 \tag{9.21}$$

For the special case of $\alpha = 0°$, the angled frame reduces to a simply supported beam of length $2L$. Accordingly, the preceding two equations reduce to

$$M_{0,cr} = \pm\pi\frac{\sqrt{EI_zGJ}}{2L} \tag{9.22}$$

which coincides exactly with the result given by Timoshenko and Gere (1961) for a simply supported beam under uniform bending.

In Figure 9.2, the critical loads $M_{0,cr}$ given in equations (9.20) and (9.21) have been plotted as a function of the inclination angle α and denoted as the "correct" solutions, in comparison with the "incorrect" solutions obtained with no consideration of the rotational properties, which were also referred to as the "conventional" approach in Section 6.8.1, solved by the finite element method. As can be seen, larger critical loads exist for the negative than for the positive bending case. This appears to be physically sound.

Of interest is the case when the bending moment ${}^2M_z$ is defined instead for the initial position of the member, i.e.,

$${}^2M_Z = EI_zv'' \tag{9.23}$$

This of course is not physically sound but has been implied in many previous finite element analyses. For the purpose of investigation, we

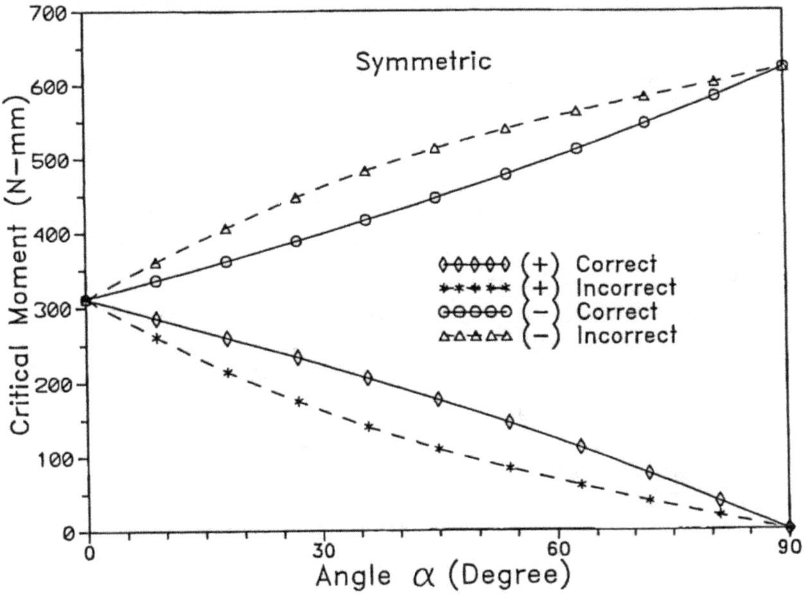

Fig. 9.2. Critical moment for symmetric frame 1 – symmetric mode.

may adopt equation (9.23) for the bending moment ${}^2M_z$, and follow the same procedure to solve for the critical loads:

$$M_{0,cr} = \frac{\sqrt{EI_zGJ}}{L}\tan^{-1}\left(\frac{1}{2}\sqrt{\frac{GJ}{EI_z}}\cot\alpha\right) \quad \text{for } M_0 > 0 \qquad (9.24)$$

$$M_{0,cr} = \frac{\sqrt{EI_zGJ}}{L}\left[\tan^{-1}\left(\frac{1}{2}\sqrt{\frac{GJ}{EI_z}}\cot\alpha\right) - \pi\right] \quad \text{for } M_0 < 0 \qquad (9.25)$$

which have been identified as the "incorrect" solutions in Figure 9.2. As can be seen, the incorrect approach tends to underestimate the positive critical moments, while overestimating the negative critical moments.

One difference between the bending moments of (9.5) and (9.23) is that the former has been defined in the buckling configuration, but the later in the initial configuration. While substituting the former in the condition of (9.18) will result in an angled joint that is in equilibrium at the buckling configuration, the same is not true with the latter.

It should be noted that the same problem has been solved by the eigenvalue approach in Chapter 6 using the space frame element that considers the joint rotational effect via incorporation of the $[k_j]$. The result was compared with the conventional approach with no knowledge of the $[k_j]$ matrix. A comparison of Figure 6.4 with Figure 9.2 indicates that the conventional finite element approach in Figure 6.4 corresponds to the "incorrect" presented in Figure 9.2. It is interesting to note that the finite element approach presented in Chapter 6 considers the joint moment matrix $[k_j]$ is equivalent to establishing the moment equation for the joint in the deformed configuration in the present buckling analysis.

- *Antisymmetric buckling*

For antisymmetric buckling, the following continuity conditions can be specified for joint B:

$$(v)_{x=L} = 0 \tag{9.26}$$

$$(\theta_x)_{x=L} \cos\alpha + (v')_{x=L} \sin\alpha = 0 \tag{9.27}$$

$$({}^2M_x)_{x=L} \sin\alpha - ({}^2M_z)_{x=L} \cos\alpha = 0 \tag{9.28}$$

The expressions given previously for the displacements in (9.10) and (9.12), and for the forces in (9.13)–(9.15) remain valid. Substituting the expressions for displacements and forces into the preceding continuity conditions, we obtain after some operations the following characteristic equation:

$$(\lambda \cot\alpha - 1)\phi \cos\phi + (\lambda^2 \cot^2\alpha - \lambda \cot\alpha + 1) \sin\phi = 0 \tag{9.29}$$

which can be solved by trial and error. The critical loads $M_{0,cr}$ obtained here have been denoted as the "correct" solutions in Figure 9.3.

On the other hand, if the inconsistent equation (9.23) has been used for the bending moment ${}^2M_Z$, the characteristic equation becomes

$$\phi \cos\phi - (1 + \lambda^2 \cot^2\alpha) \sin\phi = 0 \tag{9.30}$$

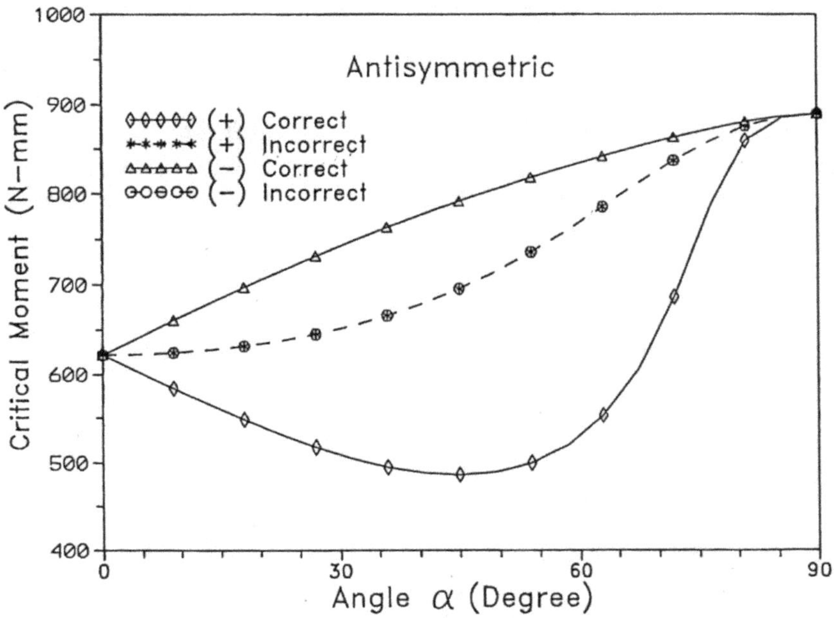

Fig. 9.3. Critical moment for symmetric frame 1 – antisymmetric mode.

which can also be solved by trial and error. From Figure 9.3, we observe that the "incorrect" solution obtained from (9.30) is independent of the sense of the applied moments, while the "correct" solution obtained from (9.29) predicts a higher resistance for the frame against negative bending than against positive bending. It should be noted that the same "correct" solution can be obtained by using the finite element presented in Chapter 6 considering the joint rotational effect via incorporation of the joint moment matrix $[k_j]$. Again, the "incorrect" solution corresponds to the one obtained by the conventional approach that has no knowledge of the joint moment matrix $[k_j]$.

9.2.2 *Symmetric frame 2 – Fixed out of plane*

The same frame as the one given in Figure 9.1 is studied here, except that it is fully restrained at the supports with respect to out-of-plane bending deformations. The following are the boundary conditions for support A:

$$(v)_{x=0} = 0, \quad (\theta_x)_{x=0} = 0, \quad (v')_{x=0} = 0 \tag{9.31}$$

Following the same procedure (i.e., the one from equation (9.7) to (9.12), we can solve the governing differential equations (9.1) and (9.2) along with the boundary conditions (9.31) for the buckling displacements as

$$v = aL\left(\cos\frac{\phi x}{L} - 1\right) + bL\left(\sin\frac{\phi x}{L} - \frac{\phi x}{L}\right) \tag{9.32}$$

$$\theta_x = a\lambda\left(\cos\frac{\phi x}{L} - 1\right) + b\lambda\sin\frac{\phi x}{L} + c\frac{x}{L} \tag{9.33}$$

Again, both the symmetric and antisymmetric buckling modes are possible for the present case.

- *Symmetric buckling*

The same continuity conditions as those given in (9.16–(9.18) can be applied for joint B. By substitution of the expressions for the forces in (9.3)–(9.5) and those for displacements in (9.32) and (9.33), from the continuity conditions we can derive the characteristic equation as follows:

$$\sin\phi = 0 \tag{9.34}$$

Correspondingly, the critical load is

$$M_{0,cr} = \pm\pi\frac{\sqrt{EI_zGJ}}{L} \tag{9.35}$$

which coincides with the result given by Argyris *et al.* (1979) and Elias (1986) using finite elements of the ST type. On the other hand, if we adopt the inconsistent expression (9.23) for the bending moment ${}^2M_z$, the following characteristic equation can be obtained:

$$\tan\frac{\phi}{2} = \left[\left(\frac{\lambda}{\phi} + \frac{\phi}{\lambda}\right) + \left(\frac{\phi}{\lambda} - \frac{\lambda}{\phi}\right)\cos 2\alpha\right]\frac{1}{2\sin 2\alpha} \tag{9.36}$$

which, though incorrect by itself, is a function of the inclination angle α. The solutions obtained from the preceding equation

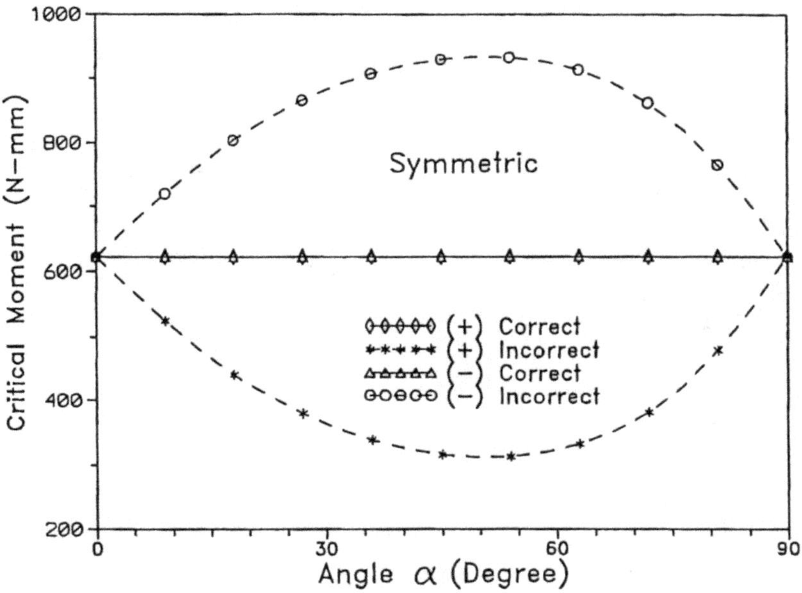

Fig. 9.4. Critical moment for symmetric frame 2 – symmetric mode.

(denoted "incorrect") have been compared with the correct solution of (9.35) for the critical moment $M_{0,cr}$ in Figure 9.4. As far as the rotational property of the bending moment is concerned, we like to mention that the ST and QT finite element solutions predicted by Argyris *et al.* (1979) correlate well with the present "correct" and "incorrect" solutions, respectively. It should be noted that the same problem was analyzed in Section 6.8.2 by the finite element method based on the eigenvalue approach, for which the results were presented in Figure 6.5.

- *Antisymmetric buckling*

The continuity conditions are the same as those given in (9.26)–(9.28). Following the same lines of derivation, we obtain for the antisymmetric buckling mode the following characteristic equation:

$$\left(\lambda^2 \cot^2 \frac{\alpha}{\phi} - \phi + \phi\lambda \cot \alpha\right) \sin \phi + (2\lambda \cot \alpha - \lambda^2 \cot^2 \alpha - 2) \cos \phi + 2(1 - \lambda \cot \alpha) = 0 \quad (9.37)$$

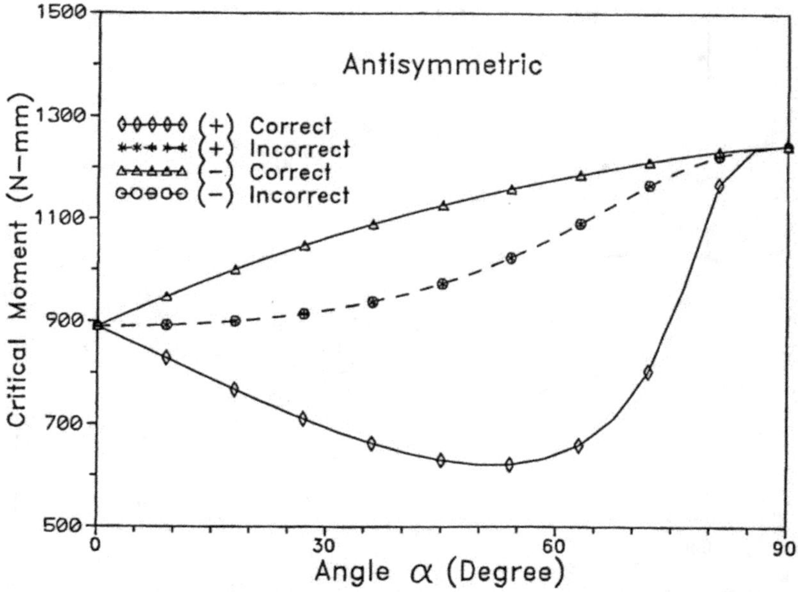

Fig. 9.5. Critical moment for symmetric frame 2 – antisymmetric mode.

In contrast, the characteristic equation derived based on the inconsistent expression (9.23) for the bending moment ${}^2M_z$ is

$$\left(\lambda^2 \cot^2 \frac{\alpha}{\phi} - \phi\right) \sin\phi - \left(\lambda^2 \cot^2 \alpha + 2\right) \cos\phi + 2 = 0 \tag{9.38}$$

The critical loads $M_{0,cr}$ solved from equations (9.37) and (9.38) have ben plotted in Figure 9.5. Similar to the previous antisymmetric buckling case of frame 1 in Section 9.2.1, the "incorrect" approach yields a critical value independent of the sense of the applied moments, while the "correct" approach indicates that the frame has a higher resistance against negative bending than against positive bending.

9.2.3 *Frame 3 – Angled frame with fixed base*

Identical cross-sections are assumed for each member of the frames shown in Figure 9.5. Since the moment M_0 applied at the free end C may be subjected to three-dimensional rotations during the buckling

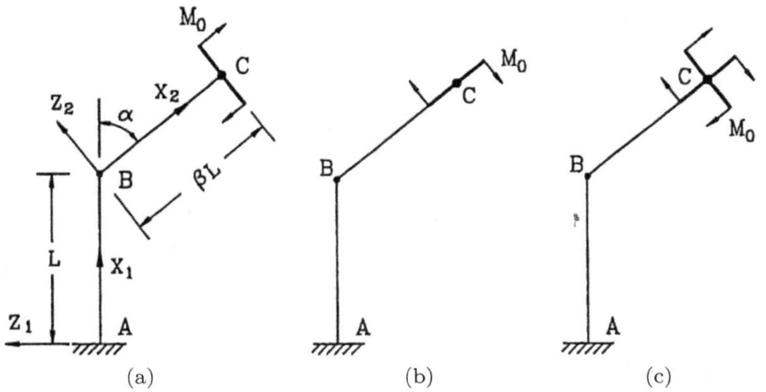

Fig. 9.6. Angled frame with fixed base subjected to end torque M_0: (a) QT-1; (b) QT-2; (c) ST.

process, the moments induced by the bending moment M_0 undergoing these rotations have to be taken into account in establishing the natural boundary conditions for joint C in the buckling position. For the present purposes, only the three types of moment mechanisms shown in Figures 9.6(a)–9.6(c) will be considered, which differ in the way by which the moments are induced upon rotations in the three-dimensional space. Except for the natural boundary conditions, other kinematic and static relations remain basically the same for the three frames shown in Figures 9.6(a)–9.6(c). For this example, right subscripts "1" and "2" will be added to quantities to signify that they are associated with members 1 and 2, respectively, of the angled frame. It should be noted that the same problem was also analyzed in Section 6.8.4 by the finite element method based on the eigenvalue approach.

- *Basic equations*

The governing differential equations for member 1 are

$$EI_z(v_1)'''' + M_0(\theta_{x1})'' = 0 \tag{9.39}$$

$$GJ(\theta_1)'' - M_0(v_1)'' = 0 \tag{9.40}$$

and those for member 2 are

$$EI_z(v_2)'''' + M_0(\theta_{x2})'' = 0 \tag{9.41}$$

$$GJ(\theta_{x2})'' - M_0(v_2)'' = 0 \tag{9.42}$$

The following are the geometric boundary conditions for the fixed base, i.e., joint A, of the frame:

$$(v_1)_{x_1=0} = 0, \quad (v_1')_{x_1=0} = 0, \quad (\theta_{x1})_{x_1=0} = 0 \tag{9.43}$$

For joint B, the conditions of equilibrium are

$$({}^2F_{y1})_{x_1=L} = ({}^2F_{y2})_{x_2=0} \tag{9.44}$$

$$({}^2M_{x1})_{x_1=L} \cos\alpha - ({}^2M_{z1})_{x_1=L} \sin\alpha = ({}^2M_{x2})_{x_2=0} \tag{9.45}$$

$$({}^2M_{x1})_{x_1=L} \sin\alpha + ({}^2M_{z1})_{x_1=L} \cos\alpha = ({}^2M_{z2})_{x_2=0} \tag{9.46}$$

and the conditions of continuity are

$$(v_1)_{x_1=L} = (v_2)_{x_2=0} \tag{9.47}$$

$$(\theta_{x1})_{x_1=L} \cos\alpha - (v_1')_{x_1=L} \sin\alpha = (\theta_{x2})_{x_2=0} \tag{9.48}$$

$$(v_1')_{x_1=L} \cos\alpha + (\theta_{x1})_{x_1=L} \sin\alpha = (v_2')_{x_2=0} \tag{9.49}$$

The general solutions to the buckling differential equations (9.39)–(9.42) can be obtained as follows:

$$v_1 = a_1 L \sin\frac{\phi x}{L} + b_1 L \cos\frac{\phi x}{L} + c_1 \phi x + d_1 L \tag{9.50}$$

$$\theta_{x1} = a_1 \lambda \sin\frac{\phi x}{L} + b_1 \lambda \cos\frac{\phi x}{L} + f_1 \frac{\phi x}{L} + g_1 \tag{9.51}$$

for member 1, and

$$v_2 = a_2 L \sin\frac{\phi x}{L} + b_2 L \cos\frac{\phi x}{L} + c_2 \phi x + d_2 L \tag{9.52}$$

$$\theta_{x2} = a_2 \lambda \sin\frac{\phi x}{L} + b_2 \lambda \cos\frac{\phi x}{L} + f_2 \frac{\phi x}{L} + g_2 \tag{9.53}$$

for member 2, where $a_1, b_1, \ldots,$ etc. are integration constants. Substitution of (9.50) and 9.51 for the buckling displacements in the geometric boundary conditions (9.43) yields: $d_1 = -b_1$, $c_1 = -\phi a_1$, and $g_1 = -\lambda b_1$. Accordingly, equations (9.50) and (9.51) reduce to

$$v_1 = a_1 L \left(\sin\frac{\phi x}{L} - \frac{\phi x}{L} \right) + b_1 L \left(\cos\frac{\phi x}{L} - 1 \right) \tag{9.54}$$

$$\theta_{x1} = a_1 \lambda \sin\frac{\phi x}{L} + b_1 \lambda \left(\cos\frac{\phi x}{L} - 1 \right) + f_1 \frac{\phi x}{L} \tag{9.55}$$

Before the buckling loads can be solved for the frames in Figures 9.6(a)–9.6(c), the natural boundary conditions for joint C have to be specified.

- *Quasitangential moment of the first kind (QT-1)*

Moment of the type depicted in Figure 9.6(a) has been referred to as the quasitangential moment of the first kind, or abbreviated as the QT-1 moment. Such a moment, initially of magnitude M_0 (about the y_2-axis), will induce a moment of magnitude $M_0\theta_{x2}$ about the z_2-axis, when subjected to rotations in the three-dimensional space. Consider the free body of joint C of the angled frame in the buckling position, as shown in Figure 9.7(a). The moments acting on the right-hand side of the joint are generated externally by the moment mechanism, and those acting on the left-hand side are generated internally by the moments treated as stress resultants of the cross-section, readily available through equations (9.4) and (9.5), or the original equations (5.111)–(5.113).

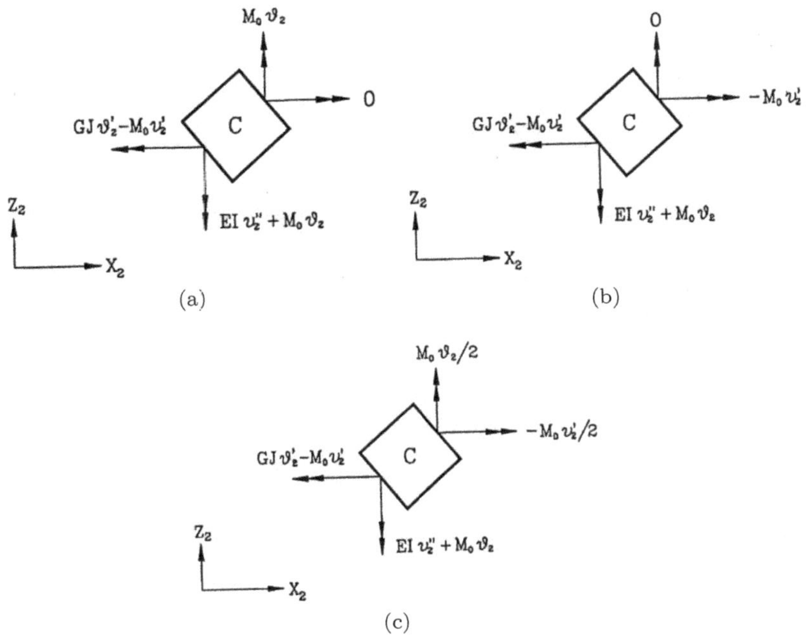

Fig. 9.7. Moment equilibrium condition for joint C under: (a) QT-1 moment; (b) QT-2 moment; (c) ST moment.

By the conditions of equilibrium for the moments acting at joint C in Figure 9.6(a) (by paying attention to the difference between the internal and external moments), the natural boundary conditions can be written from equations (5.111), (5.113), and (5.109) as follows:

$$GJ(\theta'_{x2})_{x2=\beta L} - M_0(v'_2)_{x2=\beta L} = 0 \tag{9.56}$$

$$EI_z(v''_2)_{x2=\beta L} = 0 \tag{9.57}$$

where βL denotes the length of member 2. Also, by the force equilibrium condition of joint C, the shear force ${}^2F_{y2}$ must vanish, that is,

$$EI_z(v'''_2)_{x2=\beta L} + M_0(\theta'_{x2})_{x2=\beta L} = 0 \tag{9.58}$$

Substitution of the expressions for v_2 and θ_{x2} in (9.52) and (9.53) into the moment equilibrium conditions in (9.56) and (9.57) yields the following relations: $c_2 = f_2 = 0$ and $b_2 = -a_2 \sin\beta\phi / \cos\beta\phi$. Letting $a_2 = -a_2^* \cos\beta\phi$ and $b_2 = a_2^* \sin\beta\phi$, we can rewrite equations (9.52) and (9.53) as follows:

$$v_2 = a_2^* L \sin\left[\phi\left(\beta - \frac{x}{L}\right)\right] + d_2 L \tag{9.59}$$

$$\theta_{x2} = a_2^* \lambda \sin\left[\phi\left(\beta - \frac{x}{L}\right)\right] + g_2 \tag{9.60}$$

By the use of the force–displacement relations (9.3)–(9.5) and the expressions for displacements in (9.54) and (9.55) for member 1 and (9.59) and (9.60) for member 2, we obtain the following from the equilibrium conditions (9.44)–(9.46):

$$f_1 = 0 \tag{9.61}$$

$$a_1 \phi \cos\alpha + b_1 \lambda \sin\alpha = 0 \tag{9.62}$$

$$a_1 \phi \sin\alpha - b_1 \lambda \cos\alpha = g_2 \tag{9.63}$$

Letting $g_2 = g_2^* \phi\lambda$, $a_1 = g_2^* \lambda \sin\alpha$, and $b_1 = -g_2^* \phi \cos\alpha$, we can rewrite equations (9.54), (9.55), (9.59), and (9.60) as follows:

$$v_1 = g_2^* L \left[\lambda \sin\alpha \left(\sin\frac{\phi x}{L} - \frac{\phi x}{L}\right) - \phi\cos\alpha \left(\cos\frac{\phi x}{L} - 1\right)\right] \tag{9.64}$$

$$\theta_{x1} = g_2^* \lambda \left[\lambda \sin\alpha \sin\frac{\phi x}{L} - \phi \cos\alpha \left(\cos\frac{\phi x}{L} - 1 \right) \right] \tag{9.65}$$

$$v_2 = a_2^* L \sin\left[\phi \left(\beta - \frac{x}{L} \right) \right] + d_2 L \tag{9.66}$$

$$\theta_{x2} = a_2^* \lambda \sin\left[\phi \left(\beta - \frac{x}{L} \right) \right] + g_2^* \lambda \phi \tag{9.67}$$

Furthermore, substituting the preceding expressions for displacements in the continuity conditions (9.48) and (9.49) yields the following homogeneous equations:

$$\begin{bmatrix} (\lambda^2 - \phi^2)\sin\alpha\cos\alpha\sin\phi - \lambda\phi\cos\phi & -\lambda\sin\beta\phi \\ (\lambda^2\sin^2\alpha + \phi^2\cos^2\alpha)\sin\phi & \phi\cos\beta\phi \end{bmatrix} \begin{Bmatrix} g_2^* \\ a_2^* \end{Bmatrix} = \begin{Bmatrix} 0 \\ 0 \end{Bmatrix} \tag{9.68}$$

from which the following characteristic equation can be obtained:

$$\begin{aligned} &(\lambda^3 \sin^2\alpha + \lambda\phi^2\cos^2\alpha)\sin\phi\sin\beta\phi - \lambda\phi^2\cos\phi\cos\beta\phi \\ &\quad + (\lambda^2 - \phi^2)\sin\alpha\cos\alpha(\phi\sin\phi\cos\beta\phi) = 0 \end{aligned} \tag{9.69}$$

Three special cases will be considered in the following.

Case 1 – Member 2 is of zero length: This is exactly the case shown in Figure 9.8(a). For $\beta = 0$, the characteristic equation reduces to the following:

$$2\phi\cot\phi = (\lambda - \phi)\sin 2\alpha \tag{9.70}$$

The critical loads $M_{0,cr}$ solved from this equation have been plotted in Figure 9.9. It should be noted that for the special case of $\alpha = 0°$,

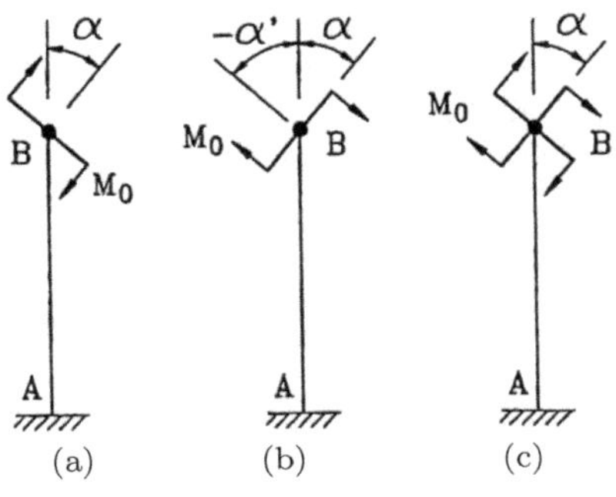

Fig. 9.8. Cantilevers subjected to various moments: (a) QT-1; (b) QT-2; (c) ST.

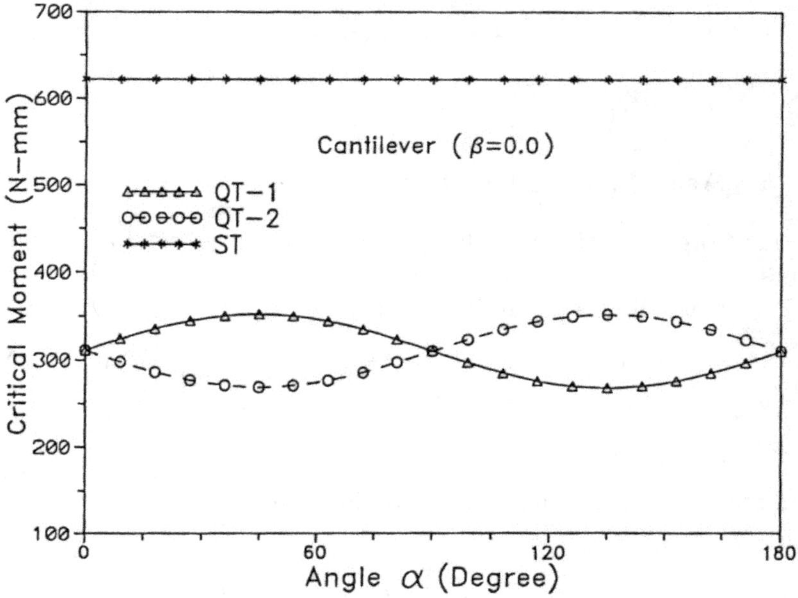

Fig. 9.9. Critical loads for cantilevers ($\beta = 0$) subjected to various moments.

the critical load become

$$M_{0,cr} = \pm\pi\frac{\sqrt{EI_zGJ}}{2L} \tag{9.71}$$

which is identical to the one given by Timoshenko and Gere (1961). Of interest is that the angled frame under the ST moment appears to have higher buckling loads, compared with the other two kinds of applied moments.

Case 2 – Equal rigidities for torsion and flexure: For this case since $\lambda = \mu$, i.e., $GJ = EI_z$ and $\phi = \sqrt{\lambda\mu}$, equation (9.69) reduces to

$$\cos[(1+\beta)\phi] = 0 \tag{9.72}$$

The critical loads are simply

$$M_{0,cr} = \pm\pi\frac{\sqrt{EI_zGJ}}{2(1+\beta)L} \tag{9.73}$$

which are independent of angle α.

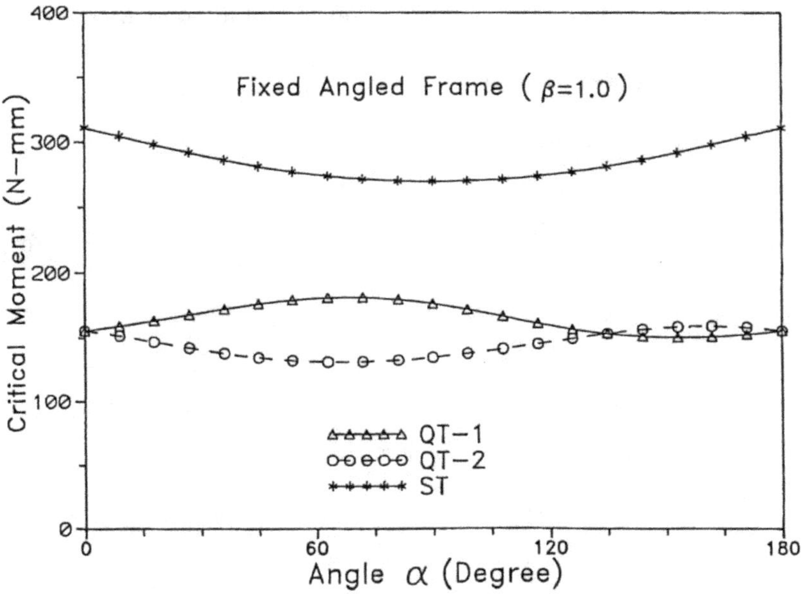

Fig. 9.10. Critical loads for angled frames of members of equal length ($\beta = 1$).

Case 3 – Equal length for members 1 and 2: For this case, $\beta = 1$ and equation (9.69) reduces to

$$2\cot^2\phi + \left(\frac{\phi}{\lambda} - \frac{\lambda}{\phi}\right)\sin 2\alpha \cot\phi - \left[\left(1 + \frac{\lambda}{\mu}\right) + \left(1 - \frac{\lambda}{\mu}\right)\cos 2\alpha\right] = 0 \tag{9.74}$$

The critical loads $M_{0,cr}$ Obtained from this equation have been plotted for various angles α in Figure 9.10. Similarly, the angled frame under the ST moment appears to have higher buckling loads, compared with the other two kinds of applied moments.

- *Quasitangential moment of the second kind (QT-2)*

External moment of the type given in Figure 9.6(b) has been referred to as the quasitangential moment of the second kind (QT-2). According to Section 5.3, such a moment initially of magnitude M_0 (about the y_2-axis) will induce a moment of magnitude $-M_0 v_2'$ about the x_2-axis at the buckling position. Considering the equilibrium of joint C in the buckling position (Figure 9.7(b)) (by paying attention to the difference between the internal and external moments), we can write the natural boundary conditions for the joint from equations (5.111), (5.113) and (5.109) as follows:

$$GJ(\theta'_{x2})_{x_2=\beta L} = 0 \tag{9.75}$$

$$EI_z(v''_2)_{x_2=\beta L} + M_0(\theta_{x2})_{x_2=\beta L} = 0 \tag{9.76}$$

$$EI_z(v'''_2)_{x_2=\beta L} + M_0(\theta'_{x2})_{x_2=\beta L} = 0 \tag{9.77}$$

Following basically the same procedure, the characteristic equation can be derived:

$$\begin{aligned}(\phi^3 \sin^2\alpha + \lambda^2\phi\cos^2\alpha)\sin\phi\sin\beta\phi - \lambda^2\phi\cos\phi\cos\beta\phi \\ - \lambda(\lambda^2 - \phi^2)\sin\alpha\cos\alpha(\sin\phi\cos\beta\phi) = 0\end{aligned} \tag{9.78}$$

In the following, three special cases will be studied.

Case 1 – Member 2 is of zero length: For this particular case, the frame reduces to the one shown in Figure 9.7(b). For $\beta = 0$, equation (9.78) can be simplified to

$$2\phi\cot\phi = -(\lambda - \phi)\sin 2\alpha \tag{9.79}$$

which can also be derived from equation (9.70) by substituting the angle $\alpha'(=\alpha - 90°)$ for a (see Figures 9.8(a) and (b)). For comparison, the critical loads $M_{0,cr}$ obtained from (9.79) have also been plotted in Figure 9.9, from which the influence of external moment mechanisms on the buckling load can be readily seen.

Case 2 – Equal rigidities for torsion and flexure: For this case, results identical to those presented in equation (9.73) can be obtained.

Case 3 – Same length for members 1 and 2: For this case ($\beta = 1$), equation (9.78) reduces to

$$2\cot^2\phi + \left(\frac{\lambda}{\phi} - \frac{\phi}{\lambda}\right)\sin 2\alpha\cot\phi - \left[\left(1 + \frac{\mu}{\lambda}\right) + \left(1 - \frac{\mu}{\lambda}\right)\cos 2\alpha\right] = 0 \tag{9.80}$$

A comparison of the critical loads $M_{0,cr}$ obtained for the QT-1 and QT-2 moments in Figure 9.10 reveals that they are very sensitive to the manner by which the external moments are applied. Among the three moment mechanisms, the ST moment shows the highest resistance to frame buckling.

- *Semitangential moment (ST)*

The semitangential moment M_0 of Figure 9.6(c) will induce a moment of magnitude $-1/2(M_0 v_2')$ about the x_2-axis and another moment of magnitude $-1/2(M_0 \theta_{x2})$ about the z_2-axis at the buckling position. From the equilibrium conditions for joint C in Figure 9.7(c) (by paying attention to the difference between the internal and external moments), we can write the natural boundary conditions for the joint from equations (5.111), (5.113) and (5.109) as follows:

$$GJ(\theta_{x2}')_{x_2=\beta L} - \frac{1}{2}M_0(v_2')_{x_2=\beta L} = 0 \tag{9.81}$$

$$EI_z(v_2'')_{x_2=\beta L} + \frac{1}{2}M_0(\theta_{x2})_{x_2=\beta L} = 0 \tag{9.82}$$

$$EI_z(v_2''')_{x_2=\beta L} + M_0(\theta_{x2}')_{x_2=\beta L} = 0 \tag{9.83}$$

Adopting the same procedure for solving the boundary-value problems, we obtain for the present case, the characteristic equation as follows:

$$2 + 2\cos\phi\cos\beta\phi - \left[2\cos^2\alpha + \left(\frac{\mu}{\lambda} + \frac{\lambda}{\mu}\right)\sin^2\alpha\right]\sin\phi\sin\beta\phi \tag{9.84}$$

Three special cases will be studied in the following.

Case 1 – Member 2 is of zero length: For this case, $\beta = 0$ and the characteristic equation reduces to the following:

$$\cos\phi = -1 \tag{9.85}$$

from which the critical load is

$$M_{0,cr} = \pm\pi\frac{\sqrt{EI_z GJ}}{L} \tag{9.86}$$

As can be seen from Figure 9.9, the critical load $M_{0,cr}$ for the ST moment, being independent of the inclination angle α (between the two members) is about twice as large as those predicted for the QT-1 and QT-2 moments.

Case 2 – Equal rigidities for torsion and flexure: For this case ($\lambda = \mu$), equation (9.84) reduces to

$$\cos[(1+\beta)\phi] = -1 \tag{9.87}$$

The critical load obtained is

$$M_{0,cr} = \pm\pi\frac{\sqrt{EI_z GJ}}{(1+\beta)L} \tag{9.88}$$

which is twice as large as that given in equation (9.73).

Case 3 – Same length for members 1 and 2: Assuming members 1 and 2 to be of the same length, i.e., with $\beta = 1$, the critical loads are found to be

$$M_{0,cr} = \pm\pi\frac{\sqrt{EI_z GJ}}{L}\sin^{-1} \times \left\{\frac{1}{1+\left[(GJ/EI_z + EI_z/GJ - 2)\sin^2\alpha\right]/4}\right\} \tag{9.89}$$

which have also been plotted in Figure 9.10. From this figure, we observe that when subjected to the ST moment, the frame tends to buckle at a critical load significantly higher than those with the QT-1 and QT-2 moments for the frame with the same inclination angle α.

In the study of the third frame in Figure 9.6, it has been demonstrated that the critical load for the angled frame to buckle laterally depends to a large extent on the way by which the moment loadings are applied. In other words, for frames subjected to loadings of the moment type, the nature of applied moments has to be considered in the solution of the critical loads. Before closing this section, we like to reiterate that one key step in the analysis of the lateral buckling of plane frames has been the establishment of all kinematic and static relations in the buckling configuration, which should include, in particular, conditions of equilibrium for structural joints, natural boundary conditions, and governing differential equations, and that in such a process, the properties of nodal moments and torques undergoing three-dimensional rotations have to be duly considered.

9.3 Part II: Out-of-Plane Buckling of Angled Frames Under Torsional Loadings

The nature of a torsional or bending moment can be classified according to the way by which it is generated. Semitangential and quasitangential moments are two types of conservative moments commonly encountered in the design of structural and machine components. They are distinguished by the fact that different moment increments may be induced about the axes perpendicular to the axis of the moment when subjected to finite rotations in the three-dimensional space (see Section 5.3).

The instability of a torsionally loaded straight bar has been studied as early as in 1883 by Greenhill, and later by numerous researchers. The buckling of a circular bar subjected to a torque was studied by Ziegler (1977), where a brief review of previous works on this subject was given. Based on a finite element approach, Barsoum and Gallagher (1970) studied the instability of a circular shaft under the action of a semitangential or quasitangential torque.

To the knowledge of the authors, very few analytical studies have been conducted on the torsionally-loaded buckling of space frames, or in the limit case, on the lateral or out-of-plane buckling of plane frames subjected to torsional loadings, based rigorously on the governing differential equations, joint equilibrium and continuity conditions, and geometric and natural boundary conditions. As part of the effort to fill this gap, a series of analytical works were conducted by Yang and co-workers in the 1980s–1990s on the buckling of various single-member or two-member frames. Concerning the torsional loadings, Yang and Yau (1989) studied the stability of pretwisted bars under various end torques; Yang and Kuo (1991b) studied the buckling of two-member frames subjected to various torsional loadings, and Yau *et al.* (1992) studied the stability of tapered bars of circular cross sections under semi- and quasi-tangential torques.

In this section, we shall proceed further to investigate the general buckling behavior of plane frames subjected to loadings not of the in-plane type, but of the torsional type, which is by nature three-dimensional. This type of loadings differs from the in-plane loadings in that prebuckling deformations are not restricted to occur in the

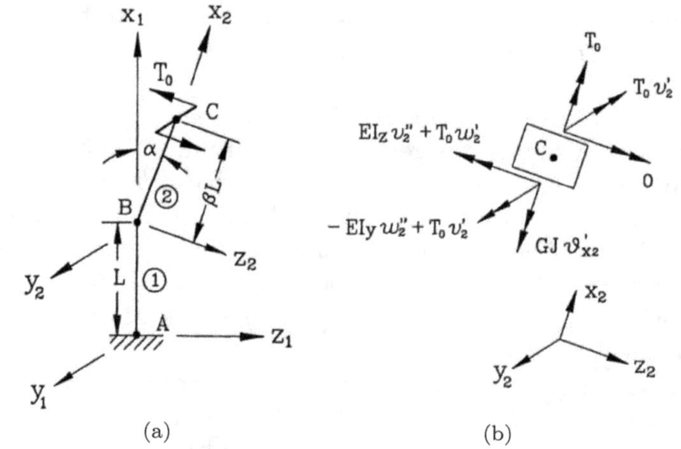

Fig. 9.11. Angled frame under QT-1 torque: (a) frame; (b) joint C.

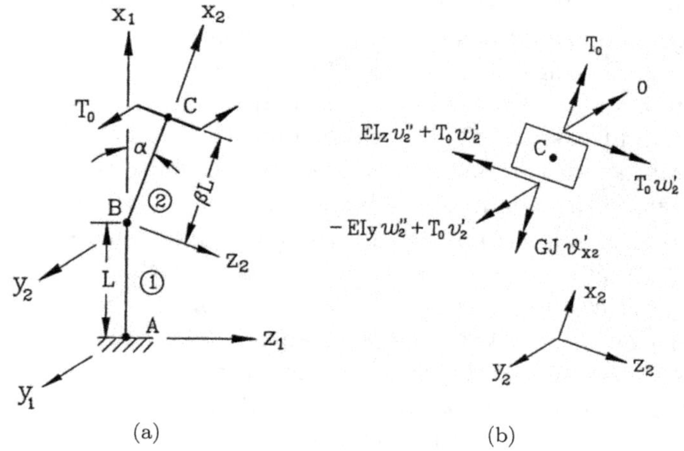

Fig. 9.12. Angled frame under QT-2 torque: (a) frame; (b) joint C.

plane of the frame. As a result, this section differs from the preceding one in that bending about the y-axis should also be taken into account.

The two-member-angled frames are shown in Figures 9.11–9.13, where L is the length of member 1 and βL the length of member 2. Particularly, the internal moments acting on the inner side of each of

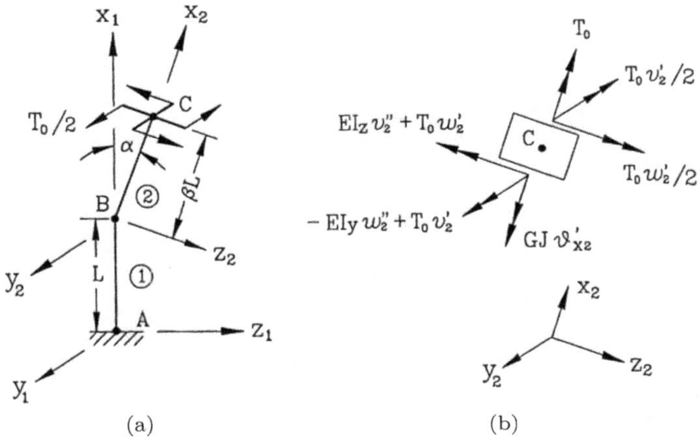

Fig. 9.13. Angled frame under ST torque: (a) frame; (b) joint C.

the free body diagrams of joint C have been obtained from the natural boundary conditions given in equations (9.95)–(9.97) or the original equations (5.111)–(5.113) in Chapter 5. Though the frames shown in Figures 9.11–9.13 are simple in geometry and loading, they serve as an effective vehicle for highlighting the concepts, procedures, and physical meanings involved in the buckling analysis of space frames. Aside from demonstrating the effects of various torsional loadings, the present solutions are by themselves benchmarks for calibrating the accuracy of various finite element procedures, as far as the rotational properties of the applied torques are concerned. It should be mentioned that the results presented here are valid in the scope of elastic linearized buckling, that is, the prebuckling deformations of structures are assumed to be so small that they can be neglected entirely from the buckling analysis. The materials to be presented herein in Part II are based mainly on the paper presented by Yang and Kuo (1991b).

9.3.1 *Fundamental equations and conditions*

All the fundamental equations and conditions to be invoked later in the analysis of the two-member angled frame will be first summarized. For simplicity, the two members of the frame are assumed to have identical cross-sections and identical material properties.

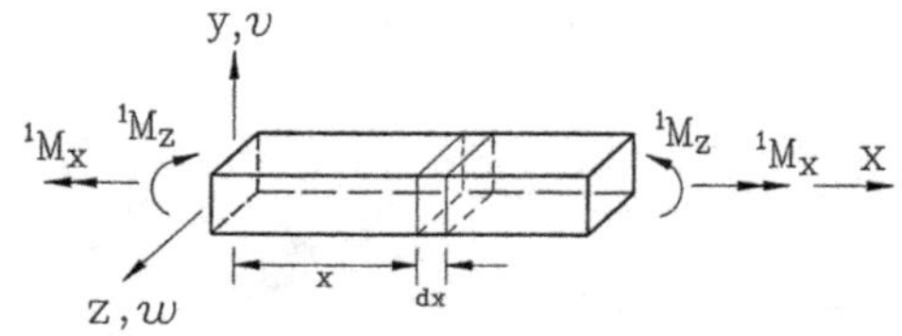

Fig. 9.14. Coordinates of solid beam under bending and torsion.

- *Governing differential equations*

Consider a solid beam acted upon by a pair of torques ${}^1M_x$ and a pair of bending moments ${}^1M_z$ in the C_1 configuration as shown in Figure 9.14. Both flexural and torsional deformations are possible with the beam when buckling occurs. Based on the theory of Section 5.5, the governing differential equations of equilibrium for a generic cross section x of the beam can be obtained from equations (5.105)–(5.107) as follows:

$$EI_y w'''' - {}^1M_x v''' + {}^1M_z \theta_x'' = 0 \tag{9.90}$$

$$EI_z v'''' + {}^1M_x w''' = 0 \tag{9.91}$$

$$GJ\theta_x'' - {}^1M_z w'' = 0 \tag{9.92}$$

where a prime ($'$) denotes differentiation with respect to the coordinate x; v and w the translations along the y-and z-axis, respectively; θ_x the angle of twist about the x-axis; E and G the moduli of elasticity and rigidity, respectively; J the torsional constant; and I_y and I_z the moments of inertia about the y-and z-axis, respectively. In arriving at equation (9.91), the higher-order term $({}^1M_z u')''$ in equation (5.106) due to axial straining has been neglected

- *Cross-sectional forces*

For problems of the bifurcation type, we can still use the updated Lagrangian formulation presented in Chapter 5 for the motion of the beam from C_1 to C_2, by simply taking the C_1 configuration equal to the C_0 configuration Hence, the shear forces (${}^2F_y$, ${}^2F_z$), torque ${}^2M_x$, and bending moment (${}^2M_y$, ${}^2M_Z$) acting on the member in the buckling configuration C_2 can be obtained from equations (5.109), (5.110), (5.111), (5.112), and (5.113), respectively, as follows:

$${}^2F_y = -EI_z v''' - {}^1M_x w'' \tag{9.93}$$

$$ {}^2F_z = -EI_y w''' + {}^1M_x v'' - {}^1M_z \theta_x' \tag{9.94} $$

$$ {}^2M_x = GJ\theta_x' - {}^1M_z w' \tag{9.95} $$

$$ {}^2M_y = -EI_y w'' + {}^1M_x v' - {}^1M_z \theta_x \tag{9.96} $$

$$ {}^2M_Z = EI_z v'' + {}^1M_x w' \tag{9.97} $$

where the member forces ${}^1M_x$ and ${}^1M_z$ should be recognized as the ones existing on the member prior to buckling, and the relation $\theta_y = -w'$ has been used in reaching equation (9.97).

- *Geometrical boundary conditions at joint A*

Throughout this section, the quantities associated with members 1 and 2 of the angled frame in Figures 9.11–9.13 will be denoted with subscripts "1" and "2", respectively. The geometric boundary conditions for the frame are

$$ (v_1)_{x1=0} = 0 \tag{9.98} $$

$$ (v_1')_{x1=0} = 0 \tag{9.99} $$

$$ (\theta_{x1})_{x1=0} = 0 \tag{9.100} $$

$$ (w_1)_{x1=0} = 0 \tag{9.101} $$

$$ (w_1')_{x1=0} = 0 \tag{9.102} $$

all of which are evaluated at $x_1 = 0$ for the fixed support.

- *Equilibrium conditions at joint B*

Let L and βL denote the lengths of members 1 and 2, respectively, of the angled frame. The equilibrium conditions of the frame at joint B should be specified for the buckling configuration C_1 as follows:

$$ ({}^2F_{y1})_{x1=L} = ({}^2F_{y2})_{x2=0} \tag{9.103} $$

$$ ({}^2F_{z1})_{x1=L} = ({}^2F_{x2})_{x2=0} \sin\alpha + ({}^2F_{z2})_{x2=0} \cos\alpha \tag{9.104} $$

$$ ({}^2M_{x1})_{x1=L} = ({}^2M_{x2})_{x2=0} \cos\alpha - ({}^2M_{z2})_{x2=0} \sin\alpha \tag{9.105} $$

$$ ({}^2M_{y1})_{x1=L} = ({}^2M_{y2})_{x2=0} \tag{9.106} $$

$$ ({}^2M_{z1})_{x1=L} = ({}^2M_{x2})_{x2=0} \sin\alpha + ({}^2M_{z2})_{x2=0} \cos\alpha \tag{9.107} $$

where α denotes the inclination angle between members 1 and 2. Note that the preceding conditions are valid for joint B in the buckling position only when the associated cross-sectional forces, i.e, ${}^2F_{y1}, {}^2F_{y2}, \ldots$, etc., are correctly established for the joint at the buckling position, such as those given in (9.93)–(9.97).

- *Slope-continuity conditions at joint B*

For the present case, only the slope-continuity conditions at joint B are essential to the calculation of the critical loads:

$$(w_1')_{x1=L} = (w_2')_{x1=0} \tag{9.108}$$

$$(v_1')_{x1=L} \cos\alpha - (\theta_{x1})_{x1=L} \sin\alpha = (v_2')_{x2=0} \tag{9.109}$$

The others will not be given here, since they are needed only when the deformed shape of the frame is desired.

- *Natural boundary conditions at joint C*

The only nonzero force for member 2 at C_1 (i.e., at the beginning of the buckling stage) is ${}^1M_{x2}$, which is equal to the applied torque T_0. Based on the force expressions in (9.93)–(9.97), the natural boundary conditions that are not related to the properties of the applied torque T_0 for joint C undergoing finite rotations can be given as follows:

$$({}^2F_{y2})_{x2=\beta L} = (-EI_z v_2''' - T_0 w_2'')_{x2=\beta L} = 0 \tag{9.110}$$

$$({}^2F_{z2})_{x2=\beta L} = (-EI_y w_2''' + T_0 v_2'')_{x2=\beta L} = 0 \tag{9.111}$$

$$({}^2M_{y2})_{x2=\beta L} = (GJ\theta_{x2}')_{x2=\beta L} = 0 \tag{9.112}$$

Other natural boundary conditions for joint C will be established once the rotational properties of the applied torque T_0 about the three axes are prescribed, as will be discussed below.

9.3.2 *Rotational properties of applied torques*

Three types of torques will be considered in this section. They can all be generated by mechanisms consisting of a couple or couples of direct forces as shown in Figures 9.14(a)–9.14(c).

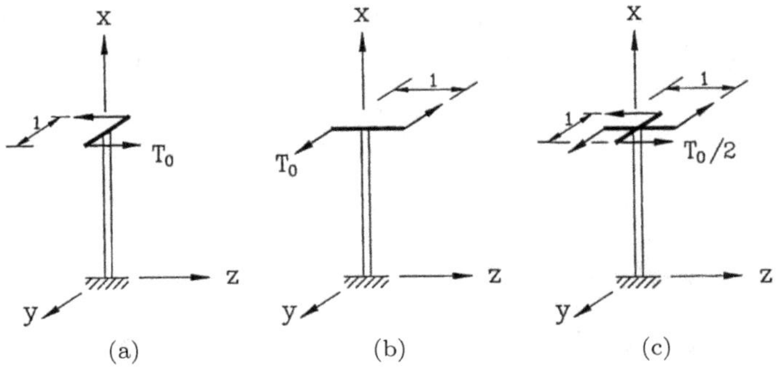

Fig. 9.15. Various types of torques: (a) QT-1; (6) QT-2; (c) ST.

- *Quasitangential torque of first kind (QT-1)*

The torque T_0 depicted in Figure 9.15(a) is generated by a couple of conservative forces acting on a rigid lever lying along the y-axis. Such a torque, referred to as quasitangential torque of the first kind (QT-1), induces moment increments of magnitude T_0v' and 0 about the y- and z-axis respectively, when subjected to rotations in the three-dimensional space (see Section 5.3). Consider the free body of joint C in the buckling position in Figure 9.11(b). The moments indicated on the top side of the joint are generated externally by the QT-1 torsion mechanism, while those on the bottom side are generated internally as the stress resultants of the cross-section, readily available through equations (9.95)–(9.97) based on the fact that ${}^1M_x = T_0$ and ${}^1M_z = 0$ for member 2. From the conditions of equilibrium for moments acting at joint C, the following can be written:

$$-EI_yw_2'' = 0 \tag{9.113}$$

$$EI_zv_2'' + T_0w_2' = 0 \tag{9.114}$$

- *Quasitangential torque of second kind (QT-2)*

External torque of the type given in Figure 9.15(b) has been referred to as quasitangential torque of the second kind (QT-2). According to Section 5.3, a QT-2 torque T_0 will induce moment increments of magnitude 0 and T_0w' about the y-and z-axis, respectively. Such moment increments have been indicated on the top side of the free body for joint C in Figure 9.12(b), in addition to those generated as stress resultants on the bottom side. Considering the equilibrium of joint C in the buckling position (with the internal moments given in equations (9.95)–(9.97), we can write the following natural boundary conditions:

$$-EI_y w_2'' + T_0 v_2' = 0 \tag{9.115}$$

$$EI_z v_2'' = 0 \tag{9.116}$$

- *Semitangential torque (ST)*

A torque generated by the mechanism shown in Figure 9.15(c) is called the semitangential (ST) torque. Such a torque will induce moments of magnitude $1/2(T_0 v')$ and $1/2(T_0 w')$ about the y-and z-axis, respectively, when subjected to three-dimensional rotations, which have been indicated on the top side of the free body for joint C in Figure 9.13(b). From the conditions of equilibrium of moments at joint C (with the internal moments given in equations (9.95)–(9.97), the following can be written:

$$-EI_y w_2'' + \frac{1}{2} T_0 v_2' = 0 \tag{9.117}$$

$$EI_z v_2'' + \frac{1}{2} T_0 w_2' = 0 \tag{9.118}$$

9.3.3 *Solutions for various torsional loads*

With the natural boundary conditions given above for each type of applied torque, the critical loads for the angled frames shown in Figures 9.11–9.13 can be determined.

- *Critical value for QT-1 torque*

Setting ${}^1M_{x1} = T_0 \cos\alpha$ and ${}^1M_{z1} = T_0 \sin\alpha$, we obtain the general solution to equations (9.90)–(9.92) for member 1:

$$\frac{v_1}{L} = \frac{\mu}{\eta} \cos\alpha \left(a_1 \cos\eta \frac{x_1}{L} - b_1 \sin\eta \frac{x_1}{L} \right) + f_1 + g_1 \frac{x_1}{L} + h_1 \left(\frac{x_1}{L} \right)^2 \tag{9.119}$$

$$\frac{w_1}{L} = a_1 \sin\eta \frac{x_1}{L} + b_1 \cos\eta \frac{x_1}{L} + c_1 + d_1 \frac{x_1}{L} + e_1 \left(\frac{x_1}{L} \right)^2 \tag{9.120}$$

$$\theta_{x1} = \lambda \sin\alpha \left(a_1 \eta \frac{x_1}{L} + b_1 \cos\eta \frac{x_1}{L} \right) + i_1 + j_1 \frac{x_1}{L} \tag{9.121}$$

where $a_1, b_1, \ldots, j_1$ are integration constants, and

$$\lambda = \frac{TL}{GJ}, \quad \mu = \frac{TL}{EI_z}, \quad \phi = \frac{TL}{\sqrt{EI_y EI_z}} \tag{9.122a-c}$$

$$\eta = \sqrt{\frac{T^2L^2}{EI_y EI_z}\cos^2\alpha + \frac{T^2L^2}{EI_y GJ}\sin^2\alpha} = \phi\sqrt{\cos^2\alpha + \frac{\lambda}{\mu}\sin^2\alpha} \tag{9.123}$$

Enforcement of the geometric boundary conditions (9.98)– (9.102) enables us to rewrite equations (9.119– (9.121) as

$$\frac{v_1}{L} = \frac{\mu}{\eta}\cos\alpha\left[a_1\left(\cos\eta\frac{x_1}{L} - 1\right) - \underline{b_1\left(\sin\eta\frac{x_1}{L} - \eta\frac{x_1}{L}\right)}\right] + h_1\left(\frac{x_1}{L}\right)^2 \tag{9.124}$$

$$\frac{w_1}{L} = a_1\left(\sin\eta\frac{x_1}{L} - \eta\frac{x_1}{L}\right) + \underline{b_1\left(\cos\eta\frac{x_1}{L} - 1\right)} + e_1\left(\frac{x_1}{L}\right)^2 \tag{9.125}$$

$$\theta_{x1} = \lambda\sin\alpha\left[a_1\sin\eta\frac{x_1}{L} + \underline{b_1\left(\cos\eta\frac{x_1}{L} - 1\right)}\right] + j_1\frac{x_1}{L} \tag{9.126}$$

Similarly, by noting that ${}^1M_{x2} = T_0$ and ${}^1M_{z2} = 0$, the general solution to equations (9.90)–(9.92) can be written for member 2 as follows:

$$\frac{v_2}{L} = \frac{\mu}{\phi}\left(a_2\cos\phi\frac{x_2}{L} - b_2\sin\phi\frac{x_2}{L}\right) + f_2 + g_2\frac{x_2}{L} + h_2\left(\frac{x_2}{L}\right)^2 \tag{9.127}$$

$$\frac{w_2}{L} = a_2\sin\phi\frac{x_2}{L} + b_2\cos\phi\frac{x_2}{L} + c_2 + d_2\frac{x_2}{L} + e_2\left(\frac{x_2}{L}\right)^2 \tag{9.128}$$

$$\theta_{x2} = i_2 + j_2\frac{x_2}{L} \tag{9.129}$$

where $a_2, b_2, \ldots, j_2$ are integration constants. By substituting the preceding displacements into the natural boundary conditions for joint C, i.e., (9.110)–(9.112), the following coefficients can be determined:

$$e_2 = h_2 = 0 \tag{9.130}$$

$$j_2 = \frac{TL}{GJ} \tag{9.131}$$

In addition, from the remaining natural boundary conditions for joint C, i.e., (9.113) and (9.114), the following can be determined:

$$a_2 \sin \beta\phi + b_2 \cos \beta\phi = 0, \quad d_2 = 0 \tag{9.132a,b}$$

Consequently, equations (9.127)–(9.129) for member 2 reduce to

$$\frac{v_2}{L} = \underline{\frac{\mu}{\phi} a^*2 \cos \phi \left(\beta - \frac{x_2}{L}\right)} + f_2 + \underline{g_2 \frac{x_2}{L}} \tag{9.133}$$

$$\frac{w_2}{L} = \underline{-a^*2 \sin \phi \left(\beta - \frac{x_2}{L}\right)} + c_2 \tag{9.134}$$

$$\theta_{x2} = i_2 + \underline{\frac{T_0 x_2}{GJ}} \tag{9.135}$$

where $a_2^* = a_2 / \cos \beta\phi = -b_2 / \sin \beta\phi$.

Now let us turn to the equilibrium conditions for joint B in equations (9.103)–(9.107). Note that for member 1, ${}^1M_{x1} = T_0 \cos \alpha$ and ${}^1M_{z1} = T_0 \sin \alpha$, and that for member 2, ${}^1M_{x2} = T_0$ and ${}^1F_{x2} = {}^1F_{y2} = {}^1F_{z2} = {}^1M_{y2} = {}^1M_{z2} = 0$, we may substitute the displacement functions (9.124)–(9.126) and (9.133)–(9.135) into the joint equilibrium conditions (9.103)–(9.107), and then solve for the following:

$$g_2 = b_1(\lambda \sin^2 \alpha + \mu \cos^2 \alpha) \tag{9.136}$$

$$e_1 = 0 \tag{9.137}$$

$$a_1 = \frac{(\lambda - \mu) \sin \alpha \cos \alpha}{\eta(\lambda \sin^2 \alpha + \mu \cos^2 \alpha)} \tag{9.138}$$

$$h_1 = \frac{\mu\lambda \sin \alpha}{2(\lambda \sin^2 \alpha + \mu \cos^2 \alpha)} \tag{9.139}$$

$$j_1 = \frac{\mu\lambda \cos \alpha}{\lambda \sin^2 \alpha + \mu \cos^2 \alpha} \tag{9.140}$$

As can be seen from the preceding expressions, the coefficients a_1, e_1, h_1, and j_1 have no correlation with the other coefficients. It follows that the underlined terms in (9.124)–(9.126) are the only terms central to the determination of the critical loads. On the other, hand, an examination of the displacement functions (9.133)–(9.135) reveals that the coefficients c_2, f_2, and i_2 represent basically the rigid body displacements at joint B, which again bear no relation to the critical loads. Those terms that are essential to the determination of the critical loads in (9.133)–(9.135) have also been underlined.

Based on the preceding observations, the slope-continuity conditions (9.108) and (9.109) for joint B, in which the rigid body displacements have been excluded, can be relied upon to calculate the critical loads. Substituting the displacement functions as given in (9.124)–(9.126) and (9.133)–(9.135) into the slope-continuity conditions yields

$$\begin{bmatrix} \phi\cos\beta\phi & \eta\sin\eta \\ \mu\sin\beta\phi & (\lambda\sin^2\alpha+\mu\cos^2\alpha)\cos\eta \end{bmatrix} \begin{Bmatrix} a_2^* \\ b_1 \end{Bmatrix} = \begin{Bmatrix} \frac{(\lambda-\mu)(\cos\eta-1)\sin\alpha\cos\alpha}{\lambda\sin^2\alpha+\mu\cos^2\alpha} \\ -\frac{(\lambda-\mu)\sin\eta\sin\alpha}{\cos\alpha}\eta \end{Bmatrix} \tag{9.141}$$

in which the relation (9.136) has been utilized. By setting the determinant of the matrix in the preceding equation equal to zero, a transcendental equation can be obtained:

$$\cos\beta\phi\cos\eta\sqrt{\cos^2\alpha+\frac{\lambda}{\mu}\sin^2\alpha}-\sin\beta\phi\sin\eta=0 \tag{9.142}$$

from which the critical load $T_{0,cr}$ can be solved.

- *Critical value for QT-1 torque*

In the following, three special cases will be considered.

Case 1 – Member 2 is of zero length: This is exactly the cantilever shown in Figure 9.16(a). For the present case, $\beta=0$, equation (9.142) reduces to $\cos\eta=0$, from which the critical load $T_{0,cr}$ is obtained as

$$T_{0,cr}=\pm\frac{\pi}{2L}\sqrt{\frac{EI_yEI_zGJ}{GJ\cos^2\alpha+EI_z\sin^2\alpha}} \tag{9.143}$$

Two useful extreme cases should be discussed here. First, for $\alpha=0°$, the cantilever is subjected to a QT-1 torque about the member axis. The critical load $T_{0,cr}$ for this case is

$$T_{0,cr}=\pm\frac{\pi}{2L}\sqrt{EI_yEI_z} \tag{9.144}$$

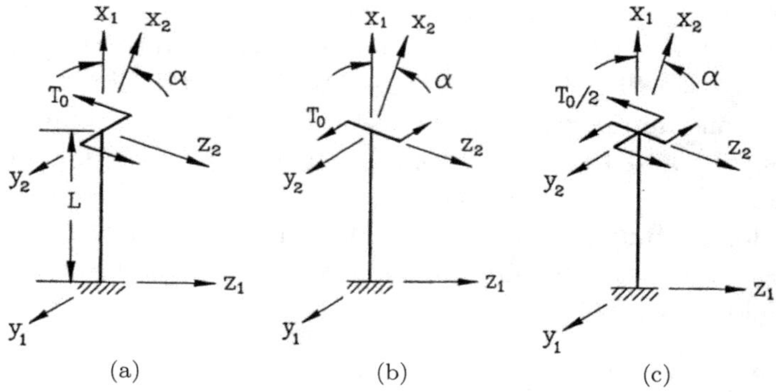

Fig. 9.16. Cantilever with various torques: (a) QT-1; (b) QT-2; (c) ST.

which reduces to that given by Ziegler (1977) for a circular bar with $I_y = I_z$.

The other case occurs for $\alpha = 90°$, which implies that the cantilever is under the action of a QT-1 bending moment with its rigid arm lying along the y-axis. The corresponding critical moment $T_{0,cr}$ is

$$T_{0,cr} = \pm\frac{\pi}{2L}\sqrt{EI_yGJ} \tag{9.145}$$

which coincides with the result given by Timoshenko and Gere (1961).

Case 2 – Equal rigidities for flexure and torsion: Since $EI_z = GJ$ (i.e., $\lambda = \mu$) and $\eta = \phi$, equation (9.142) reduces to

$$\cos(1+\beta)\phi = 0 \tag{9.146}$$

from which the critical load $T_{0,cr}$ can be obtained as

$$T_{0,cr} = \pm\frac{\pi}{2(1+\beta)L}\sqrt{EI_yEI_z} = \pm\frac{\pi}{2(1+\beta)L}\sqrt{EI_yGJ} \tag{9.147}$$

Case 3 – Members 1 and 2 are of equal length: For $\beta = 1$, equation (9.142) reduces to

$$\frac{\eta}{\phi}\cos\phi\cos\eta - \sin\phi\sin\eta = 0 \tag{9.148}$$

which can be solved by numerical approaches.

- *Critical value for QT-2 torque*

All the equations and conditions employed here are the same as those of the preceding loading case, except that the natural boundary conditions of (9.113) and (9.114) should be replaced by those given in (9.115) and (9.116) to stand for the rotational effects of the QT-2 torque. Following the same procedure, we can derive the characteristic equation as

$$\sin\beta\phi\sin\eta\sqrt{\cos^2\alpha+\frac{\lambda}{\mu}\sin^2\alpha}-\cos\beta\phi\cos\eta=0 \tag{9.149}$$

Again, three special cases will be considered.

Case 1 – Member 2 is of zero length: This case is exactly the one shown in Figure 9.16(b). For $\beta = 0$, equation (9.149) reduces to $\cos\eta = 0$. The critical load $T_{0,cr}$ obtained here is identical to that given in (9.143).

Case 2 – Equal rigidities for flexure and torsion: For $EI_z = GJ$ (i.e., $\lambda = \mu$) and $\eta = \phi$, equation (9.149) reduces identically to (9.146). Consequently, the same critical load $T_{0,cr}$ as that given in (9.147) can be obtained.

Case 3 – Members 1 and 2 are of equal length: For $\beta = 1$, equation (9.149) reduces to the following:

$$\frac{\eta}{\phi}\sin\phi\sin\eta-\cos\phi\cos\eta=0 \tag{9.150}$$

which can be solved numerically.

- *Critical value for ST torque*

Using the natural boundary conditions of (9.117) and (9.118), and following the same lines of derivation, we can derive the characteristic equation for the present loading case:

$$2(1+\cos\beta\phi\cos\eta)-\sin\beta\phi\sin\eta\left(\frac{\eta}{\phi}+\frac{\phi}{\eta}\right)=0 \tag{9.151}$$

Solutions for the three special cases are presented as follows.

Case 1 – Member 2 is of zero length: This is the case shown in Figure 9.16(c). For this case, $\beta = 0$ and the characteristic equation (9.151) reduces to $\cos\eta = -1$, from which the critical load $T_{0,cr}$ is

$$T_{0,cr} = \pm\frac{\pi}{L}\sqrt{\frac{EI_y EI_z GJ}{GJ\cos^2\alpha + EI_z\sin^2\alpha}} \tag{9.152}$$

For the special case of $\alpha = 0°$, the cantilever is subjected to an ST torque about the member axis. This case has a critical value of

$$T_{0,cr} = \pm\frac{\pi}{L}\sqrt{EI_y EI_z} \tag{9.153}$$

which again agrees with Ziegler's (1977) result for the special case of circular bars with $I_y = I_z$. The other special case assumes an angle of $\alpha = 90°$, implying that the cantilever is acted upon by an ST bending moment at the free end. The critical moment obtained from equation (9.152) is

$$T_{0,cr} = \pm\frac{\pi}{L}\sqrt{EI_y GJ} \tag{9.154}$$

which equals twice the value given in (9.145) for the QT torques.

Case 2 – Equal rigidities for flexure and torsion: For $EI_z = GJ$ (i.e., $\lambda = \mu$) and $\eta = \phi$, equation (9.151) reduces to

$$\cos(1+\beta)\phi = -1 \tag{9.155}$$

from which the critical load $T_{0,cr}$ is obtained as

$$T_{0,cr} = \pm\frac{\pi}{(1+\beta)L}\sqrt{EI_y EI_z} = \pm\frac{\pi}{(1+\beta)L}\sqrt{EI_y GJ} \tag{9.156}$$

which is twice the value given in (9.147) for the frame under the QT-1 or QT-2 loadings.

Case 3 – Members 1 and 2 are of equal length: For $\beta = 1$, equation (9.151) reduces to

$$2(1 + \cos\phi\cos\eta) - \sin\phi\sin\eta\left(\frac{\eta}{\phi} + \frac{\phi}{\eta}\right) = 0 \tag{9.157}$$

which can also be solved by numerical means.

9.4 Numerical Examples

To demonstrate the effect of various torsional loadings, three frames made up of members with different orientations of cross-sectional axes will be studied in this section.

9.4.1 *Example 1: Cross-sections with* $I_y \gg I_z$

This is exactly the case shown in Figure 9.17(a). The following data are assumed for both members 1 and 2: $E = 71{,}240\,\text{N/mm}^2$; $G = 27{,}190\,\text{N/mm}^2$; $L = 240\,\text{mm}$; $I_y = 50\,\text{mm}^4$; $I_z = 0.18\,\text{mm}^4$; and $J = 0.72\,\text{mm}^4$. The critical torque $T_{0,cr}$ has been plotted against angle α in Figures 9.18(a)–9.18(c) for three length ratios: $\beta = 0$ (cantilever); $\beta = 1$; and $\beta = 2$. An observation of these figures reveals the following: (1) The ability of the frame to resist torsional loadings, whether the QT or ST torques, decreases significantly as the length ratio β increases; (2) the critical value of the ST torque is about twice that of the QT torque; (3) in general, no significant difference can be observed between the critical values of the QT-1 and QT-2 torques.

9.4.2 *Example 2: Cross-sections with* $I_y \ll I_z$

This is the case shown in Figure 9.17(b). The following data are assumed: $E = 71,240\,\text{N/mm}^2$; $G = 27,190\,\text{N/mm}^2$; $L = 240\,\text{mm}$; $I_y = 0.18\,\text{mm}^4$; $I_z = 50\,\text{mm}^4$; and $J = 0.72\,\text{mm}^4$. The analysis results have been plotted in Figures 9.19(a)–9.19(c) for three length

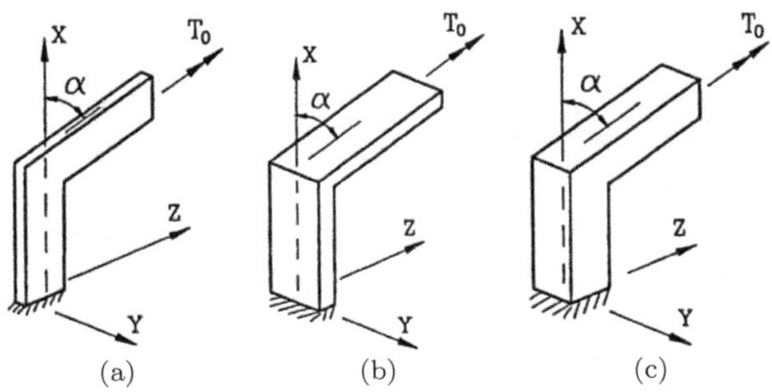

Fig. 9.17. Example frames: (a) $I_y \gg I_z$; (b) $I_y \ll I_z$; (c) $I_y = I_z$.

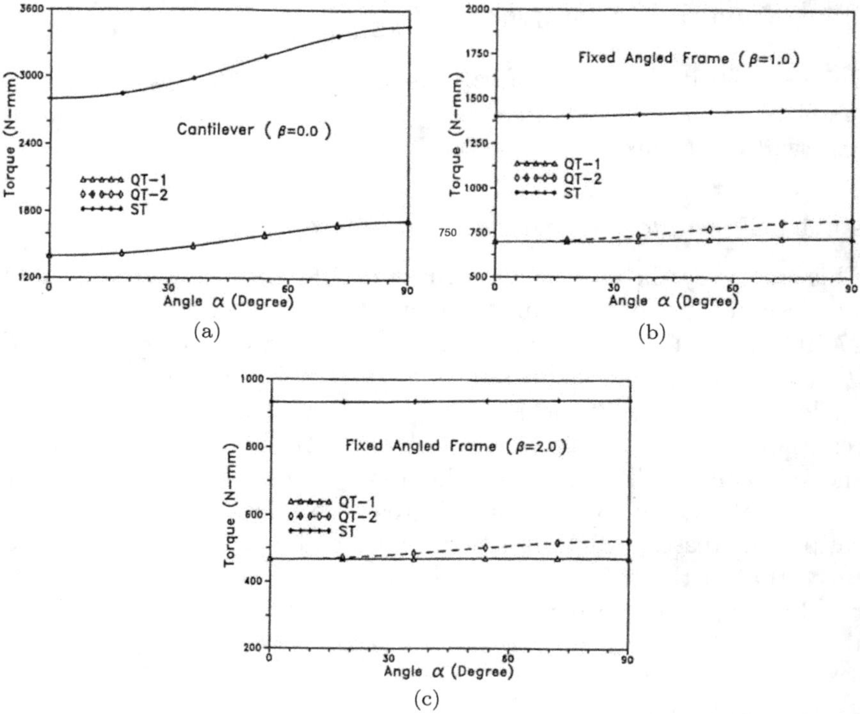

Fig. 9.18. Critical torques for frames with $I_y \gg I_z$: (a) $\beta = 0$; (b) $\beta = 1$; (c) $\beta = 2$.

ratios: $\beta = 0$ (cantilever); $\beta = 1$; and $\beta = 2$. A comparison of Figures 9.19(a)–9.19(c) with Figures 9.18(a)–9.18(c) indicates that the critical value for Example 1 either increases or remains rather constant, but never decreases, as angle α increases. However, the critical value for the present example decreases dramatically as angle α increases. This can be attributed to the fact that in Example 1, member 1 is bent about the weak axis of the cross-section prior to buckling, while in this example it is bent initially about the strong axis; the latter has a stronger tendency to buckle.

It remains true that the critical values for the ST and QT torques decrease as the length ratio β increases. But the ability of the frame to resist TS torques is only marginally higher than that to resist

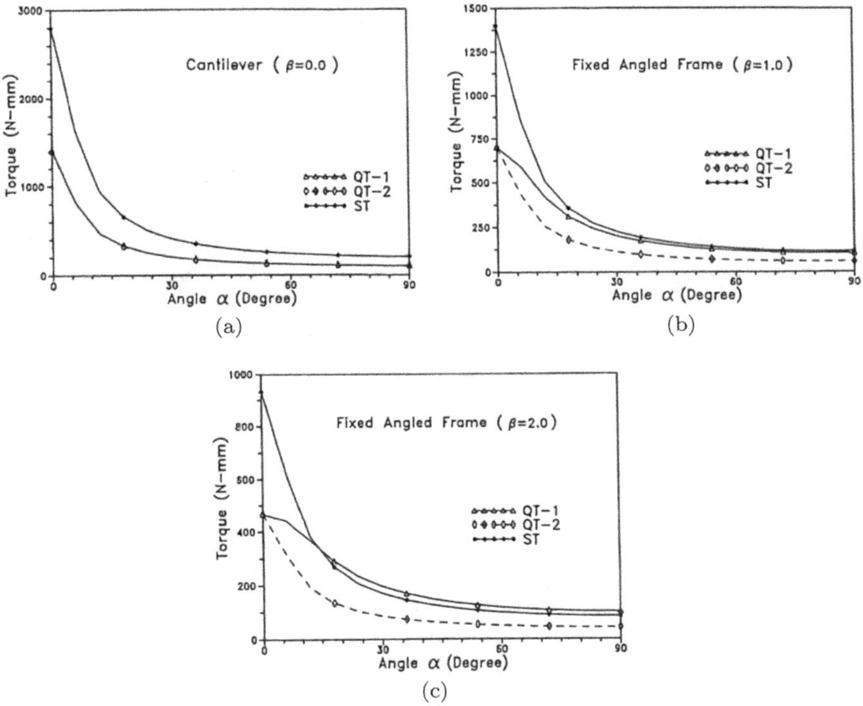

Fig. 9.19. Critical torques for frames with $I_y \ll I_z$: (a) $\beta = 0$; (b) $\beta = 1$; (c) $\beta = 2$.

the QT torques. Such a margin even vanishes as the length ratio β increases, say, reaching the value of 2.

9.4.3 *Example 3: Cross-sections with $I_y = I_z$*

Except for the use of $I_y = I_z = 12.56\,\text{mm}^4$ and $J = 25.12\,\text{mm}^4$, as in Figure 9.17(c), the other data are the same as those used in the preceding two examples. From Figures 9.20(a)–9.20(c), we observe that the three points made previously for Example 1 remain valid here. One difference is that for the current example, there is a slight decrease in the resistance against the ST and QT-2 torques as the inclination angle α becomes larger.

Before closing this section, we like to comment that the fundamental aspects in the analysis of the buckling of a three-dimensional

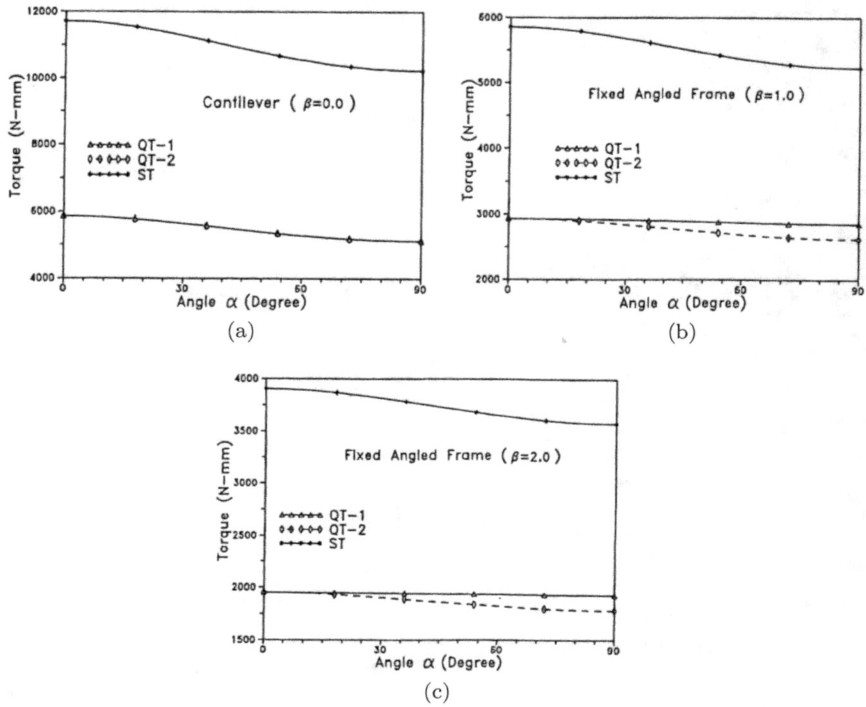

Fig. 9.20. Critical torques for frames with $I_y = I_z$: (a) $\beta = 0$; (b) $\beta = 1$; (c) $\beta = 2$.

frame have been highlighted through the study of an angled frame subjected to various types of torsional loadings, which remains a special case of the general space frames. Central to the present derivation is the construction of all physical relations in the buckling configuration for the frame considered, which include, in particular, the expressions for internal moments appearing as stress resultants and external moments generated by various mechanical devices. From the analysis results, it is observed that the buckling loads of an angled frame are very sensitive to the way by which the torsional loadings are applied. For the limited number of special cases for which previous solutions are available, the validity of the present results has been confirmed. The numerical examples presented in this section, in addition to those of Sections 9.2 and 9.3, can be used as benchmarks for calibrating various finite element procedures.

References

Argyris, J.H., Hilbert, O., Malejannakis, G.A., and Scharpf, D.W. (1979). On the geometric stiffness of a beam in space: a consistent V.W. approach, *Comp. Meth. Appl. Mech. Eng.*, 20, 105–131.

Barsoum, R.S. and Gallagher, R.H. (1970). Finite element analysis of torsional-flexural stability problems, *Int. J. Numer. Meth. Eng.*, 2, 335–352.

Chajes, A. (1974). *Principles of Structural Stability Theory*, Prentice Hall, Englewood Cliffs, N.J.

Chen, W.F. and Lui, E.M. (1987). *Structural Stability-Theory and Implementation*, Elsevier, New York, NY.

Elias, Z.M. (1986). *Theory and Methods of Structural Analysis*, John Wiley, New York, NY.

Greenhill, A.G. (1883). On the strength of shafting when exposed to torsion and to end thrust, *Proc., Instit. Mech. Eng.*, 182–225.

Simitses, G.J. (1976). *An Introduction to the Elastic Stability of Structures*, Prentice Hall, Englewood Cliffs, N.J.

Timoshenko, S.P. and Gere, J.M. (1961). *Theory of Elastic Stability*, 2nd edn, McGraw-Hill, New York, NY.

Vlasov, V.Z. (1961). *Thin-Waled Elastic Beams*, Israel Program for Scientific Translation, Jerusalem, Israel.

Yang, Y.B. and Kuo, S.R. (1986). Static stability of curved thin-walled beams, *J. Eng. Mech.*, ASCE, 112(8), 821–841.

Yang, Y.B. and Kuo, S.R. (1987). Effect of curvature on stability of curved beams, *J. Struct. Eng.*, ASCE, 113(6), 1185–1202.

Yang, Y.B. and Kuo, S.R. (1991a). Out-of-plane buckling of angled frames, *Int. J. Mech. Sci.*, 33(1), 55–67.

Yang, Y.B. and Kuo, S. R. (1991b). Buckling of frames under various torsional loadings, *J. Eng. Mech.*, ASCE, 117(8), 1681–1697.

Yang, Y.B. and Kuo, S.R. (1991c). Consistent frame buckling analysis by finite element method, *J. Struct. Eng.*, ASCE, 117(4), 1053–1069.

Yang, Y.B. and Yau, J.D. (1987). Stability of beams with tapered I-sections, *J. Eng. Mech.*, ASCE, 113(9), 1337–1357.

Yang, Y.B. and Yau, J.D. (1989). Stability of pretwisted bars with various end torques, *J. Eng. Mech.*, ASCE, 115(4), 671–688.

Yau, J.D., Yang, Y.B., and Kuo, S.R. (1992). Stability of tapered bars of circular cross sections under semi- and quasi-tangential torques, *Int. J. Mech. Sci.*, 34(1), 31–40.

Ziegler, H. (1977). *Principles of Structural Stability*, 2nd edn, Birkhäuser, Stuttgart, Germany.

Appendix A

List of Integral Matrices

In Chapter 2, we have defined the following notation for the integral matrices in equation (2.44) that appear in the formation of various stiffness matrices:

$$[K_{gh}^{stv}] = \int_0^1 i^v \{n_g^s\}\{n_h^t\}^T di \tag{A.1}$$

where the subscripts "g" and "h" denote the degrees of the interpolation functions $\{n\}$, the superscripts "s" and "t" the order of differentiation, and "v" the exponent of the multiplying factor i. Various integral matrices based on this definition have been used extensively in the calculation of the stiffness matrices throughout the book. The following is a list of the most commonly used integral matrices:

$$[K_{33}^{021}] = \frac{1}{30}\begin{bmatrix} -3 & -6 & 3 & 3 \\ 0 & -1 & 0 & 1 \\ 33 & 6 & -33 & 27 \\ -3 & 0 & 3 & -3 \end{bmatrix}$$

$$[K_{33}^{110}] = \frac{1}{30}\begin{bmatrix} 36 & 3 & -36 & 3 \\ 3 & 4 & -3 & -1 \\ -36 & -3 & 36 & -3 \\ 3 & -1 & -3 & 4 \end{bmatrix}$$

$$[K_{33}^{100}] = \frac{1}{30}\begin{bmatrix} -15 & -3 & -15 & 3 \\ 3 & 0 & -3 & 0.5 \\ 15 & 3 & 15 & -3 \\ -3 & -0.5 & 3 & 0 \end{bmatrix}$$

$$[K_{33}^{200} = \frac{1}{30}\begin{bmatrix} -36 & -3 & 36 & -3 \\ -33 & -4 & 3 & 1 \\ 36 & 3 & -36 & 3 \\ -3 & 1 & 33 & -4 \end{bmatrix}$$

$$[K_{33}^{111}] = \frac{1}{30}\begin{bmatrix} 18 & 3 & -18 & 0 \\ 3 & 1 & -3 & -0.5 \\ -18 & -3 & 18 & 0 \\ 0 & -0.5 & 0 & 3 \end{bmatrix}$$

$$[K_{33}^{121}] = \frac{1}{30}\begin{bmatrix} -18 & 6 & 18 & -24 \\ -9 & -2 & 9 & -7 \\ 18 & -6 & -18 & 24 \\ 21 & 8 & -21 & 13 \end{bmatrix}$$

$$[K_{33}^{220}] = \begin{bmatrix} 12 & 6 & -12 & 6 \\ 6 & 4 & -6 & 2 \\ -12 & -6 & 12 & -6 \\ 6 & 2 & -6 & 4 \end{bmatrix}$$

$$[K_{33}^{112}] = \frac{1}{210}\begin{bmatrix} 72 & 15 & -72 & -6 \\ 15 & 4 & -15 & -3 \\ -72 & -15 & 72 & 6 \\ -6 & -3 & 6 & 18 \end{bmatrix}$$

$$[K_{33}^{120}] = \begin{bmatrix} 0 & 1 & 0 & -1 \\ -1 & -0.5 & 1 & -0.5 \\ 0 & -1 & 0 & 1 \\ 1 & 0.5 & -1 & 0.5 \end{bmatrix}$$

$$[K_{33}^{101}] = \frac{1}{420}\begin{bmatrix} -78 & -18 & -132 & 24 \\ -4 & -2 & -31 & 5 \\ 78 & 18 & 132 & -24 \\ -11 & -2 & 46 & -2 \end{bmatrix}$$

$$[K_{13}^{110}] = \begin{bmatrix} 1 & 0 & -1 & 0 \\ -1 & 0 & 1 & 0 \end{bmatrix} \quad [K_{13}^{010}] = \frac{1}{12}\begin{bmatrix} -6 & 1 & 6 & -1 \\ -6 & -1 & 6 & 1 \end{bmatrix}$$

$$[K_{13}^{020}] = \begin{bmatrix} -1 & -1 & 1 & 0 \\ 1 & 0 & -1 & 1 \end{bmatrix} \quad [K_{13}^{111}] = \frac{1}{12}\begin{bmatrix} 6 & 1 & -6 & -1 \\ -6 & -1 & 6 & 1 \end{bmatrix}$$

$$[K_{13}^{120}] = \begin{bmatrix} 0 & 1 & 0 & -1 \\ 0 & -1 & 0 & 1 \end{bmatrix} \quad [K_{13}^{021}] = \frac{1}{6}\begin{bmatrix} 0 & -1 & 0 & 1 \\ 6 & 1 & -6 & 5 \end{bmatrix}$$

$$[K_{13}^{121}] = \begin{bmatrix} -1 & 0 & 1 & -1 \\ 1 & 0 & -1 & 1 \end{bmatrix} \quad [K_{11}^{110}] = \begin{bmatrix} 1 & -1 \\ -1 & 1 \end{bmatrix}$$

It should be noted that many matrices not listed herein can be obtained merely by the formula: $[K_{hg}^{tsv}]^T$.

Appendix B

Geometry Updating of Space Frame Element Involving Finite Rotations and Calculation of Element Forces

B.1 Introduction

The analysis of *space frames* is much more complicated than any of the structures mentioned in Chapter 7 due to the involvement of three rotational degrees of freedom at each end of the frame element. Conventionally, the end rotations of each element of the space frame are assumed to be small (Connor *et al.*, 1968; Tezcan and Mahapatra, 1969; Oran, 1973; Remseth, 1979; Chan and Kitipornchai, 1987). Since the *law of commutativity* remains valid for small rotations about the three axes, the end rotations of each element can be calculated simply by adding the rotation increments generated at each incremental step to those accumulated up to the previous step. Thus, with the assumption of small rotations, the procedure for updating the end orientations is not that difficult, if the concept of *convected coordinates* (Belytschko and Hsieh, 1974) is adopted for updating the geometry of each element.

However, for the cases where the incremental end rotations are of *finite* magnitudes, the law of commutativity does not apply. The procedure for updating the end orientations of each frame element in the three-dimensional sense has to be developed based on the theory of *finite rotations*, in order to account for the effect of *noncommutativity*. In this Appendix, we shall first talk about the updating of element axes and section axes. After this, we shall proceed to calculate the natural deformations for the space frame element and further to calculate the element forces, as necessary in the incremental-iterative analysis. The materials to be presented below are based mainly on the paper by Kuo *et al.* (1993).

B.2 Reference Axes of Element Nodes

As was stated previously, in an incremental nonlinear analysis, we may denote the initial undeformed configuration of a solid body by C_0, the last calculated configuration by C_1, and the current deformed configuration by C_2. For a space frame element of solid cross-section, six degrees of freedom are needed at each end (node) of the element, which consists of three translations and three rotations, as shown in Figure 5.11. In this Appendix, as well as in previous chapters, only connections of the rigid type will be assumed for the framed structures. Throughout this text, a square matrix is denoted by brackets [ ], a column vector by braces { }, and a row vector by $\langle\ \ \rangle$.

For the present purposes, we may arbitrarily attach a set of orthogonal reference axes of each node of the frame element. As shown in Figure B.1, the reference axes for node A of the frame element at configuration C_0 are denoted by three unit vectors ${}^0\vec{\xi}_a, {}^0\vec{\eta}_a, {}^0\vec{\zeta}_a$, those at configuration C_1 by ${}^1\vec{\xi}_a, {}^1\vec{\eta}_a, {}^1\vec{\zeta}_a$, and those at configuration C_2 by ${}^2\vec{\xi}_a, {}^2\vec{\eta}_a, {}^2\vec{\zeta}_a$. The coordinates of node A at the three configurations are denoted by $({}^0X_a, {}^0Y_a, {}^0Z_a)$, $({}^1X_a, {}^1Y_a, {}^1Z_a)$, and $({}^2X_a, {}^2Y_a, {}^2Z_a)$. Note that at configuration C_0, the reference axes for each node are parallel to the global coordinate axes XYZ. In the following, the procedure for updating the end rotations of each element based on the theory of finite rotations will be described.

As shown in Figure B.2, the plane OSP will move to OSP' when subjected to a rigid rotation ϕ about the axis OS, which is accompanied by the movement of the position vector $\vec{r}$ to $\vec{r}'$. According to Rodriguez' (Euler's) finite rotation formula (Cheng and Gupta, 1989),

$$\vec{r}' = \cos\phi\vec{r} + \sin\phi\,(\vec{n}\times\vec{r}) + (1-\cos\phi)(\vec{n}\cdot\vec{r})\vec{n} \tag{B.1}$$

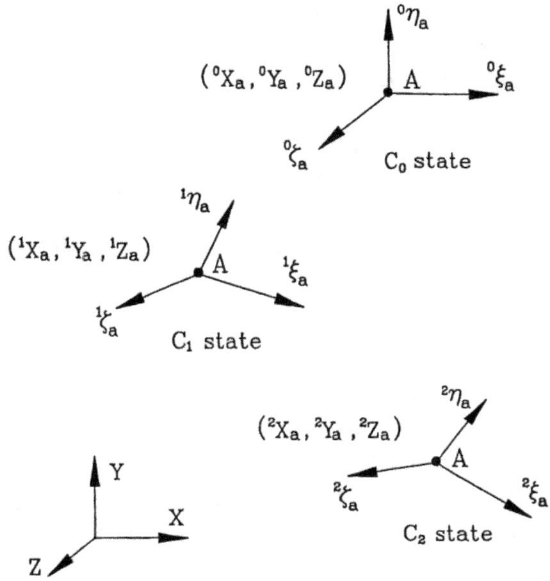

Fig. B.1. Nodal coordinates of frame element.

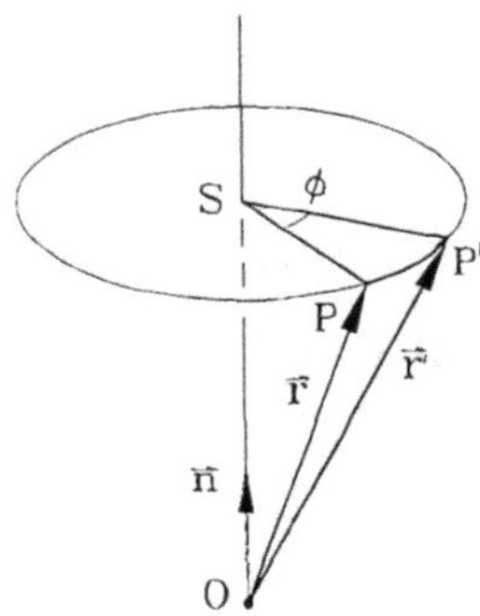

Fig. B.2. Concept of finite rotation.

where "×" indicates the cross product, "·" the dot product, and $\vec{n}$ a unit vector along the axis OS,

$$\vec{n} = \frac{\overrightarrow{OS}}{|\overrightarrow{OS}|} \tag{B.2}$$

During the movement of the element from configuration C_1 to C_2, the displacement increment $\Delta\vec{u}_a$ and rotation increment $\Delta\vec{\theta}_a$ experienced by node A can be denoted as

$$\Delta\vec{u}_a = \Delta u_x\vec{i} + \Delta u_y\vec{j} + \Delta u_z\vec{k} \tag{B.3}$$

$$\Delta\vec{\theta}_a = \Delta\theta_x\vec{i} + \Delta\theta_y\vec{j} + \Delta\theta_z\vec{k} \tag{B.4}$$

The magnitude ϕ_a and unit direction vector $\vec{n}_a$ of the rotation increment $\Delta\vec{\theta}_a$ are as follows:

$$\phi_a = |\Delta\vec{\theta}_a| = \sqrt{\Delta\theta_x^2 + \Delta\theta_y^2 + \Delta\theta_z^2} \tag{B.5}$$

$$\vec{n}_a = \frac{\Delta\vec{\theta}_a}{\phi_a} \tag{B.6}$$

According to equations (B.1)–(B.6), the unit vectors for the reference axes of node A at configuration C_2 can be written,

$${}^2\vec{\xi}_a = \cos\phi_a\,{}^1\vec{\xi}_a + \sin\phi_a(\vec{n}_a \times {}^1\vec{\xi}_a) + (1-\cos\phi_a)(\vec{n}_a \cdot {}^1\vec{\xi}_a)\vec{n}_a \tag{B.7}$$

$${}^2\vec{\eta}_a = \cos\phi_a\,{}^1\vec{\eta}_a + \sin\phi_a(\vec{n}_a \times {}^1\vec{\eta}_a) + (1-\cos\phi_a)(\vec{n}_a \cdot {}^1\vec{\eta}_a)\vec{n}_a \tag{B.8}$$

$${}^2\vec{\zeta}_a = \cos\phi_a\,{}^1\vec{\zeta}_a + \sin\phi_a(\vec{n}_a \times {}^1\vec{\zeta}_a) + (1-\cos\phi_a)(\vec{n}_a \cdot {}^1\vec{\zeta}_a)\vec{n}_a \tag{B.9}$$

On the other hand, the coordinates of node A at C_2 are simply

$$({}^2X_a, {}^2Y_a, {}^2Z_a) = ({}^2X_a + \Delta u_x, {}^2Y_a + \Delta u_y, {}^2Z_a + \Delta u_z) \tag{B.10}$$

The expressions as presented above in (B.3)–(B.10) can be easily extended to node B of the element.

B.3 Element Axes and Nodal Section Axes

In this section, the variation of the element axes and nodal section axes in the process of member deformation will be discussed. As shown in Figure B.3(a), the section axes for node A of the frame element at configuration C_0 are denoted by vectors ${}^0\vec{\alpha}_a$, ${}^0\vec{\beta}_a$, and ${}^0\vec{\gamma}_a$, where ${}^0\vec{\alpha}_a$ is taken to be coincident with the member axis passing through the centroid of the cross-section, and ${}^0\vec{\beta}_a$ and ${}^0\vec{\gamma}_a$ are the major and minor principal directions, respectively, of the cross-section. Similarly, another set of orthogonal axes ${}^0\vec{\alpha}_b$, ${}^0\vec{\beta}_b$, and ${}^0\vec{\gamma}_b$

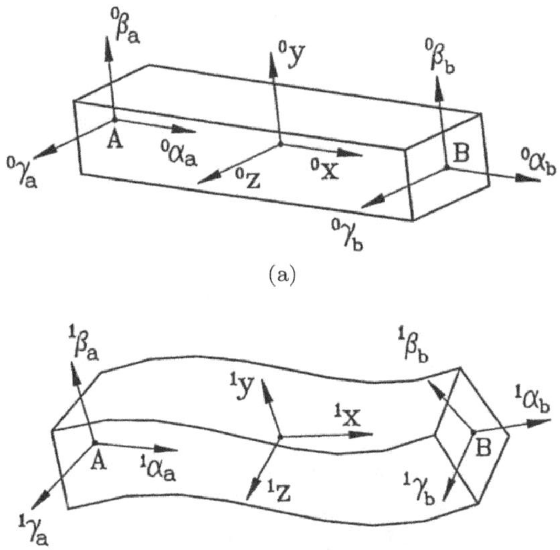

Fig. B.3. Element coordinates and nodal coordinates: (a) C_0 configuration; (b) C_1 configuration.

can be established for node B at C_0. Likewise, it is necessary to introduce a third set of orthogonal axes ${}^0\vec{x}$, ${}^0\vec{y}$, and ${}^0\vec{z}$ to the frame element at configuration C_0, of which ${}^0\vec{x}$ is taken to be the centroidal axes of the member. From Figure B.3(a), it can be seen that at configuration C_0 the above three sets of orthogonal axes are parallel to each other.

When the element moves from C_0 to the C_1 configuration (Figure B.3(b)), the section axes $({}^0\vec{\alpha}_a, {}^0\vec{\beta}_a, {}^0\vec{\gamma}_a)$ and $({}^0\vec{\alpha}_b, {}^0\vec{\beta}_b, {}^0\vec{\gamma}_b)$ for the two ends will change as the normal and principal directions of the end sections rotate. Since the centroidal axis of the element becomes curved, the two end sections in general will not remain parallel to each other. Based on the concept of *convected coordinates* (Belytschko and Hsieh, 1974), we may define the element axis ${}^1\vec{x}$ at configuration C_1 as the axis passing through the centroids of the end sections. Moreover, by assuming a plane ${}^1S$ perpendicular to the ${}^1\vec{x}$-axis, we may define ${}^1\vec{y}$ and ${}^1\vec{z}$ at the same configuration as the average of the projections of the major and minor principal directions, respectively, of the two end sections on the ${}^1S$ plane. In the following, the procedure for determining the element axes and nodal section axes based on the positions and reference axes of element nodes is presented.

The section axes of node A at configuration C_0 can be related to the reference axes as

$$\{{}^0p_a\} = [{}^0R]\{{}^0q_a\} \quad (p = \alpha, \beta, \gamma; \quad q = \xi, \eta, \zeta) \tag{B.11}$$

where the left superscript "0" denotes the C_0 configuration, $\{{}^0p_a\}$ the section axes of node A for the frame element, and $\{{}^0q_a\}$ the reference axes of node A. Since the reference axes are parallel to the global axes XYZ at C_0, the transformation matrix $[{}^0R]$ can therefore be written as

$$[{}^0R] = [\{{}^0\alpha_a\} \quad \{{}^0\beta_a\} \quad \{{}^0\gamma_a\}] \tag{B.12}$$

in which $\{{}^0\alpha_a\}, \{{}^0\beta_a\}$, and $\{{}^0\gamma_a\}$ denote the components of the section axes of node A in the global coordinates XYZ. The matrix $[{}^0R]$ can also be interpreted as the transformation matrix between the element axes and the global coordinate axes for the frame element at configuration C_0, since at this configuration, the element is undeformed and the section axes of the element nodes are parallel to the element axes.

Based on the assumption of *rigid joints*, the transformation between the reference axes and section axes of the element nodes will not be influenced by deformations of the element. Hence,

$$\{{}^1p_a\} = [{}^0R]\{{}^1q_a\} \quad (p = \alpha, \beta, \gamma; \; q = \xi, \eta, \zeta) \tag{B.13}$$

$$\{{}^2p_a\} = [{}^0R]\{{}^2q_a\} \quad (p = \alpha, \beta, \gamma; \; q = \xi, \eta, \zeta) \tag{B.14}$$

where $\{{}^1p_a\}$ and $\{{}^2p_a\}$ represent the section axes of node A at C_1 and C_2, respectively. Similarly, the following can be written for node B:

$$\{{}^1p_b\} = [{}^0R]\{{}^1q_b\} \quad (p = \alpha, \beta, \gamma; \; q = \xi, \eta, \zeta) \tag{B.15}$$

$$\{{}^2p_b\} = [{}^0R]\{{}^2q_b\} \quad (p = \alpha, \beta, \gamma; \; q = \xi, \eta, \zeta) \tag{B.16}$$

In equations (B.11)–(B.16), all vectors have been referred to the global coordinates XYZ.

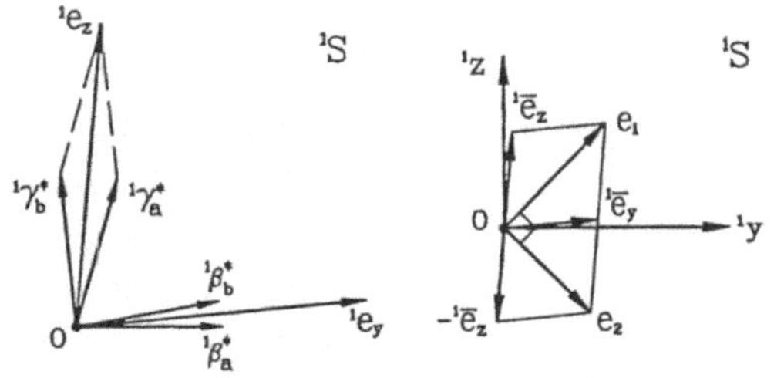

(a) (b)

Fig. B.4. Calculation of element coordinates: (a) averaged projections; (b) new element axes.

The axis ${}^1\vec{x}$ of the frame element at C_1 can be determined by the position vector formed by the two nodes A and B,

$$\{{}^1x\} = \frac{1}{{}^1L}\left\langle {}^1X_b - {}^1X_a \quad {}^1Y_b - {}^1Y_a \quad {}^1Z_b - {}^1Z_a \right\rangle^T \tag{B.17}$$

where ${}^1L$ denotes the length of the element at C_1,

$${}^1L = \sqrt{({}^1X_b - {}^1X_a)^2 + ({}^1Y_b - {}^1Y_a)^2 + ({}^1Z_b - {}^1Z_a)^2} \tag{B.18}$$

By defining ${}^1S$ as a plane normal to the ${}^1\vec{x}$-axis, the projections of the section axes ${}^1\vec{\beta}$ and ${}^1\vec{\gamma}$ for nodes A and B on the ${}^1S$ plane can be written (Figure B.4(a)

$$\{{}^1p_j^*\} = \{{}^1p_j\} - \overline{\lambda}_{pj}\{{}^1x\} \quad (j = a, b;\ \ p = \beta, \gamma) \tag{B.19}$$

where

$$\overline{\lambda}_{pj} = \{{}^1p_j\}^T\{{}^1x\} \quad (j = a, b;\ \ p = \beta, \gamma) \tag{B.20}$$

Normalizing,

$$\{{}^1\overline{p}_j\} = \frac{1}{\mu_{pj}}\{{}^1p_j^*\} \quad (j = a, b;\ \ p = \beta, \gamma) \tag{B.21}$$

where

$$\mu_{pj} = \sqrt{\{{}^1p_j^*\}^T\{{}^1p_j^*\}} \quad (j = a, b;\ \ p = \beta, \gamma) \tag{B.22}$$

In general, the section axes ${}^1\vec{\beta}$ and ${}^1\vec{\gamma}$ for the two nodes will not remain parallel to each other after the deformation of the element. Consequently, the projections of the section axes of the two nodes will not be coincident with each other. By taking their average (Figure B.4(a)), we have

$$\{{}^1e_s\} = \{{}^1\overline{p}_a\} + \{{}^1\overline{p}_b\} \quad (s = y, z;\ p = \beta, \gamma) \tag{B.23}$$

which can be normalized to yield

$$\{{}^1\overline{e}_s\} = \frac{1}{\mu_s}\{{}^1e_s\} \quad (s = y, z) \tag{B.24}$$

where

$$\mu_s = \sqrt{\{{}^1e_s\}^T\{{}^1e_s\}} \quad (s = y, z) \tag{B.25}$$

In general, the unit vectors $\{{}^1\overline{e}_y\}$ and $\{{}^1\overline{e}_z\}$, which represent the averaged directions of the projections of the section axes at the two nodes, will not be perpendicular to each other. In order to establish an orthogonal axis system, further modifications have to be made.

Based on the property that the diagonals of a rhombus are perpendicular to each other and that they divide the interior angles of the rhombus into two equal parts, the diagonals of the rhombus formed by the vectors $\{{}^1\overline{e}_y\}$ and $\{{}^1\overline{e}_z\}$ (Figure B.4(b)) can be expressed as

$$\{e_1\} = \{{}^1\overline{e}_y\} + \{{}^1\overline{e}_z\} \tag{B.26}$$

$$\{e_2\} = \{{}^1\overline{e}_y\} - \{{}^1\overline{e}_z\} \tag{B.27}$$

which can be normalized as

$$\{\overline{e}_k\} = \frac{1}{\mu_k}\{e_k\} \quad (k = 1, 2) \tag{B.28}$$

along with

$$\mu_k = \sqrt{\{e_k\}^T\{e_k\}} \quad (k = 1, 2) \tag{B.29}$$

Rotating the unit vectors $\{\overline{e}_y\}$ and $\{\overline{e}_z\}$ counterclockwise by 45° (Figure B.4(b)), we obtain the section axes for the frame element as follows:

$$\{^1y\} = \frac{1}{\sqrt{2}}(\{\bar{e}_1\} + \{\bar{e}_2\}) \tag{B.30}$$

$$\{^1z\} = \frac{1}{\sqrt{2}}(\{\bar{e}_1\} - \{\bar{e}_2\}) \tag{B.31}$$

The rotation matrix for the frame element at configuration C_1 can be formed by combining the axes $\{^1x\}$ of (B.17), $\{^1y\}$ of (B.30), and $\{^1z\}$ of (B.31), i.e.,

$$[^1R] = \left[\{^1x\} \quad \{^1y\} \quad \{^1z\}\right] \tag{B.32}$$

Similarly, the section axes for the element at configuration C_2 can be formulated based on the positions and reference axes of the element nodes at the said configuration.

B.4 Calculation of Natural Deformations

We do not advocate very much the concept of natural deformations in this book, since the displacements $\{u\}$ solved in the incremental-iterative analysis can be directly used in computing the force increment $\{\Delta f\}$ together with the stiffness matrix $[k]$, i.e., $\{\Delta f\} = [k]\{u\}$. Furthermore, according to the rigid body rule, the initial forces $\{^1f\}$ acting at the beginning of an incremental (or iterative) step can be directly treated as those acting at the end of the incremental (or iterative) step. Summing up the above two effects yields the total forces $\{^2f\}$ acting at the end of the incremental (or iterative) step.

There are cases where it is convenient if the displacement increments $\{u\}$ solved from the incremental-iterative step can be conceptually decomposed into two parts as the *rigid displacements* $\{u\}_r$ and the *natural (member) deformations* $\{u\}_n$. As such, we may use the natural deformations $\{u\}_n$ to calculate the element force increments $\{\Delta f\}$. The following is a presentation for such an approach.

The displacement increments $\{u\}$ can be decomposed into two parts as the rigid displacements $\{u\}_r$ and the natural deformations $\{u\}_n$:

$$\{u\} = \{u\}_r + \{u\}_n \tag{B.33}$$

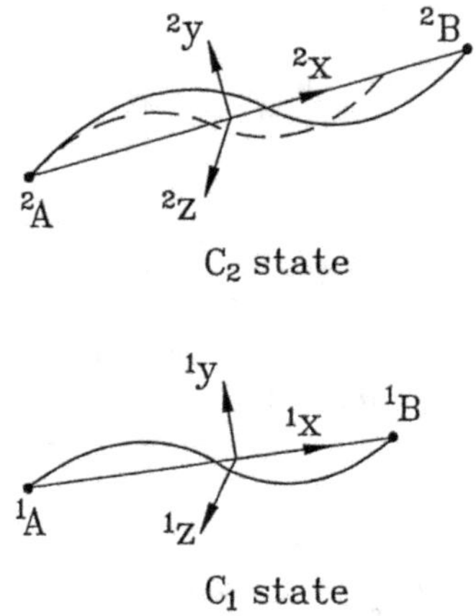

Fig. B.5. Rigid body displacements and natural deformations.

According to the rigid body rule, the effect of rigid body displacements is to rotate all the initial forces acting on a finite element by an angle equal to the rigid body rotation. The result is the preservation of equilibrium of the finite element in the displaced configuration, with no change in the magnitudes of the acting forces (see Section 3.5 or Yang and Chiou, 1987).

By the rigid displacements, the axes ${}^1\vec{x}$, ${}^1\vec{y}$, and ${}^1\vec{z}$ of the member at C_1 will be converted into the axes ${}^2\vec{x}$, ${}^2\vec{y}$, and ${}^2\vec{z}$ at C_2 as indicated by the dashed curve in Figure B.5. In this figure, the deformation of the element from the dashed to the solid curve at C_2 has been referred to as the *natural* or *member deformation.* For a space frame element with six degrees of freedom at each node, the natural deformation vector $\{u\}_n$, referred to the C_2 axes ${}^2\vec{x}$, ${}^2\vec{y}$, and ${}^2\vec{z}$, can be written as

$$\{u\}_n = \langle 0 \quad 0 \quad 0 \quad \Theta_{xa} \quad \Theta_{ya} \quad \Theta_{za} \quad U_b \quad 0 \quad 0 \quad \Theta_{xb} \quad \Theta_{yb} \quad \Theta_{zb}\rangle^T \tag{B.34}$$

where U_b denotes the axial elongation; Θ_{xa}, Θ_{ya} and Θ_{za} the rotations of the frame element at node A about the three axes ${}^2\vec{x}$, ${}^2\vec{y}$, and ${}^2\vec{z}$, respectively; and similar definitions hold for Θ_{xb}, Θ_{yb} and Θ_{zb}. By definition,

$$U_b = {}^2L - {}^1L \tag{B.35}$$

where ${}^2L$ and ${}^1L$ represent the chord length of the element at C_1 and C_2, respectively [see equation (B.18) for ${}^1L$].

The other components of the natural deformation vector $\{u\}_n$ in (B.34) can be determined from the nodal section axes and element axes at C_1 and C_2. First, the direction vectors of the section axes in terms of the global coordinate axes in (B.13) and (B.14) at configuration C_k (where $k = 1, 2$) can be transformed into those in terms of the element axes ${}^k\vec{x}, {}^k\vec{y}$, and ${}^k\vec{z}$ as

$$\{{}^k_0p_a\} = [{}^kR]^T\{{}^kp_a\} \quad (k = 1, 2;\; p = \alpha, \beta, \gamma) \tag{B.36}$$

where the subscript "0" indicates that the reference configuration is C_0 and $[{}^kR]$ is the transformation matrix for the element at the C_k configuration:

$$[{}^kR] = \left[\{{}^kx\} \quad \{{}^ky\} \quad \{{}^kz\}\right] \tag{B.37}$$

As shown in Figures B.6(a) and B.6(b), $\{{}^1_0p_a\}$ and $\{{}^2_0p_a\}$ denote the natural deformations (rotations) of the element at node A from configuration C_0 to C_1 and C_2, respectively. Consequently, the deformational increments of the element at node A during the incremental step can be expressed with reference to the C_1 section axes as

$$\{{}^2_1p_a\} = \left[\{{}^1_0\alpha_a\} \quad \{{}^1_0\beta_a\} \quad \{{}^1_0\gamma_a\}\right]\{{}^2_0p_a\} \quad (p = \alpha, \beta, \gamma) \tag{B.38}$$

where $\{{}^1_0\alpha_a\}, \{{}^1_0\beta_a\}$ and $\{{}^1_0\gamma_a\}$ are the base vectors for the section axes of node A at C_1.

Thus, we have shown that as the element moves from C_1 to C_2, the three axes $\{{}^2_0\alpha_a\}, \{{}^2_0\beta_a\}$ and $\{{}^2_0\gamma_a\}$ embedded at node A will be rotated and become the axes denoted as $\{{}^2_1\alpha_a\}, \{{}^2_1\beta_a\}$ and $\{{}^2_1\gamma_a\}$ in Figure B.6(c). In this connection, the amount of rotation ϕ_a at

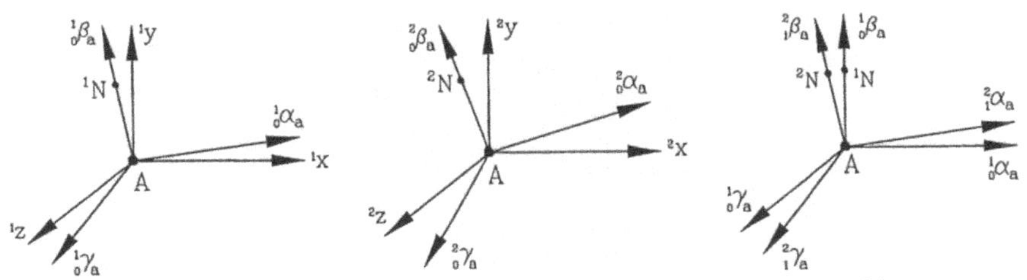

Fig. B.6. Variation of section coordinates: (a) C_1 state; (b) C_2 state; (c) rotational increments.

node A occurring during the incremental step from C_1 to C_2 and its axis of rotation $\{_1n_a\}$ can be solved. Let

$$\{_1n_a\} = \langle n_1 \quad n_2 \quad n_3\rangle^T \tag{B.39}$$

subject to the condition

$$n_1^2 + n_2^2 + n_3^2 = 1 \tag{B.40}$$

Further, by letting

$$\vec{r} = \{^1_0\alpha_a\} = \langle 1 \quad 0 \quad 0\rangle^T \tag{B.41}$$

$$\vec{r}' = \{^2_1\alpha_a\} = \langle \alpha_1 \quad \alpha_2 \quad \alpha_3\rangle^T \tag{B.42}$$

we can obtain the following relations for the ${}^2_1\vec{\alpha}_a$-axis from Rodriguez' rotation formula (B.1):

$$\alpha_1 = \cos\phi_a + (1 - \cos\phi_a)n_1^2 \tag{B.43}$$

$$\alpha_2 = n_3 \sin\phi_a + (1 - \cos\phi_a)n_1 n_2 \tag{B.44}$$

$$\alpha_3 = -n_2 \sin\phi_a + (1 - \cos\phi_a)n_1 n_3 \tag{B.45}$$

Similarly, by letting

$$\vec{r} = \{^1_0\beta_a\} = \langle 0 \quad 1 \quad 0\rangle^T \tag{B.46}$$

$$\vec{r}' = \{^2_1\beta_a\} = \langle \beta_1 \quad \beta_2 \quad \beta_3\rangle^T \tag{B.47}$$

the following relation can be derived for the ${}^2_1\vec{\beta}_a$-axis:

$$\beta_1 = -n_3 \sin\phi_a + (1 - \cos\phi_a)n_1 n_2 \tag{B.48}$$

Subtracting equation (B.48) from (B.44) yields

$$2n_3 \sin\phi_a = -\beta_1 + \alpha_2 \tag{B.49}$$

By defining

$$\{^1_0\gamma_a\} = \langle 0 \quad 0 \quad 1\rangle^T \tag{B.50}$$

$$\{^2_1\gamma_a\} = \langle \gamma_1 \quad \gamma_2 \quad \gamma_3\rangle^T \tag{B.51}$$

we may show that, in addition to equation (B.49), the following relations are valid:

$$2n_2 \sin\phi_a = -\alpha_3 + \gamma_1 \tag{B.52}$$

$$2n_1 \sin\phi_a = -\gamma_2 + \beta_3 \tag{B.53}$$

From equations (B.40), (B.49), (B.52), and (B.53), the following components can be solved for the axis of rotation:

$$n_1 = -\frac{\gamma_2 - \beta_3}{\lambda} \tag{B.54}$$

$$n_2 = -\frac{\alpha_3 - \gamma_1}{\lambda} \tag{B.55}$$

$$n_3 = -\frac{\beta_1 - \alpha_2}{\lambda} \tag{B.56}$$

along with the angle of rotation ϕ_a as

$$\phi_a = \sin^{-1}\left(\frac{\lambda}{2}\right) \tag{B.57}$$

where

$$\lambda = \sqrt{(\gamma_2 - \beta_3)^2 + (\alpha_3 - \gamma_1)^2 + (\beta_1 - \alpha_2)}^{2} \tag{B.58}$$

The vector $\{{}_1n_a\}$ as derived above for the axis of rotation has been referred to as the section axes $\{{}^1_0p_a\}$, where $p = \alpha, \beta, \gamma$. It can also be transformed into the element axes as follows:

$$\{n_a\} = \left[\{{}^1_0\alpha_a\} \quad \{{}^1_0\beta_a\} \quad \{{}^1_0\gamma_a\}\right]\{{}_1n_a\} \tag{B.59}$$

With the expressions of (B.57) and (B.59), the natural rotations associated with node A of the element during the incremental step from C_1 to C_2 can be calculated simply as

$$\langle \Theta_{xa} \quad \Theta_{ya} \quad \Theta_{za}\rangle = \phi_a\{n_a\}^T \tag{B.60}$$

Similar procedures can be followed to calculate the natural rotations of the element at node B. The procedure presented in the sections of this appendix is valid for the case when the rotations contained element displacements $\{u\}$ are of finite magnitudes.

B.5 Calculation of Element Forces

As long as the natural deformations are calculated, the nodal forces acting on the space frame element at configuration C_2 can be determined. Based on the updated Lagrangian formulation of Chapters 4–6, the equations of equilibrium for the truss or frame elements can be established at the current deformed configuration C_2, but referred to the last calculated configuration C_1 as the one given in equation (6.7), i.e.,

$$[k]\{u\} = \{{}_1^2f\} - \{{}_1^1f\} \tag{B.61}$$

where $[k]$ is the *tangent stiffness matrix*, $\{u\}$ the displacements generated during the incremental step from C_1 to C_2, and $\{{}_1^2f\}$ and $\{{}_1^1f\}$ are the forces acting on the element at C_2 and C_1, respectively. The subscripts "1" have been added to signify that the quantities are expressed with reference to C_1.

For the space frame element, the stiffness matrix $[k]$ is composed of three components:

$$[k] = [k_e] + [k_g] + [k_i] \tag{B.62}$$

where $[k_e], [k_g]$ and $[k_i]$ are the elastic, geometric and induced moment matrices, respectively, as described in Chapter 6. Specifically, the induced moment matrix $[k_i]$ is needed to account for the rotational properties of the nodal bending moments and torques, or for the space frame element to pass the rigid body test. For the plane frame element, the stiffness matrix $[k]$ reduces to

$$[k] = [k_e] + [k_g] \tag{B.63}$$

since there is no out-of-plane rotations for plane frames.

For the truss element, the stiffness matrix $[k]$ can be given in the following general form:

$$[k] = [k_e] + [k_g] + [s_1] + [s_2] + [s_3] \tag{B.64}$$

where the matrices $[s_1]$, $[s_2]$, and $[s_3]$ have been given in Section 4.2 to account for various stretching and rigid rotation effects.

As was stated, the element displacement increment $\{u\}$ can be decomposed into the rigid body displacements $\{u\}_r$ and natural deformations $\{u\}_n$. By the rigid body rule described in Chapter 3, the effect of rigid body rotations (and displacements) is to rotate all the initial forces $\{^1_1f\}$ acting on the finite element by an angle equal to the rigid body rotation. To account for such a rigid rotation effect in computer programming, we can simply regard the initial forces $\{^1_1f\}$ (acting on the element at C_1 and referred to C_1 prior to the rigid rotation) as the forces $\{^2_2f\}$ acting on the element at C_2 and referred to C_2 after the rigid body rotation. Meanwhile, the force increment $\{\Delta f\}$ generated by the natural deformations during the incremental step is simply equal to $[k]\{u\}_n$, to the accuracy of the first order of the displacement increments. By adding up these two effects, we can obtain the total element forces $\{^2_2f\}$ acting on the element at C_2 as

$$\{^2_2f\} = \{^1_1f\} + [k]\{u\}_n \tag{B.65}$$

where the left subscript "2" has been added to signify that the total element forces have been referred to configuration C_2.

B.6 Reduction of the Procedure to Two-Dimensional Case

The procedures presented in the preceding sections have been developed mainly for space frames concerning three-dimensional rotations. They can be easily reduced to the two-dimensional case for planar frames as will be shown in this section. Such a reduction serves not only as a useful check on the formulas derived in preceding sections but also as an illustration of how they can be applied to practical problems. Here, we shall assume that the plane element is located on the $^1x - {}^1y$ plane at the C_1 configuration, with the axis of the element directed along the ^1x-axis, as shown in Figure 3.2 or 3.3.

For the purpose of illustration, we shall concentrate on the calculation of the natural deformations $\{u\}_n$, i.e., (B.34), for the plane

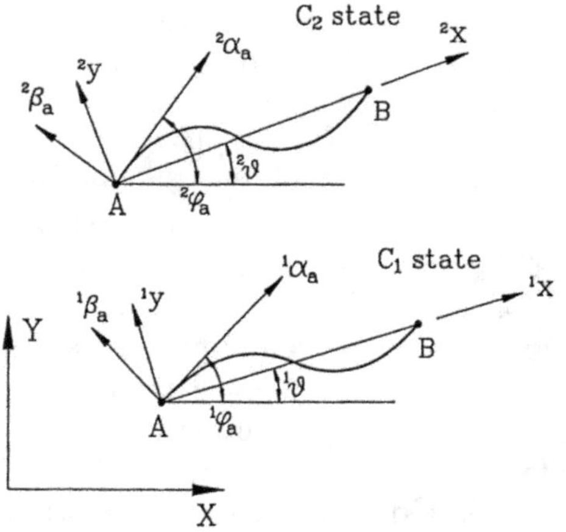

Fig. B.7. Natural deformations of plane frame element.

frame element. First of all, the axial elongation U_b of the plane frame élement can be calculated using equation (B.35) if the following expression has been adopted for the length ${}^1L$ of the element at C_1 (see equation (B.18)):

$$ {}^1L = \sqrt{({}^1X_b - {}^1X_a)^2 + ({}^1Y_b - {}^1Y_a)}^2 \tag{B.66} $$

and a similar definition for ${}^2L$.

Next, let us proceed with the calculation of natural rotations at the two ends of the element. According to Figure B.7, the section coordinates $({}^1\vec{\alpha}_a, {}^1\vec{\beta}_a, {}^1\vec{\gamma}_a)$ and $({}^2\vec{\alpha}_a, {}^2\vec{\beta}_a, {}^2\vec{\gamma}_a)$ for nodes A at C_1 and C_2 are

$$ \{{}^1\alpha_a\} = \begin{Bmatrix} \cos{}^1\phi_a \\ \sin{}^1\phi_a \\ 0 \end{Bmatrix}, \quad \{{}^1\beta_a\} = \begin{Bmatrix} -\sin{}^1\phi_a \\ \cos{}^1\phi_a \\ 0 \end{Bmatrix}, \quad \{{}^1\gamma_a\} = \begin{Bmatrix} 0 \\ 0 \\ 1 \end{Bmatrix} \tag{B.67a-c} $$

$$ \{{}^2\alpha_a\} = \begin{Bmatrix} \cos{}^2\phi_a \\ \sin{}^2\phi_a \\ 0 \end{Bmatrix}, \quad \{{}^2\beta_a\} = \begin{Bmatrix} -\sin{}^2\phi_a \\ \cos{}^2\phi_a \\ 0 \end{Bmatrix}, \quad \{{}^2\gamma_a\} = \begin{Bmatrix} 0 \\ 0 \\ 1 \end{Bmatrix} \tag{B.68a-c} $$

where ${}^1\phi_a$ and ${}^2\phi_a$ denote the total angles of rotation of node A at C_1 and C_2, respectively.

As shown in Figure B.7, the element axes at the C_1 and C_2 configurations, i.e., $({}^1\vec{x}, {}^1\vec{y}, {}^1\vec{z})$ and $({}^2\vec{x}, {}^2\vec{y}, {}^2\vec{z})$, can be expressed with respect to the XYZ-axes as follows:

$$\{{}^1x\} = \begin{Bmatrix} \cos{}^1\theta \\ \sin{}^1\theta \\ 0 \end{Bmatrix}, \quad \{{}^1y\} = \begin{Bmatrix} -\sin{}^1\theta \\ \cos{}^1\theta \\ 0 \end{Bmatrix}, \quad \{{}^1z\} = \begin{Bmatrix} 0 \\ 0 \\ 1 \end{Bmatrix} \tag{B.69a-c}$$

$$\{{}^2x\} = \begin{Bmatrix} \cos{}^2\theta \\ \sin{}^2\theta \\ 0 \end{Bmatrix}, \quad \{{}^2y\} = \begin{Bmatrix} -\sin{}^2\theta \\ \cos{}^2\theta \\ 0 \end{Bmatrix}, \quad \{{}^2z\} = \begin{Bmatrix} 0 \\ 0 \\ 1 \end{Bmatrix} \tag{B.70a-c}$$

where ${}^1\theta$ and ${}^2\theta$ denote the inclination angles of the element axes ${}^1\vec{x}$ and ${}^2\vec{x}$ with respect to the horizontal axis at C_1 and C_2, respectively. According to equation (B.37), the transformation matrices $[{}^1R]$ and $[{}^2R]$ can be written as

$$[{}^1R]^T = \begin{bmatrix} \cos{}^1\theta & \sin{}^1\theta & 0 \\ -\sin{}^1\theta & \cos{}^1\theta & 0 \\ 0 & 0 & 1 \end{bmatrix} \tag{B.71}$$

$$[{}^2R]^T = \begin{bmatrix} \cos{}^2\theta & \sin{}^2\theta & 0 \\ -\sin{}^2\theta & \cos{}^2\theta & 0 \\ 0 & 0 & 1 \end{bmatrix} \tag{B.72}$$

which, along with equations (B.67a-c) and (B.68a-c), can be substituted back into equation (B.36) to yield

$$\{{}^1_0\alpha_a\} = \begin{Bmatrix} \cos({}^1\phi_a - {}^1\theta) \\ \sin({}^1\phi_a - {}^1\theta) \\ 0 \end{Bmatrix}, \quad \{{}^1_0\beta_a\} = \begin{Bmatrix} -\sin({}^1\phi_a - {}^1\theta) \\ \cos({}^1\phi_a - {}^1\theta) \\ 0 \end{Bmatrix},$$

$$\{{}^1_0\gamma_a\} = \begin{Bmatrix} 0 \\ 0 \\ 1 \end{Bmatrix} \tag{B.73a-c}$$

$$\{{}_0^2\alpha_a\} = \begin{Bmatrix} \cos({}^2\phi_a - {}^2\theta) \\ \sin({}^2\phi_a - {}^2\theta) \\ 0 \end{Bmatrix}, \quad \{{}_0^2\beta_a\} = \begin{Bmatrix} -\sin({}^2\phi_a - {}^2\theta) \\ \cos({}^2\phi_a - {}^2\theta) \\ 0 \end{Bmatrix},$$

$$\{{}_0^1\gamma_a\} = \begin{Bmatrix} 0 \\ 0 \\ 1 \end{Bmatrix} \tag{B.74a-c}$$

From equation (B.38), we have

$$\{{}_1^2\alpha_a\} = \begin{Bmatrix} \cos\left[({}^2\phi_a - {}^2\theta) - ({}^1\phi_a - {}^1\theta)\right] \\ \sin\left[({}^2\phi_a - {}^2\theta) - ({}^1\phi_a - {}^1\theta)\right] \\ 0 \end{Bmatrix} \tag{B.75}$$

$$\{{}_1^2\beta_a\} = \begin{Bmatrix} -\sin\left[({}^2\phi_a - {}^2\theta) - ({}^1\phi_a - {}^1\theta)\right] \\ \cos\left[({}^2\phi_a - {}^2\theta) - ({}^1\phi_a - {}^1\theta)\right] \\ 0 \end{Bmatrix} \tag{B.76}$$

$$\{{}_1^2\gamma_a\} = \begin{Bmatrix} 0 \\ 0 \\ 1 \end{Bmatrix} \tag{B.77}$$

which imply that

$$\alpha_1 = \beta_2 = \cos\left[({}^2\phi_a - {}^2\theta) - ({}^1\phi_a - {}^1\theta)\right] \tag{B.78}$$

$$\alpha_2 = -\beta_1 = \sin\left[({}^2\phi_a - {}^2\theta) - ({}^1\phi_a - {}^1\theta)\right] \tag{B.79}$$

$$\alpha_3 = \beta_3 = \gamma_1 = \gamma_2 = 0 \tag{B.80}$$

$$\gamma_3 = 1 \tag{B.81}$$

based on the definitions of (B.42), (B.47), and (B.51). From equation (B.58), the quantity λ is

$$\lambda = -2\sin\left[({}^2\phi_a - {}^2\theta) - ({}^1\phi_a - {}^1\theta)\right] \tag{B.82}$$

Also, from equation (B.57), the angle of rotation ϕ_a at node A is

$$\phi_a = -[({}^2\phi_a - {}^2\theta) - ({}^1\phi_a - {}^1\theta)] \tag{B.83}$$

Substituting (B.78)–(B.82) into (B.54)–(B.56) yields the following components for the axis of rotation $\{_1n_a\}$ at node A:

$$n_1 = 0, \quad n_2 = 0, \quad n_3 = -1 \tag{B.84}$$

According to equation (B.60), the natural rotations at node A are

$$\langle \Theta_{xa} \quad \Theta_{ya} \quad \Theta_{za} \rangle = \langle 0 \quad 0 \quad ({}^2\phi_a - {}^2\theta) - ({}^1\phi_a - {}^1\theta) \rangle \tag{B.85}$$

which agrees completely with the formula given in Yang and Chiou (1987). Obviously, the same procedure can be applied to the calculation of the natural rotations for node B.

References

Belytschko, T. and Hsieh, B.J. (1974). Nonlinear transient analysis of shells and solids of revolution by convected elements, *AIAA J.*, 12(8), 1031–1035.

Chan, S.L. and Kitipornchai, S. (1987). Geometric nonlinear analysis of asymmetric thin-walled beam-columns, *Eng. Struct.*, 9, 243–254.

Cheng, H. and Gupta, K.C. (1989). A historical note on finite rotations, *J. Appl. Mech.*, 56, 139–145.

Connor, J.J., Logcher, D., and Chan, S.C. (1968). Nonlinear analysis of elastic framed structures, *J. Struct. Div*, ASCE, 94(6), 1525–1547.

Kuo, S.R., Yang, Y.B., and Chou, J.H. (1993). Nonlinear analysis of space frames with finite rotations," *J. Struct. Eng.*, ASCE, 119(1), 1–15.

Oran, C. (1973). Tangent stiffness in space frames, *J. Struct. Div.*, ASCE, 99(6), 987–1001.

Remseth, S.N. (1979). Nonlinear static and dynamic analysis of framed structures, *Comput. Struct.*, 10, 879–897.

Tezcan, S.S. and Mahapatra, B.C. (1969). Tangent stiffness matrix for space frame members, *J. Struct. Div.*, ASCE, 95(6), 1257–1270.

Yang, Y.B. and Chiou, H.T. (1987). Rigid body motion test for nonlinear analysis with beam elements, *J. Eng. Mech.*, ASCE, 113(9), 1404–1419.

Appendix C

Geometric Stiffness Matrix for the Rigid TPE

The geometric stiffness matrix derived in equation (8.73) of Chapter 8 for the triangular plate element (TPE) is

$$[k_g]_{TPE} = \begin{bmatrix} [E_1] & [F_1] & -[E_{12}] & [G_{12}] & -[E_{31}] & -[H_{31}] \\ [F_1]^T & [I_1] & -[H_{12}]^T & [0] & [G_{31}]^T & [0] \\ -[E_{12}]^T & -[H_{12}] & [E_2] & [F_2] & -[E_{23}] & [G_{23}] \\ [G_{12}]^T & [0] & [F_2]^T & [I_2] & -[H_{23}]^T & [0] \\ -[E_{31}]^T & [G_{31}] & -[E_{23}]^T & -[H_{23}] & [E_3] & [F_3] \\ -[H_{31}]^T & [0] & [G_{23}]^T & [0] & [F_3]^T & [I_3] \end{bmatrix} \tag{C.1}$$

where

$$[E_{ij}] = \begin{bmatrix} a^{ij} & c^{ij} & d^{ij} \\ c^{ij} & b^{ij} & e^{ij} \\ d^{ij} & e^{ij} & f^{ij} \end{bmatrix} \tag{C.2}$$

$$[G_{ij}] = \begin{bmatrix} n^{ij} & p^{ij} & 0 \\ q^{ij} & o^{ij} & 0 \\ r^{ij} & s^{ij} & -m^{ij} \end{bmatrix} \tag{C.3}$$

$$[H_{ij}] = \begin{bmatrix} g^{ij} & i^{ij} & 0 \\ j^{ij} & h^{ij} & 0 \\ k^{ij} & l^{ij} & m^{ij} \end{bmatrix} \tag{C.4}$$

$$[I_1] = \begin{bmatrix} t_1^{12} + t_2^{31} & u_1^{12} + u_2^{31} & w_1^{12} + w_2^{31} \\ v_1^{12} + v_2^{31} & -(t_1^{12} + t_2^{31}) & y_1^{12} + y_2^{31} \\ x_1^{12} + x_2^{31} & z_1^{12} + z_2^{31} & 0 \end{bmatrix} \tag{C.5}$$

$$[I_2] = \begin{bmatrix} t_1^{23} + t_2^{12} & u_1^{23} + u_2^{12} & w_1^{23} + w_2^{12} \\ v_1^{23} + v_2^{12} & -(t_1^{23} + t_2^{12}) & y_1^{23} + y_2^{12} \\ x_1^{23} + x_2^{12} & z_1^{23} + z_2^{12} & 0 \end{bmatrix} \tag{C.6}$$

$$[I_3] = \begin{bmatrix} t_1^{31} + t_2^{23} & u_1^{31} + u_2^{23} & w_1^{31} + w_2^{23} \\ v_1^{31} + v_2^{23} & -(t_1^{31} + t_2^{23}) & y_1^{31} + y_2^{23} \\ x_1^{31} + x_2^{23} & z_1^{31} + z_2^{23} & 0 \end{bmatrix} \tag{C.7}$$

$$[E_1] = [E_{12}] + [E_{31}] \tag{C.8}$$

$$[E_2] = [E_{23}] + [E_{12}] \tag{C.9}$$

$$[E_3] = [E_{31}] + [E_{23}] \tag{C.10}$$

$$[F_1] = [H_{12}] - [G_{31}] \tag{C.11}$$

$$[F_2] = [H_{23}] - [G_{12}] \tag{C.12}$$

$$[F_3] = [H_{31}] - [G_{23}] \tag{C.13}$$

In the above expressions, the parameters are given as follows:

$$a^{ij} = -\frac{1}{L_{ij}^4}(X_{ij}Y_{ij}^2F_{xa}^{ij} - X_{ij}^2Y_{ij}F_{ya}^{ij} - L_{ij}^2Y_{ij}F_{ya}^{ij}) \tag{C.14}$$

$$b^{ij} = \frac{1}{L_{ij}^4}(X_{ij}Y_{ij}^2F_{xa}^{ij} - X_{ij}^2Y_{ij}F_{ya}^{ij} + L_{ij}^2X_{ij}F_{xa}^{ij}) \tag{C.15}$$

$$c^{ij} = -\frac{1}{L_{ij}^4}(Y_{ij}^3F_{xa}^{ij} + X_{ij}^3F_{ya}^{ij}) \tag{C.16}$$

$$d^{ij} = -\frac{1}{L_{ij}^2}X_{ij}F_{za}^{ij} \tag{C.17}$$

$$e^{ij} = -\frac{1}{L_{ij}^2} Y_{ij} F_{za}^{ij} \tag{C.18}$$

$$f^{ij} = \frac{1}{L_{ij}^2} (X_{ij} F_{xa}^{ij} + Y_{ij} F_{ya}^{ij}) \tag{C.19}$$

$$g^{ij} = -\frac{1}{2L_{ij}^4} (X_{ij} Y_{ij}^2 M_{xa}^{ij} - X_{ij}^2 Y_{ij} M_{ya}^{ij} - L_{ij}^2 Y_{ij} M_{ya}^{ij}) \tag{C.20}$$

$$h^{ij} = \frac{1}{2L_{ij}^4} (X_{ij} Y_{ij}^2 M_{xa}^{ij} - X_{ij}^2 Y_{ij} M_{ya}^{ij} + L_{ij}^2 X_{ij} M_{xa}^{ij}) \tag{C.21}$$

$$i^{ij} = -\frac{1}{2L_{ij}^4} (Y_{ij}^3 M_{xa}^{ij} - X_{ij} Y_{ij}^2 M_{ya}^{ij} + L_{ij}^2 Y_{ij} M_{xa}^{ij}) \tag{C.22}$$

$$j^{ij} = \frac{1}{2L_{ij}^4} (X_{ij}^2 Y_{ij} M_{xa}^{ij} - X_{ij}^3 M_{ya}^{ij} - L_{ij}^2 X_{ij} M_{ya}^{ij}) \tag{C.23}$$

$$k^{ij} = -\frac{1}{2L_{ij}^2} X_{ij} M_{za}^{ij} \tag{C.24}$$

$$l^{ij} = -\frac{1}{2L_{ij}^2} Y_{ij} M_{za}^{ij} \tag{C.25}$$

$$m^{ij} = \frac{1}{2L_{ij}^2} (X_{ij} M_{xa}^{ij} + Y_{ij} M_{ya}^{ij}) \tag{C.26}$$

$$n^{ij} = \frac{1}{L_{ij}^2} X_{ij} Y_{ij} F_{za}^{ij} + \frac{1}{2L_{ij}^4} \left(X_{ij} Y_{ij}^2 M_{xa}^{ij} - X_{ij}^2 Y_{ij} M_{ya}^{ij} - L_{ij}^2 Y_{ij} M_{ya}^{ij} \right) \tag{C.27}$$

$$\begin{aligned} o^{ij} = &-\frac{1}{L_{ij}^2} X_{ij} Y_{ij} F_{za}^{ij} \\ &- \frac{1}{2L_{ij}^4} \left(X_{ij} Y_{ij}^2 M_{xa}^{ij} - X_{ij}^2 Y_{ij} M_{ya}^{ij} + L_{ij}^2 X_{ij} M_{xa}^{ij} \right) \end{aligned} \tag{C.28}$$

$$p^{ij} = \frac{1}{L_{ij}^2} Y_{ij}^2 F_{za}^{ij} + \frac{1}{2L_{ij}^4} \left(Y_{ij}^3 M_{xa}^{ij} - X_{ij} Y_{ij}^2 M_{ya}^{ij} + L_{ij}^2 Y_{ij} M_{xa}^{ij} \right) \tag{C.29}$$

$$q^{ij} = -\frac{1}{L_{ij}^2} X_{ij}^2 F_{za}^{ij} - \frac{1}{2L_{ij}^4} \left(X_{ij}^2 Y_{ij} M_{xa}^{ij} - X_{ij}^3 M_{ya}^{ij} - L_{ij}^2 X_{ij} M_{ya}^{ij} \right) \tag{C.30}$$

$$r^{ij} = \frac{1}{L_{ij}^2} X_{ij} M_{za}^{ij} - \frac{1}{L_{ij}^2}(X_{ij} Y_{ij} F_{xa}^{ij} - X_{ij}^2 F_{ya}^{ij}) \tag{C.31}$$

$$s^{ij} = \frac{1}{L_{ij}^2} Y_{ij} M_{za}^{ij} - \frac{1}{L_{ij}^2}(Y_{ij}^2 F_{xa}^{ij} - X_{ij} Y_{ij} F_{ya}^{ij}) \tag{C.32}$$

$$t_1^{ij} = \frac{1}{L_{ij}^2} X_{ij} Y_{ij} M_{za}^{ij} \tag{C.33}$$

$$u_1^{ij} = \frac{1}{L_{ij}^2} Y_{ij}^2 M_{za}^{ij} \tag{C.34}$$

$$v_1^{ij} = -\frac{1}{L_{ij}^2} X_{ij}^2 M_{za}^{ij} \tag{C.35}$$

$$w_1^{ij} = -\frac{1}{2L_{ij}^2}(X_{ij} Y_{ij} M_{xa}^{ij} + Y_{ij}^2 M_{ya}^{ij}) \tag{C.36}$$

$$x_1^{ij} = -\frac{1}{2L_{ij}^2}(X_{ij} Y_{ij} M_{xa}^{ij} - X_{ij}^2 M_{ya}^{ij} - L_{ij}^2 M_{ya}^{ij}) \tag{C.37}$$

$$y_1^{ij} = \frac{1}{2L_{ij}^2}(X_{ij}^2 M_{xa}^{ij} + X_{ij} Y_{ij} M_{ya}^{ij}) \tag{C.38}$$

$$z_1^{ij} = -\frac{1}{2L_{ij}^2}(Y_{ij}^2 M_{xa}^{ij} - X_{ij} Y_{ij} M_{ya}^{ij} + L_{ij}^2 M_{xa}^{ij}) \tag{C.39}$$

$$t_2^{ij} = -\frac{1}{L_{ij}^2} X_{ij} Y_{ij} M_{za}^{ij} + \frac{1}{L_{ij}^2}(X_{ij} Y_{ij}^2 F_{xa}^{ij} - X_{ij}^2 Y_{ij} F_{ya}^{ij}) \tag{C.40}$$

$$u_2^{ij} = -\frac{1}{L_{ij}^2} Y_{ij}^2 M_{za}^{ij} + \frac{1}{L_{ij}^2}(Y_{ij}^3 F_{xa}^{ij} - X_{ij} Y_{ij}^2 F_{ya}^{ij}) \tag{C.41}$$

$$v_2^{ij} = \frac{1}{L_{ij}^2} X_{ij}^2 M_{za}^{ij} - \frac{1}{L_{ij}^2}(X_{ij}^2 Y_{ij} F_{xa}^{ij} - X_{ij}^3 F_{ya}^{ij}) \tag{C.42}$$

$$w_2^{ij} = \frac{1}{2L_{ij}^2}(X_{ij} Y_{ij} M_{xa}^{ij} + Y_{ij}^2 M_{ya}^{ij}) \tag{C.43}$$

$$x_2^{ij} = X_{ij} F_{za}^{ij} + \frac{1}{2L_{ij}^2}(X_{ij} Y_{ij} M_{xa}^{ij} - X_{ij}^2 M_{ya}^{ij} - L_{ij}^2 M_{ya}^{ij}) \tag{C.44}$$

$$y_2^{ij} = -\frac{1}{2L_{ij}^2}(X_{ij}^2 M_{xa}^{ij} + X_{ij}Y_{ij}M_{ya}^{ij}) \tag{C.45}$$

$$z_2^{ij} = Y_{ij}F_{za}^{ij} + \frac{1}{2L_{ij}^2}(Y_{ij}^2 M_{xa}^{ij} - X_{ij}Y_{ij}M_{ya}^{ij} + L_{ij}^2 M_{xa}^{ij}) \tag{C.46}$$

In the above, the nodal forces for element ij, that is, $F_{xa}^{ij}, F_{ya}^{ij}, \ldots,$ $M_{xa}^{ij}, M_{ya}^{ij}, \ldots,$ with the left superscript "1" omitted, have been given in equations (8.66) and (8.67), $X_{ij} = X_i - X_j, Y_{ij} = Y_i - Y_j$, and L_{ij} is the length of element ij.

Author Index

Subject Index

D

E

J

K

L

M

N

O

T

U

www.ingramcontent.com/pod-product-compliance
Lightning Source LLC
LaVergne TN
LVHW011658200926
848484LV00004B/980

* 9 7 8 9 8 1 9 8 0 8 2 1 2 *